有的城市
Cities with Heart

（美）潘德明(Thomas M.Paine) 著

中国建筑工业出版社

鸣谢

本书的问世是源于中国某设计公司的创始人相信：设计公司有责任推动对重要设计问题的思考，尤其是对于我们现今快速城市化的时代来说。迄今为止，鲜有公司能够在推广自身作品之外，从客观的角度整体看待这些问题。2009年，本书作者潘德明加入意格国际两年后，意格创始人马晓暐先生萌生了出版这本书的想法。潘德明加入意格时已积累了丰富的公共开放空间设计导则编写经验，他热爱历史，同时，由于他的两个儿子分别在苹果和 Adobe 公司工作，促使他对新材料与技术怀有浓厚的兴趣。马先生坚信，这本专著将对我们这个时代的设计对话产生影响，并引发一系列专著的问世。

在此特别感谢冉未名和赵娇阳对整本书的精确翻译。本书的编著需要大量人员的协助，包括：马友友，王石，王培娟，毛玲玲，孙志华，冯林，刘冰洁，陆潇潇，沈欣宇，吴天煜，杨天棋，罗青，俞孔坚，侯亚凤，高旻，徐心怡，夏金梅，黄仕倩，Gary Anderson, Nina Brown, Joe Brown, Albert Chun, Ann Clifford, Nick Danforth, Benny Djohan, Eric Douglas, Barnaby Evans, James Fallows, Ronald Lee Fleming, Valerie Fletcher, Craig Halvorson, Joanne Hiromura, Hsu Wensheng, Claire Kim, Robert Krim, Rachel Lee, Jessica Leete, Christian Lemon, Julie Lucier, Kathy Madden, Roy Mann, Julie Messervy, Robert O'Brien, Lydia Paine, Lynn Paine, Mallory Paine, Sumner Paine, Alison Richardson, Clarissa Rowe, Olivia Saw, Lewis Stackpole, Armando Tancio, Robert Tullis, Arthur Waldron, Jack Weber, Arthur Weissman, Sebastian Yeow 等等（中文名按姓氏笔画排序，英文名按姓氏字母排序）。在要旨与风格上，作者从哈佛大学及设计研究生院的一些教授那里得益甚多，包括 Charles Harris, Peter Hornbeck, J. B. Jackson, Clare Cooper Marcus, Norman T. Newton, Eduard F. Sekler 和 Carl Steinitz。

作者尤其想感谢公共土地信托的华盛顿特区办公室城市公园卓越中心的 Peter Harnik 和他的同事。

大部分有关设计的书籍强调平面图与效果图，但我们倾向于尽可能地以图解的方式展示人们实际使用公共空间的情景。除了作者自己的摄影作品，还使用了一些通过公共网络搜集的免版税图片，主要是 wikimedia.org，此外还有 123rf.com 和 dreamstime.com。此外，还包括美国国家公园服务部门的弗雷德里克·劳·奥姆斯特德国家历史遗址和马晓暐提供的历史档案图片，在此深表谢意。

由于篇幅有限，我们无法收纳所有优秀的设计案例。略去的作品并不代表达不到我们的评判标准。在未来更多的新版书籍中，我们希望增加更多的案例并进行必要的修改。所有错误均由作者承担责任。

谨以此书献给马晓暐先生，是他的热情支持对这本书的出版起到了极大的推动作用，相信这也会对中国公共开放空间的设计质量产生影响。

Acknowledgments

This book came about because the founder of a China-based design firm believed that design firms have a responsibility to advance thinking on the important design issues brought about in an age of rapid urbanization such as our own. Until now, few if any firms have gone beyond promoting their own work to look at the problems objectively and holistically. Xiaowei Ma, Founder of AGER, originally conceived the project in 2009, two years after the author, Thomas M. Paine ASLA, joined the firm. Paine came to AGER with considerable experience in writing public open space design guidelines, a love of history, and a fascination with new materials and technologies, driven no doubt because his two sons worked at Apple and Adobe. Mr. Ma spared nothing to be sure that the resulting monograph would make a difference in the design dialogue of our time, indeed inaugurate a series of subsequent monographs.

Special thanks are due Ran Weiming and Zhao Jiaoyang for their extraordinary efforts to assure a faithful translation throughout this book. Covering as much ground as this book does required many helping hands and inspiring words all along the way: Gary Anderson, Joe Brown, Nina Brown, Albert Chun, Ann Clifford, Nick Danforth, Benny Djohan, Eric Douglas, Barnaby Evans, James Fallows, Eunice Feng, Ronald Lee Fleming, Valerie Fletcher, Sophia Gao, Joanne Hiromura, Hsu Wensheng, Craig Halvorson, Molly Huang, Claire Kim, Robert Krim, Rachel Lee, Jessica Leete, Christian Lemon, Xiaoxiao Lu, Julie Lucier, Luo Qing, Yo-Yo Ma, Kathy Madden, Roy Mann, Julie Messervy, Robert O'Brien, Lydia Paine, Lynn Paine, Mallory Paine, Sumner Paine, Ran Weiming, Alison Richardson, Clarissa Rowe, Olivia Saw, Jonathan Shen, Shen Xuyao, Lewis Stackpole, Sandy Sun, Armando Tancio, Robert Tullis, Charles Waldheim, Arthur Waldron, Wang Shi, Jack Weber, Arthur Weissman, Arthur Wu, Shelly Xu, Sebastian Yeow, and Kongjian Yu, to name a few. In substance and style the author owes much to a core group of professors at Harvard College and the Graduate School of Design: Charles Harris, Peter Hornbeck, J. B. Jackson, Clare Cooper Marcus, Norman T. Newton, Eduard F. Sekler, and Carl Steinitz.

The author particularly wishes to recognize the work of Peter Harnik and his colleagues at the Center for City Park Excellence in the Washington DC office of The Trust for Public Land.

While most books on design emphasize plans and renderings, our preference is for illustrations showing people actually using public space, if at all possible. To supplement his own photographs, the author has relied on the power of the internet to provide royalty-free images in the public domain, predominantly from wikimedia.org, but also from 123rf.com and dreamstime.com. In addition he is most grateful for permission to include images at Frederick Law Olmsted National Historic Site of the U. S. National Park Service and in the historic archive of Xiaowei Ma.

Due to space limitations we could not include all the examples of excellent design, and omission is in no way indicative of our judgment. In future editions we hope to augment examples as well as make all necessary corrections. Errors are solely the responsibility of the author.

This book is dedicated to Xiaowei Ma, whose enthusiastic support for this project made all the difference in this project, and will make a difference in the quality of public open space design in China.

序言

城市是人类聚居所产生的形态，城市为高密度的人群提供了丰富的物质生活条件（让他们舒适幸福地生活）。城市中的开放空间为人们提供了相互交往的场所，人们在此交换信息，获取知识，产生归属感。城市开放空间的品质直接影响了城市对于个人的吸引力，从而决定了城市自身的定位和属性。

Thomas Paine 先生集数十年的（逾40年的）景观设计与地产开发专业从业感悟，汇集成了《有心的城市》一书，通过对于城市和个人关系的细致观察，针对城市开放空间在东西方城市发展演变过程中所扮演的角色进行了系统化的剖析，将开放空间在城市中的地位上升到"心"的高度，生动地再现了开放空间在城市中扮演的重要角色。对于社会公众、专业规划设计人员以及城市发展的决策者了解和改进城市开放空间品质将具有重要的启迪作用。

在人类发展的历史长河中，东、西方的城市形态除了自然地理因素和防卫需要以外，社会形态（社会结构）是主导城市形态的根本因素。西方以宗教信仰对社会大众的控制下的城市与中国的君王统治下的世俗社会城市之间具有明显的差异。人们对城市开放空间的解读也有明显的差异。早在汉代中国城市中就已经大量出现为帝王服务的皇家园林、为商人巨贾服务的私家园林以及面向社会大众的寺庙园林。佛教崇尚的超尘脱俗与道教提倡的旷达随性都助长了寄情山林、道法自然之风，"不出城郭而获山林之怡，身居闹市而有林泉之乐"成为上至皇亲国戚下至城市平民的共同愿望。中国传统城市里的开放空间一直以寺庙、集市、山林这三者为要素构成，人们的生活和精神需求在这样的开放空间中得到最大限度的满足，类似《西厢记》的爱情故事也发生在这类场所便是顺理成章的事情了。到了1905年在无锡产生的中国第一个城市公园，也是将庙宇和几处私园结合起来而形成的。20世纪30年代民国时期许多新式公园在各地的兴建，其基本目的是出于对国民性的改造，用运动设施来倡导市民通过健身培养良好的生活习惯，这是当时刚刚从列强手中夺回城市发展主导权，百废待兴的中国社会背景所驱动的必然结果，城市开放空间被赋予了新的定义。

19世纪以来美国的城市发展给城市开放空间提供了前所未有的舞台，这种高强度的发展方式到了21世纪的中国被以更高的速度复制着。在社会大众对于城市开放空间到底是什么还没有完全明了的时候，成百上千座新城市便已经建立了起来。基于对美式生活方式的向往和对美式城市形态的简单理解，这些以极短的时间规划和建立起来的城市既来不及消化西方城市规划的教训和学习优秀的经验，又没有仔细聆听自己的内心，明了自己的文化传承和生活需求，在迅速改变公众生活方式的同时，大量的永久性的错误也在快速被固化下来。千城一面，交通拥堵，雾霾大面积发生，大量的人们在被城市化的同时永远地失去了故乡。未能够良好设计的城市开放空间没有将积极友善的信息传达给使用者，非人尺度的空间使一夜之间进入现代城市生活的人们更加感受到迷失和渺小。传统上以寺庙、集市、山林为主而构成的城市活动中心被政府广场、购物中心和大型绿地所替代，穿插着喧闹而宽阔的城市干道。似曾相识的商店里叫卖着相同的商品，城市变成了超大型的连锁商店，就在 20 年前还曾存在过的故乡转瞬间成了一种模糊的记忆。

潘德明先生的努力促使我们重新审视我们的城市，让我们把关注的视角重新聚焦在城市里面的关键要素，那就是生活在城市里面的每一个人（他们在城市里生活与工作，也需要享受城市开放空间）。一个城市的活力、魅力、竞争力均来自生活在其中的人们，在大规模城市基础设施快速到位，城市规模和形象快速成型的今天，决策者和设计师的关注点应当迅速转移到使用者的角度，对城市开放空间的品质进行再次的梳理。西方大量的有关城市公共空间的建设经验都是在快速发展期过后的城市更新过程中所产生的，在西方进行更新的对象往往是百年以上的老街区和公园，而在中国则很可能变成只建好不到 10 年的新城。潘德明先生的研究用大量鲜活的实例验证城市公共空间的关键构成要素以及其对于构建城市性格的重要性，并指出了明确而翔实的操作步骤用以改善城市对人的吸引力。我们今天所面对的城市已经很难再用所谓东西方的概念来进行概括了。当上海内环线内生活着数以十万计的西方人的时候，伦敦新售的近 1/3 公寓的消费者则来自于中国大陆。当西方设计师设计的摩天楼在浦东建造的同时，纽约历史上最大规模的商业开发地块正转手到中国大陆来的开发商手中。我们生活在一个快速变"小"的地球之上，各地的差异性正快速被共性所覆盖，相同的问题在各文化各地域显现，大量的经验比以往任何时候都具备相互学习的必要性和可行性。潘德明先生的研究必将为今天大量的城市空间塑造提供极有价值的思考借鉴。《有心的城市》为快速城市化的当今世界，特别是为当今中国，就如何达成卓越的设计，搭建了一个宝贵的构架。

愿明天的人们在回忆起今天规划建设的城市的时候能感受到我们怀念故乡时的那种亲切和温馨。我们这一代人是幸运的，因为我们至少还有回忆中的故乡。

（意格国际总裁兼首席设计师）

Preface

City form derives from the group habitation of human beings. In principle, the high density of the city efficiently provides a wide range of services to allow its many inhabitants to live happily and comfortably. The public open spaces in a city are essential places for providing many of those services, where people can interact with each other, exchange information, draw knowledge, and cultivate a sense of belonging. The quality of urban public open space fundamentally determines the attractiveness of the city, indeed may be said to define urban identity and urban character.

Thomas Paine's perception and insight gleaned over four decades practicing landscape architecture and in real estate converge in *Cities with Heart*. The author examines in depth the role that urban public open space has played over the course of urban development in both the East and the West. To sum up all the important roles that urban open spaces must now play in the life of urban residents, Mr. Paine uses the term "heart", elevating urban public open space to the level of importance that it deserves. This book will inspire anyone who is concerned about improving the quality of urban public open space, whether concerned citizens, professional planners or urban authorities.

In the course of human civilization, social structure has been an essential factor in city formation, along with other factors such as environment, geology and defense. Over the centuries, as the form of cities in the East and the West exhibited certain differences, so did the role of urban public open space. In China, as early as in Han Dynasty, there were royal gardens for emperors, private gardens for the rich and temple gardens open to the public. Buddhism advocated a detachment from the trivial life, and the Taoism encouraged people to follow their convictions while cultivating serenity and broadness of mind. Both cultivated people's love of mountains and woods and knowledge of the laws of nature. Individuals regardless of their social status shared a desire to "enjoy mountain and wood scenery without the trouble of leaving the city, interact with spring water while in the middle of busy streets". Temple grounds, fair grounds and mountain woods were traditional Chinese public open space archetypes in which people's social and spiritual needs were accommodated to the full. It is appropriate that the famous Chinese love story *Romance of the Western Chamber* took place in public open space. The first truly Chinese urban park was a renovation and restructuring of a temple garden and several nearby private gardens in Wuxi, in 1905. During the 1930s many modern parks appeared in China. The Republic of China authorities of that era hoped to revive a devastated nation by providing exercise facilities to cultivate the habit of exercise so that the people could lead healthier lives. It was an inevitable reaction to the colonial period after the Chinese regained autonomy in city development. Urban public open space was henceforth embedded with new and profound meaning.

American urbanization in the 19th century brought an unprecedented era of rapid development of urban public open space. This was repeated even more intensively in 21st century China. Thousands of new towns and cities have been built with only superficial knowledge about urban public open space. Most of the development has been based on the people's yearning for an American life-style and the most rudimentary understanding of American urban form. On the one hand, we Chinese were planning and building cities in such haste that we had no time time to digest advanced Western urban planning theories or learn from their experience, and mistakes. On the other hand, we didn't listen to our own hearts or take to heart our need to perpetuate our extraordinary cultural legacy. Given the fast pace of change in peoples' life-styles, a lot of mistakes with long-lasting consequences have been solidified on the ground. New cities all look alike, suffer from traffic jams, and are shrouded in a haze of air pollution. Poorly designed urban public open spaces fail to impart a positive message to people. Vastly overscaled spaces are so forbidding that people feel lost and insignificant. The traditional urban activity areas of temple, fair and mountain woods have been replaced by the municipal plaza, shopping mall and gigantic green space divided and surrounded by noisy and broad city boulevards. Cities have become big chain stores where the retail products are all déjà vu. The hometown which was so dear to our heart twenty years ago has become a distant memory.

Thomas Paine's book inspires us to examine the public open space in our cities all over again, and focuses our attention on the most essential factor of the city—each and every individual who lives and works there, and needs open space to go to. The vigor, charm and competitiveness of a city are bestowed by its people. Today large-scale infrastructure can be constructed quickly, city scale can expand rapidly and the city image can be transformed overnight. Decision makers and designers should take a much closer look at the quality of urban public open space and shift their focus to the end users. In the Western experience with urban development, a period of fast-paced development was followed by a period of urban renovation. While Western cities may see the renovation of blocks and parks a hundred years old, we may have to renovate blocks and parks after less than one decade. Thomas Paine includes enough examples of recent work and emerging best practice to show vividly the key components of urban public open space and makes a compelling case for the importance of urban public open space in creating urban character, and its potential for advancement beyond anything we have seen. He elaborates clear and detailed guidelines on how to improve the design quality of different urban public open space types. Cities in our era cannot be simply categorized as Eastern or Western cities. There are a hundred thousand Westerners living in the inner ring area of Shanghai; one third of the condo buyers in London are from mainland China; the skyscrapers built in Pudong Shanghai are designed by Western designers; the largest commercial development ever in New York history was recently transferred to a Chinese developer. We live on a globe, which suddenly has become very small. The local character of various regions is replaced by global homogeneity. Similar problems arise in different cultures and regions. It is more urgent than ever before, and more feasible, to learn from our mutual experience. *Cities with Heart* provides a most valuable framework for understanding how to achieve design excellence in our era of rapid global urbanization, nowhere more concentrated than in China.

I sincerely hope that when future generations recall our current urban planning many years from now, they too can feel the intimacy and warmth that we feel when we recall our hometown. We are a lucky generation, for at least we have our hometown in memories.

Xiaowei Ma
(Founder and president of AGER Group)

目录 | Contents

鸣谢
Acknowledgments

序言
Preface

导言
Introduction

012 第一章 | 人民公园的起源
    Chapter One | Origins of the People's Parks

068 第二章 | 为什么城市开放空间比以往更重要
    Chapter Two | Why Urban Public Open Space Matters More Than Ever

096 第三章 Chapter Three | 规划原则 Planning Principles

118 第四章 Chapter Four | 设计原则 Design Principles

152 第五章 Chapter Five | 城市广场、市区公园、大型公园、绿道及社区公园的设计指南
Design Guidelines for Civic Plazas, Downtown Parks, Large Parks, Greenways and Neighborhood Parks

196 第六章 Chapter Six | 齐心协力展望空间的未来发展
Taking Heart, a Vision for the Next Level of Excellence

230 附录 A Appendix A

235 附录 B Appendix B

导言

中国和西方国家都把市中心比作"城市的心脏"——这个我们常常提及的词语，听起来似乎很严肃，却不见得有人真正意识到其重要性。也许，是时候该我们去认真呵护城市的"心脏"了。"心脏"本意为人体主要器官之一，而后其意被引申至其他领域，包括"城市心脏"，意为市中心。但无论其意义如何推演，都保留了它另一个主要的意义："情感"。

"城市心脏"的开放空间相当珍贵，它备受日常访客、附近居民、商务人士、游客、老人、孩子以及每个人的青睐。因此，我们甚至可以说，市中心开放空间牵动着"人们"的心。调查显示，人们对设计优秀且维护良好的公共空间情有独钟，他们与这些地方形成了一种美好的情感纽带，并将其视为生活的一部分，对这种体验恋恋不舍。能为人们提供这些设施的城市就是"有心之城"，这样城市的领导者已经将此当作自己庄严的使命，为公众创造免费开放的城市公共空间，打造一种独特的体验，以丰富市民生活，而这种体验在家中或办公场所是无法实现的。现代城市甚至作出更多努力，建造完整的绿色基建体系，包括绿色走廊、湿地和自然保护区、邻里公园系统，这些形成了城市的"肺"、"肾脏"和"心脏"，赋予城市生命力。那些竭诚为市民提供优质城市空间的政府官员，都是凭着对人民的感情，在用心地做事情。

面对全球气候变化及珍贵资源可持续使用等挑战，从真正意义上说，提供充足的城市开放空间是问题的核心和关键。

意格团队已在中国及全球其他地区的城市设计了不少优秀的公共空间，在解决市民忧虑的同时，也维护了他们对公共空间的使用权。我们相信，无论是政府官员、专业设计师或是来自其他行业对美好城市生活的倡导者，都会通过本书重新认识各地城市领导者该如何慎重管理城市的资源，不负众望地担起责任，为市民提供其应得的优质城市公共开放空间，并以此而自豪。本书内容虽总体而言是以讨论城市空间为中心，但也着眼宏观，从省级和全国角度思考，虽重点讨论公共开放空间，不是开发商项目，但是有意为民众提供公共开放空间的开发商一定会在书中得到启示。

通观本书，"设计"特指为人类的活动对空间的塑造。关于地下的工程系统和结构设计，例如：雨洪排水、土木工程等其他工程学科，虽然对宜居城市的建立至关重要，却不在本书的讨论范围内。工程技术是必要的，但它只解决了技术层面的问题，不能补充充满生活气息的元素，这也就不足以保证城市开放空间的优质。

全球信息化革命，特别是地理学的信息革命，使得全球范围内的衡量和比较成为可能，这种趋势意味着，优秀的设计将以更快的速度扩张，大大节省了当地的珍贵资源、时间和金钱。不同于一般的手册和导则，也有别于那些一味堆砌与真实世界里真实人物真实照片不符的效果图的设计类书籍，本书只在最后一章使用了效果图。且有着极为鲜明的观点和立场，丰富的信息，极强的可读性，可帮助读者更真切地评定现实中、PPT 汇报文件及网站上接收到的相关信息，也能更好地去了解城市规划和设计过程更为乐观的发展方向。规划和设计的细节步骤：从数据采集到场地分析、备选方案设计、敲定方案方向、深化设计、材料陈设研究、与工程师和其他专业人士的协作、施工图阶段、投标、施工、后期施工进程及管理等对于行业手册或导则的编撰而言，都是颇具价值的题材，书中提供了一些相关内容细则阐释，旨在帮助读者透过树木体会森林，见微知著，不论他们在各自旅途的哪一站，都能受到鼓舞和启发。

Introduction

People in China and the West each use the same meaning-laden word to describe downtown: heart. We all talk about the heart of the city without even thinking about what we are saying, and yet we should, now more than ever. "Heart" is a word that took the original meaning of an essential bodily organ and extended it to other realms, including cities, but never losing its other key sense, compassion.

The heart of the city is where open space is most precious, and most appreciated by daily visitors, residents, business people, and tourists, seniors, children, virtually everyone, so we may simply say "the people". It has been shown that people form a positive emotional bond to places where the urban public open space is so abundant and so well designed and maintained that people gravitate to it and make it a part of their life, and want to return to it again and again. Cities which provide such amenities are cities with heart. In such cities, the leaders have made it their solemn duty to provide for the people urban open space open to all, free of charge, to enrich their lives in ways impossible at home or in the workplace. Cities now go further, and provide greenway corridors, protected wetlands and nature preserves and neighborhood parks in a system of green infrastructure that is the lungs, kidney and heart of the city combined. Public officials who deliver the people excellent urban public open space are also, whether they like to admit it or not, acting with compassion.

As we confront the challenges of global climate change and sustainable use of precious resources, in a very real sense, providing sufficient urban public open space is truly the heart of the matter.

AGER is proud to offer concerned citizens in China and globally its case for the people's right to excellent open space design in the cities where they live. We trust that readers—whether urban officials, design professionals, or advocates for better urban life from across all fields—will be challenged and inspired to think anew about how urban leaders everywhere should be careful stewards of the resources entrusted to them, fully live up to the responsibility that comes with providing the people with the open space that they deserve, and take pride in delivering excellent urban public open space design. There may be lessons here for open space thinking at the provincial and national level, but that is not the focus of this very much city-centered book. Likewise, developers who provide private open space accessible to the public may find our approach useful, though development projects are not the focus of this very much public-space focused book.

Throughout this book, "design" refers specifically to the shaping of spaces for human activity. The design of invisible underground systems and structures, such as storm sewers or tunnels, though absolutely essential to making cities livable, are outside the scope of this book. So too are many other disciplines within engineering besides civil engineering. A world designed as solutions to engineering problems and nothing more misses much of what makes life worth living. Engineering is necessary, but it is not sufficient to guarantee excellent urban public open space design.

The revolution in access to global information, not least geographical, ushers in the ability to measure performance and compare results globally. The trend also suggests that best practices in design will spread more quickly, saving precious local resources, time and money. Rather than a handbook or a manual, this book plays an advocacy role. It is designed to stand apart from the many books overwhelmingly illustrated by renderings at the expense of real photographs of real people in the real world. We use renderings only in the final chapter. Armed with this informative and readable book, readers will be able to evaluate more substantively what they see on the ground, in Powerpoint presentations and web sites, and be better informed about what the process of planning and design in their communities ought to aspire to. The detailed steps in that process, from data collection to site analysis, generation of schematic design alternatives, selection of a preferred design alternative, further development of that design, research into materials and furnishings, coordination with engineering and other disciplines, construction documentation, bidding, construction, and post-construction operation and management, are all the worthy subject of a laborious manual or handbook. This book provides the context for making sense of their details. It is intended to help readers see the forest through the trees, to inspire and encourage them, wherever they are on their journey.

人民公园的起源
Origins of the People's Parks

Chapter One
第一章

我不得不承认，民主的美国最无可比拟的就是人民的公园……并且这所有景致宜人的游乐场地都毫无保留地供人民享用。
——弗雷德里克·劳·奥姆斯特德，论伯金海德公园，1852年
I was ready to admit that in democratic America there was nothing to be thought of as comparable with this People's Garden...And all this magnificent pleasure-ground is entirely, unreservedly and for ever the people's own.
—Frederick Law Olmsted on Birkenhead Park, 1852

依我之见，每个城镇都应该有一个公园……一个永久的共同财产，供教育和娱乐之用……瓦尔登的树林作为公园永远保留下来，而瓦尔登湖则镶嵌其中。
——亨利·戴维德·梭罗，1859年
I think that each town should have a park...a common possession forever, for instruction and recreation...All Walden wood might have been preserved for our park forever, with Walden [Pond] in its midst.
—Henry David Thoreau, 1859

图 1.1
雅典广场（阿果拉），雅典
Agora, Athens

雅典广场一侧的遮阴柱廊为人们会面提供了庇荫场所。会面的人们也包括斯多葛学派哲学家和他们的学生。纪念柱是阿塔罗斯二世赠予城市以表对于当时在那里接受教育的感激之情。柱廊园在1953-1956年的时候再经改造，现在已经成为博物馆。
A covered portico along one side of the Greek marketplace (or agora) of ancient Athens, the Stoa provided cool shade for people to meet, including "Stoic" philosophers and their students. The monumental portico was the gift of Attalus II of Pergamon (ca.159-138 BC) to the city in gratitude for the education he had received there. Reconstructed in 1953-6, the Stoa is now a museum. (Credit: Ken Russell Salvador)

在世界城市发展的数百年历史里，或许没有什么比可自由进出的公共开放空间的质量和数量更能作为城市生活衍变的度量衡了。开放社会和开放空间齐头并进，共同进化。当今中国正在经历文化转型，这非常值得研究并从多方面效仿。在中国的文化转型中，城市公共开放空间的开发将占据相应的地位，且会为世界城市发展史树立新的里程碑。

我们皆活在当下，所以很容易忽视这些辉煌的里程碑，一直以来，是它们承载记录了城市衍变的进程。我们终日忙碌，鲜有时间审视历史发展的轨迹；这是何其不幸。过去的成功以经历无数质疑而得到印证的永恒原则激励着我们。受到激励的人们甘冒风险追随自己的愿景，这转而激励了今天的梦想家，使他们创造出了现今的人居环境，打造了我们向往的城市公共开放空间。

开放空间的第一个里程碑要追溯到2500年以前，其间的历史遗珍浩如烟海，在此限于篇幅，只能浮光掠影、断难一一尽述。但本章亦不会只关注个别知名私家别墅或皇家园林的景观设计——无论是西方的凡尔赛宫还是东方的颐和园或西安旧址的大明宫，都不是此章的重点。准确来讲，本章旨在探究城市规划和景观设计领域与*城市公共开放空间*的交集。

中国城市公共开放空间的历史应该与西方一样源远流长。毋庸置疑，杭州西湖风景区的形成就是其历史悠久的最佳佐证之一。在西方，回溯至古典文明时代，古希腊集市和古罗马广场的出现都是因为欧洲城市人口集聚，从而催生了对城市中心开放空间的

Over the centuries of global urban history, there may be no better metric for the progress in urban living than the quality and quantity of freely accessible urban public open space. Open societies and "open space" have evolved hand in hand. Now that China is truly undergoing a cultural transformation that deserves to be studied and in many ways emulated, the dedication to urban public open space will take its rightful place in China's cultural transformation and add new milestones to global urban history.

We live so much in the present that we too easily lose sight of the extraordinary milestones that have marked this progress so far. We are usually so busy that we do not take the time to review the course of history. That is unfortunate. Past triumphs can inspire us with their introduction of timeless principles that originally attracted their full measure of skeptics. And the inspired individuals who risked much to follow through on their vision can in turn inspire today's visionaries to provide urban public open space for people the way we live now or would choose to live if only we had the chance.

Consider this chapter suggestive of a much larger legacy of spaces for people than it has room to cover adequately, since the first milestone occurred 2,500 years ago. Nor is this the history of landscape design reserved for the few: the private villa and palace gardens, whether Versailles in the West or the Summer Palace or the Xian Daming Palace

图 1.2
罗马广场,罗马(根据推测重建)
Forum, Rome (conjectural restoration)

罗马广场起初是一个位于市中心并以帝国名称命名的集市。在公元前 600 年 - 公元 300 年间逐步地发展成一个被纪念碑装饰、公共建筑环绕的市民空间。作为罗马帝国和城市本身的庆典、政治、审判和经济生活中心,罗马广场一度被称为世界上最著名的集会场所。
Originally a marketplace in the heart of the ancient city that gave its name to an empire, the Forum developed piecemeal from 600 BC to 300 AD into a civic space ornamented with monuments and surrounded by public buildings. As the center of ceremonial, political, judicial, and commercial life of the city and the Roman Empire, the Forum has been called the most celebrated meeting place in the world. (Credit: Amadscientist, model by Lasha Tskhondia)

图 1.3
庞贝广场,意大利
Forum, Pompei, Italy

庞贝广场位于那不勒斯海湾附近罗马度假城镇中心的主要十字路口,广场被设计成矩形的形状,周围围绕着公共建筑,和维苏威火山连成直线。这个火山在公元 76 年的时候爆发了,不幸地吞没了这个城市的第一层楼。
At the main intersection in the heart of the Roman resort town near the Bay of Naples, the Forum was designed as a rectangle surrounded by public buildings and aligned with Mount Vesuvius, whose eruption in 76 AD preserved most of the ground floor of the ill-fated city. (Credit: GuidoB)

in China, are not our focus. Rather, this chapter examines what we now call the disciplines of urban design and landscape architecture where they intersect with *urban public open space*.

It may well be that the history of urban public open space in China extends back as far as it does in the West. Certainly we see evidence as far back as the creation of West Lake in Hangzhou. In the West, as far back as the Classical Age of the Greek agora and Roman forum, the concentration of people in European cities demanded open spaces in the heart of the city distinct from sacred enclosures (*temenos*) where people could meet, trade goods and ideas, witness public events, celebrate, even participate in democracy. Cleisthenes introduced democracy in the agora in Athens in 505 BCE (1.1). Provided by the government, free of charge and open to all citizens, these paved spaces surrounded by public buildings in the heart of the city forged in people, whether they could vote or not, a sense of belonging, a common destiny and a bond across all classes that we call *communitas*. Providing public settings for *communitas* remains a worthy goal to this day. The conception of the ideal form of the agora and the forum made great strides toward ever-increasing order. The original Forum of Rome (1.2) owes its irregularity to its slow evolution and early start. The importance of place, including the landscape within which a city grew up, was fully acknowledged. The Greeks, and occasionally even the Romans, who called it *genius loci* (spirit of place), seemed to feel the urgency of aligning architecture within a sacred landscape, especially sacred mountains, in its way like Chinese *fengshui*. In the rivalry among the Greek city states, the idea that the form of the sacred architecture of individual temples could be perfected by ever more subtle variations of proportion and line to manipulate optical perception soon spread, along with the spread of empire and new city building, to the design of outdoor urban space. The agora was

需求。与围合的宗教聚集点截然不同,城市中心的开放空间能为人们提供集会、货物交易及思想交流的场地,让他们见证公共事件、参加盛会庆典甚至参与民主事务。公元前 505 年,克利斯提尼在雅典集市内向人们宣讲民主(图 1.1)。这些被公共建筑围绕的铺装空间是由政府提供的,免费向所有市民开放;它们位于城市的中心,在众人(无论他们是否有选举的权利)的心里铸就了一种归属感、打造了一个命运共同体,并形成了一个维系各阶层的纽带,我们称之为*市民共同体*。为市民共同体提供公共环境至今仍是一个值得追寻的目标。对古希腊集市及古罗马广场的理想形式的构思经历了前所未有的发展(图 1.2)。古罗马广场的不规则性缘于其年代的久远以及演变的缓慢。人们充分认识到场所,包括影响城市发展的景观环境,都扮演着重要的角色。希腊人似乎意识到将建筑与神圣景观相结合的必要性,尤其是与神圣山地景观的整合,甚至连罗马人也会偶尔意识到这点,并称之为找寻"场地之灵",这恰似中国的风水。古希腊城邦在相互角逐之中总结出建筑比例和线条的微妙变化可以作用于人们的视觉感知,从而使神庙建筑单体以更为完美的姿态呈现出来。这一理论迅速传开,得到广泛运用,并随着帝国的扩张和新城的建立,被应用到了户外城市空间的设计上。有时古希腊集市周边会有带顶的柱廊,为人们遮阴挡雨,比如雅典的阿塔罗斯柱廊。公元前 5 世纪,首位著名城市规划师希波丹姆在米利都的城市规划中开创性地使用规整矩形来设计集市,使其成为新式开放空间的代表,便于民众集会、交易及举办节日庆典。我们可以肯定,当时的使用者跟现在一样,把这些封闭式的开放空间视作房间。古罗马工程师兼建筑师维特鲁威(公元前 80 年 - 公元 15 年)撰写了一本设计守则——《建筑十书》,该书共分十卷,第一卷描述了城市规划与景观设计的感知联系。随着罗马帝国向更多新城的迅速扩张,罗马人在石砖铺砌的城市公共开放空间的设计上延续并发扬其一贯采用的欧几里得几何图形风格,比如与神圣威严的维苏威火山成中轴线的矩形庞贝古城(图 1.3),或是位于约旦杰拉什

图 1.4
杰拉什古城广场，约旦
Forum, Jerash, Jordan

罗马人对于城市公共开放空间几何学设计的探索创新遍布整个帝国，例如这个椭圆形状的被整齐的柱廊围绕的广场，它的历史可追溯到公元 100 年。
The Romans explored innovative geometry in the design of urban public open space across the Roman Empire such as this oval shaped forum surrounded by a uniform colonnade, dating from around 100 AD. (Credit: Bernard Gagnon)

的柱廊围合的椭圆形圣地庭院（图 1.4）。

在古典时代，用公共艺术装饰城市空间已然成为一种传统，英雄人物的纪念雕像尤为典型，比如铸于公元 176 年的古罗马皇帝马可·奥略流斯骑马铜像。这件精美的旷世之作至今仍完整无缺地屹立于罗马（图 1.5）。极为难得的是雕像的神情里没有一丝傲慢，而是透着仁治的威仪。除公共艺术装饰外，城市空间内几乎没有其他便利设施。有时某些集市内会种植遮阴树，但这在古集市和广场里仍属罕见。

在罗马帝国崩溃之后，西方和中东的物质文明便衰退了，这是因为整个社会更专注于精神层面的宗教信仰，并忙于应对当前

图 1.5
马克·奥略流斯皇帝骑马雕像，罗马
Equestrian Statue of Emperor Marcus Aurelius, Rome

在城市公共开放空间放置纪念性的艺术品是一个至少追溯到罗马帝国的传统。这座稀有的从那个时代幸存下来的雕像纪念着这位伟大的哲学家和帝王。雕像在 1538 年从坎皮多里奥广场移到罗马广场，并被移入博物馆室内存放以防止空气污染。1981 年，它的复制品被放置在室外展出。
Commemorative art in urban public open space is a tradition going back at least to the Roman Empire. This rare survival from that era honoring the great philosopher-emperor was moved from the Roman Forum to the Campidoglio in 1538 and was only placed inside a museum, to protect it from air pollution, in 1981, replaced outdoors by the replica shown here (Credit: Radomil)

sometimes edged with a covered, colonnaded porch (*stoa*) sheltering the public from sun and rain, like the Stoa of Attalus in Athens. The first named city planner, Hippodamus introduced the first formal rectangular agora at Miletus (5th C BCE), representative of the new kind of open place designed for people to meet, trade, and celebrate special days and festivals. We can be sure that people then, as now, perceived such enclosed outdoor spaces as rooms. The Roman engineer/architect Vitruvius (ca. 80 BCE-after 15 AD) wrote *De Architectura*, a handbook of design whose first of ten books expresses a contextual sensibility anticipating landscape architecture. As their Roman Empire rapidly expanded in many new cities, the Romans continued the march toward a more powerful Euclidean geometrical order in the stone paved urban public open spaces, (1.3) like the rectangular Forum of Pompeii, aligned with the revered and feared volcano Vesuvius, or the oval *peribolos* (colonnade enclosed court) in Jerash, Jordan. (1.4)

Public art ornamenting urban space was already a strong tradition in the classical world, typically in the form of monumental sculpture of the heroic human figure, like the exquisite bronze equestrian statue of the emperor Marcus Aurelius (176 AD) that has survived intact in Rome, unusual in its air not of arrogance but of compassion. (1.5) Except for public art, amenities were few. Sometimes an agora was planted with shade trees, but trees seem to have been a rarity in agoras and forums as well.

After the collapse of the Roman Empire, material progress in the West and Middle East regressed, as society focused on a religious faith in the next world, while fending off conflict, famine and disease on earth. During the squalor of those hard times in Europe, early medieval cities walled themselves in like their Roman predecessors and struggled as economic centers lacking the resources, and perhaps even the vision, to tidy up their limited urban public open spaces and make them inviting for people beyond providing a

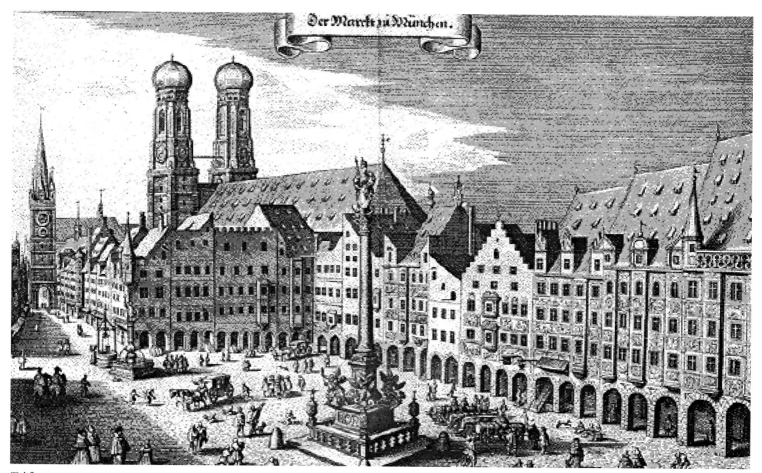

图 1.6
慕尼黑集市广场，德国，1642 年
Market Square, Munich, Germany, in 1642

广场雕刻体现出人们的自我陶醉，同时从拱形柱廊到墙上的绘画、雕塑、喷泉和铺装，也完美地展现了中世纪德国城市高品质的空间设计。
This engraving depicts people enjoying themselves and beautifully captures the high quality of open space design in medieval German cities, from the arcades to the wall art to the statues and fountains and paving. (Credit: Matthaus Merian, Topographia Germaniae, 1642)

utilitarian public fountain and wall-to-wall stone paving. The church square and town hall square provided gathering spaces in front of the architecturally ambitious sacred and secular buildings, and were usually separate from the market square, which was sometimes ornamented with an open-sided public market building, like those in the agora. The form of these spaces is sometimes merely a widening of narrow streets rather than actually rectangular or square. Typically the perimeter of the Medieval square (apart from the iconic cathedral or town hall), is enclosed by picturesque facades of attached buildings, with architectural detail framing the entrance and the windows. The peripheral facades are stylistically similar, as if the whole district is an interconnected, slowly evolving organic living thing. (1.6) Conscious beautification of Western outdoor space accelerated from the late Medieval into the early Renaissance period. By the time of the Renaissance in Eastern European cities and even small towns, peripheral walls were sometimes even ornamented with murals, as in Freiburg im Breisgau (1.7) and Wangen im Allgäu (1.8) in Germany.

的冲突、饥荒和疾病等危机。在欧洲最混乱最艰苦的时代，中世纪早期城市仿效古罗马的先驱以城墙包围自己，但却苦苦挣扎，这是因为其作为经济中心却缺乏资源，甚至是远见。他们不会想到将有限的城市公共开放空间整理妥当，吸引人群，而仅仅只是设立一座实用的喷泉，并以石头铺装城墙间的间隙而已。教堂广场和市政广场在恢宏的宗教或市政建筑前为人们提供了集会空间，它们往往跟集市广场分开。如同古希腊集市，集市广场上有时装点着一座半开敞的公共市场建筑。这些空间有时仅仅是狭窄街道的拓宽，而非实际意义上的长方形或正方形设计。典型的中世纪广场（除了地标式的大教堂或市政厅）都由美轮美奂的附属建筑物外立面围合起来，入口和窗棂饰以建筑细节，周边建筑物的外立面风格类似，仿佛整个区域就是相互关联、缓慢进化的一个有机生命（图 1.6）。从中世纪后期到文艺复兴早期，西方日渐加速对户外空间的蓄意美化。直至文艺复兴时期，东欧的城市乃至小镇，有时甚至用壁画装饰边墙，例如德国布赖斯高地区的弗赖堡（图 1.7）和旺根地区的阿尔高（图 1.8）。

在奥地利的萨尔茨堡（图 1.9）、瑞士的卢塞恩（图 1.10）及其他一些欧洲城市，中世纪时期的街景及户外空间得以保留下来，让世人研究模仿；由于保持了干净整洁的状态，这些户外活动空间给人感觉像是舒适怡人的户外房间一般（图1.11 - 图1.16）。后世欣赏它们在整体一致布局中的多样性，赞叹在前汽车时代宜人尺度下街景展开的舒缓节奏和清新风格及其带来的空间惊喜。正是这种由衷的欣赏和赞叹引发了西方20世纪后期的城镇景观运动，保留了历史街区，发展出节日市集，并带动了最近的新城市主义运动。根植于所有这些运动潮流中的是对一种理念的尊崇，即景观历史学家约翰·布尔克霍夫·杰克逊所称的"乡土"形式、尺度和格调——这是指被广泛接受的本地建筑传统。在汽车和摩托车禁止通行的地方，比如如今的波士顿昆西市场和上海新天地这些节日市集，人们摩肩接踵，其繁荣程度胜似任何一座欧洲城市从中世纪继承下来的广场。

The streetscapes and outdoor spaces surviving from the Medieval period in so many European cities like Salzburg, Austria, (1.9) and Luzern, Switzerland, (1.10) are today preserved, studied and imitated, because in their tidied-up state the outdoor spaces feel like comfortable and inviting outdoor rooms. (1.11-1.16) It is the appreciation of later generations for their variety within overall uniformity, and for the unhurried and unsullied progression of vistas and spatial surprise on an intimate pre-automotive scale, that gave rise in the latter 20th century in the West to the townscape movement, historic districts, festival marketplaces and lately the New Urbanism movement. Deeply embedded in all these trends is the veneration of what landscape historian J. B. Jackson called "vernacular" form, scale, and style—commonly accepted local traditions of building. When the cars and motorcycles are banished, today's festival marketplaces like Quincy Market in Boston and Xintiandi in Shanghai can flourish as people spaces that are as lively as in any European city that is blessed with spaces from the medieval period.

图1.7
布赖斯高地区弗赖堡市政厅广场，德国
Town Hall Square, Freiburg im Breisgau, Germany

从14世纪开始，这个城市广场就像是一个巨大的磁铁吸引着人们。弗赖堡在1120年建成，意为自由之城。
The form of this civic plaza that is still a magnet for people can be traced back to the 1300s. Founded in 1120, Freiburg means Free City. (Credit: Luidger)

图1.9
卜拉茨尔，萨尔茨堡市，奥地利
Platzl, Salzburg, Austria

这个"小广场"是典型的被良好保存下来的中世纪欧洲街景，这个小广场有一个低调的喷泉，以及给室外咖啡店的可移动的座椅。
This "small square" is typical of medieval streetscapes across Europe that have been lovingly preserved and updated, here with an understated splash fountain and movable chairs for eating outside cafés. (Credit: Andrew Bossi)

图1.8
海伦街，旺根地区阿尔高，德国
Herrenstraße, Wangen im Allgäu, Germany

海伦街是德国南部中世纪小镇的中心街道，作为公共开放空间两侧是绘画立面，并有一个喷泉雕塑。
The main street in the heart of this medieval town in southern Germany functions as the public open space and is ornamented with painted facades and a fountain statue. (Credit: Franzfoto)

图 1.10
葡萄酒广场，卢塞恩，瑞士
Weinmarkt, Luzern, Switzerland

满街飘着咖啡香，油漆刷过的立面，没有机动车的干扰，这些构成了市中心中世纪广场和谐的空间。
Café culture, painted facades, and the exclusion of vehicles together create a friendly space for people in a medieval square in the heart of the city.
(Credit: Andrew Boss)

图 1.11
巴哈拉赫，德国
Bacharach, Germany

这个位于德国西部莱茵河畔陡峭山脚下的中世纪小镇开放空间所告诉人们的是一个小镇广场，不同于城市广场，可以与花园，甚至是公共花园相结合，而不是为了人群而全部铺设铺装。
The lesson of this open space in a medieval town below a steep hillside on the Rhein River in western Germany is that a town square, unlike a city square, can accommodate a garden, even communal gardens, rather than be paved to accommodate crowds. (Credit: Manfred Heyde)

图 1.12 - 图 1.13
罗滕贝格，德国
Herrngasse, Galgenstrasse, Rothenberg ob der Tauber, Germany

在这个被良好保存下来的中世纪古城里，如画的街景被私人房屋的装饰性立面所包围，并且街道有足够的宽度同时让当地的车辆、行人和坐在户外咖啡店的人共享。
In this well preserved medieval town the highly picturesque streetscape enclosed by ornamental facades of private houses is wide enough to accommodate local vehicular traffic, pedestrians, and people sitting in outdoor café space. (Credit: Werner Berthold)

图 1.14
乌尔姆市政厅，德国
Town Hall, Ulm, Germany

可追溯到 600 年前的墙上艺术品为这个重要的中世纪纺织之城广场增加了节日气氛。
The wall art dating back over 600 years creates a festive mood for the adjacent square in this important medieval textile city. (Credit: Szeder László)

图 1.15
彼得·勃鲁盖尔，《狂欢节与四旬斋的争斗》，1559 年
Pieter Bruegel the Elder (1526/30-1569), The Fight Between Carnival and Lent, 1559

秩序在建筑回归到市民设计之前，没有一个画家能够如此生动地描述中世纪欧洲城市公共开放空间，画中展现的故事捕捉到了人们聚集在这社区中心的欢乐之情。
No painter has more vividly depicted the character of medieval European urban open space before architectural order returned to civic design. The allegory taking place in this imagined townscape captures the excitement of people coming together in the heart of the community. (Credit: The Yorck Project:10.000 Meisterwerke der Malerei. DVD-ROM, 2002. ISBN 3936122202. Distributed by DIRECTMEDIA Publishing GmbH)

图 1.16
布鲁塞尔大广场，比利时
Grand Place, Brussels, Belgium

矩形集市广场的出现可以追溯到公元 8 世纪或者更早，它们一直延续着当地居民和来访游客聚集地的繁华。这个空间被左侧集会厅富有装饰性的连续立面三面环抱，而另一面的右侧市政厅和它的高塔则是国家的标志性象征。
The rectangular market square dating back eight centuries or more continues to thrive as a venue for local residents and visitors. This space enclosed by continuous and highly ornamental historic guildhall facades on three sides and the ancient City Hall with its tall tower on the fourth is a national icon. (Credit: Vase Petrovski and Amaninto)

图 1.17
突尼斯市 Bab Suika-Suker 广场，突尼斯，约 1899 年
Bab Suika-Suker Square, Tunis, Tunisia, ca. 1899

中世纪中东的城市空间和欧洲一样为人们提供了社交场所，周围会有宗教和商业空间，但它们缺乏了欧洲地区所拥有的丰富艺术元素。图上的这张早期"麦地那"是当时这个城市的中心。
Urban public space in the Middle East in the medieval period accommodated commerce and adjoined religious and commercial space, as in Europe, but lacked the artistic exuberance of European examples, as can be inferred from this early photograph of the "medina," the ancient quarter that is the heart of the city. (Credit: Detroit Publishing Co. Collection at the U.S. Library of Congress.)

图 1.18
撒哈拉沙漠城市，马拉喀什，摩洛哥
Djemaa el Fna, Marrakesh, Morocco

随着白天的炎热逐渐退去，人们依旧会聚集在这个朴素的公共空间里，就如许多世纪以前一样。
As the harsh heat of daytime dissipates, people still gather in this unadorned public space as they have for many centuries. (Credit: Donar Reiskoffer)

图 1.19 - 图 1.20
阿契美尼德国王广场，伊斯法罕，伊朗，1839 年（右图为 2010 年）
Naqsh-e Jahan Square, Isfahan, Iran, in 1839 (and 2010)

左边这幅图是法国建筑师泽维尔·帕斯卡尔·科斯特对当时景象的描绘，画中展现了如沙漠般荒芜的景象，这与右图中如今看到的绿洲景色和熙熙攘攘的人流形成了鲜明的对比。
The left is the depiction of the scene by French architect Xavier Pascal Coste. It displays the desert-looking barrenness which contrasts strongly with the green scenery and bustling of the right image. (Credit: Arad Mojtahedi and eiaharg)

In the Middle East until perhaps a century ago urban public open space faced limitations similar to those in the West in the medieval period. (1.17) The circular plan of the city of Baghdad (762 AD) in the heart of the Cradle of Civilization predated mosques and had a central pedestrians-only open space that most likely was as devoid of public amenity as were contemporaneous cities in Europe. After the arrival of the mosque in the Islamic city, public space open to the faithful was limited to the courtyard in front of the mosque or educational center (*madrasa*). The typically treeless stone-paved square or rectangular space featured a central fountain with large vaulted and decorated entrances (*iwan*) each side. People could sit on the pavement to talk. Cities were compartmentalized into dense concentrations of modular private space, each residential unit with its own garden, but offered no collective shared park-like space. True, the challenges of a harsh climate were formidable.(1.18)

Inspired by the traditional Persian garden, the plan of Isfahan which came eight centuries later (1598) was an exception. A major urban corridor open to all people was planted with rows of chenar (*platanus orientalis*), the tree also popular in Chinese cities since the 19th century. The imperial square (*maidan*) was typically surrounded by an arcade not unlike spaces in southern Europe. (1.19-1.20)

Independent of any Western influence, for centuries China had led the East in developing in its own architectural style and tradition of urban public open space, which spread to other countries like Japan, Korea and Vietnam. Perhaps the largest city in the world in the 11th century, the Song Dynasty capital of Kaifeng, visited by Nicolo Polo, a Venetian merchant, in 1265, is recorded in the scroll painting by Zhang Zeduan as a city with a wide central open space teeming with life. (1.21) If this example is typical, the haphazard organic design of the open space inside the walled cities of China resembled that of Europe,

在中东，可能在一个世纪以前，城市公共开放空间仍面临着和中世纪欧洲相似的窘境（图1.17）。公元762年的巴格达位于人类文明的发源地中心，在清真寺传入之前，其运用的是环形都市平面设计；古城中仅有一个中央步行空间作为开放空间，且该空间极可能与同时期的欧洲城市一样缺乏公共设施。在清真寺传入这座伊斯兰教城市以后，对信徒开放的公共空间就只局限于清真寺或宗教学校（*madrasa*）前的庭院。这些庭院通常以石砖铺地，没有种植任何树木，其平面设计一般是正方形或长方形，中央有一个特色喷泉，侧面是大拱顶的装饰门洞入口（*iwan*）。人们可以坐在路面上高谈阔论。城市被密密麻麻分隔成模块式的私人空间，每栋住宅都有自己的花园，但是缺乏像公园般的集体空间。的确，难以克服的恶劣气候或许是开放空间难以成形的原因吧（图1.18）。

深受传统波斯花园布局的启发，八个世纪后（1598年），伊朗城市伊斯法罕的城市规划独辟蹊径：在一条对大众开放的主要城市廊道上种植了成排的*法国梧桐*。这一树种其实从19世纪开始也在中国城市中流行了起来。伊斯法罕的皇家广场通常被连拱廊围绕，这与南欧的空间设计不谋而合（图1.19 - 图1.20）。

几千年来，在没有受到任何西方影响的情况下，中国一直引领东方发展自己独有的城市公共开放空间的建筑风格和传统，并且传入了其他国家，比如日本、朝鲜和越南。威尼斯商人尼科洛·波罗于公元1265年造访的北宋国都开封也许是11世纪世界上最大的城市。北宋画家张择端的长卷风俗画《清明上河图》为我们描绘了这个城市里宽阔的开放空间的市井生活（图1.21）。如果这是一个典型案例的话，那么中式城墙内看似杂乱无章的有机开放空间的设计就与欧洲的设计风格不谋而合，但中心焦点并不在庙宇和衙门上。集市或者广场通常在城门里面，其熙熙攘攘的景象是当时任何欧洲城市都无法比拟的。的确，在西方的字典里面没有"热闹"一词。中文"热闹"意味着生气勃勃、宛如过节、快乐欢腾，恰如其分地表现了人群聚在一起的活跃场面。中国文化里传统节日的民俗活动或仪式都是在公共空间里举行的，

图 1.21
张择端，清明上河图，开封，清朝
Zhang Zeduan (1085-1145), *Along the River during the Qingming Festival* (detail), Kaifeng, Qing Dynasty

在中国宋朝的首都，这个 11 世纪世界上最大的城市的市中心，是一个生机勃勃的开敞城市公共开放空间。
In the heart of the Song Dynasty capital of China, the largest city in the world in the 11th C, is a wide urban public open space teeming with life. (Credit: National Palace Museum Collection)

图 1.22
西湖，杭州
West Lake, Hangzhou

这张 20 世纪初明信片所拍摄的漫步西湖的景色，也许这是世界上最早的绿道走廊了，西湖是中国最有名的景点（见图 5.38 - 图 5.39）。
The promenade captured in this early 20th C. postcard captures perhaps the first greenway corridor in the world, showcasing China's most famous scenic spot (see also 5.38-5.39). (Credit: postcard by unidentified photographer, Xiaowei Ma Collection)

图 1.23
桑德罗·波提切利（1445-1510 年），《卢克雷蒂娅的故事》，1496-1504 年
Sandro Botticelli (1445-1510), *Story of Lucretia*, 1496-1504

作为 1200 年以前的一个殉难故事的背景，这位意大利文艺复兴艺术家描绘了一个经典比例下的理想罗马城市开放空间。一些同时代的建筑师甚至尝试去把它建造出来。
As the setting for a martyrdom story from 1200 years earlier, the Italian Renaissance artist imagined an idealized Roman urban space of classical proportions even as his contemporary architects tried to build the same. (Credit: Isabella Stewart Gardner Museum Collection, Boston)

比如元宵灯会等等，这比每周有休息日的概念要早好几百年。公元 1280 年，尼科洛的儿子马可·波罗从南宋都城杭州回到故乡意大利，仍感叹于杭州的美景，特别是西湖，其景观设计及水文管理远远超越了西方国家当时的技术（图 1.22）。两位曾在杭州为官的大文豪，唐代的白居易和北宋的苏轼都为西湖留下了不朽的诗篇，为其增色不少。西湖成了人们的必游景点，频繁出现在民间故事和话本里。其湖畔层峦叠翠，山上宝塔倒映湖中，远

but without the central focus on a place of worship or town hall. Plazas or squares (广场 or *guang3chang3*, "spacious place") often lay inside the city gates, teeming with life beyond anything comparable in the West. Indeed, the West lacks the term 热闹 (*re4nao4*, literally "hot and noisy"), meaning lively, festive, and happy, exactly the mood of throngs out to have a good time. Generally, Chinese culture provided periodic festivals in which people shared timeless rituals in public space, like

the Lantern Festival, centuries before the notion of weekly leisure. In 1280 Marco Polo, son of Nicolo, returned to Italy from Hangzhou, the other Song Dynasty capital, impressed by its beauty and in particular West Lake, an act of landscape design and water management well ahead of anything similar in the West. (1.22) Owing much to two enlightened poet-governors—Bai Juyi（白居易）in the Tang Dynasty ca. 800 and Su Shi（苏轼）in the Song Dynasty ca. 1090, West Lake became a must-see scenic spot, storied in folklore, the landscape offering views of pagodas in hills near and far reflected on the expanse of water animated by ducks and hundreds of pleasure boats for hire. What Polo recorded was in essence a public park, perhaps the first such description anywhere.

> *Near the central part are two islands, upon each of which stands a superb building with an incredible number of apartments and separate pavilions…When the inhabitants of this city…resort to one of these islands…they find ready for their purpose every article that can be required…provided and kept there at the common expense of the citizens, by whom also the buildings were erected.* [1]

Travelers from all over China and beyond enjoyed boat rides and contemplative strolls along its causeways and perimeter paths. The impulse to visit "scenic spots" in China anticipates by centuries the later parks movement and the delight in "scenery" in a future world of urban public open space and national parks. After the creation of the Republic of China, one of the first parks was established on the West Lake, which rightly deserves recognition for its seminal contribution to the history of people's open space.

近错落，水面上群鸭嬉戏，游船画舫熙来攘往，摇曳灵动。马可•波罗游记里所描述的本质上是一座公园，这可能是此类描述最早的史料记载：

> "靠近湖心处有两座岛，每座岛上都有一座美丽华贵的建筑物，其中有无数房间，辅以亭台楼阁……当本城的居民举行婚礼或豪华的宴会时，就来到其中一座岛上……凡他们所需要的东西，如器皿、桌巾台布等这里都已预备齐全。这些东西是以市民公共费用备置的，岛上的建筑也都是市民修筑的。" [1]

来自中国各地和海外的游客都喜欢泛舟湖上，或徜徉于静谧的跨湖堤道和湖边小径上。早在公园运动和标榜自然景观的城市开放空间及国家公园兴起的几个世纪以前，游览风景名胜的欲望在中国民间就已经普遍存在了。中华民国创立以后，中国最早的公园之一就建在西湖。西湖在公众开放空间发展史上留下了辉煌篇章，被世界认可乃实至名归。

北京陶然亭公园的历史可追溯至元朝（1271-1368 年）。园内的慈悲庵始建于元代，又称观音庵，当时是供大众休闲及善男信女参拜观音菩萨的一座城市园林。清康熙三十四年（1695年），监管窑厂的工部侍郎江藻在慈悲庵内建亭，并取字于唐代诗人白居易抒发在自然中休憩的畅然之情的诗句，为亭题名"陶然"。这座小亭多年来一直成为封建士大夫、文人墨客聚会游览的地方，至今留下很多诗文作品。陶然亭公园于 1952 年改建成集古典名亭建筑和现代造园艺术于一体的市政公园，湖岸垂柳成行，亭台隔湖相望。此公园至今仍是北京市内被游览次数最多的城市公共开放空间之一。

而欧洲开放空间的发展历程，就像科学和发明一样，要等到中世纪黑暗时代结束以后、文艺复兴之初，西方各国的中产阶级兴起后才迎来了新生。随着贸易条件的改善，私人财富的积累增加了对世俗性文化艺术，如绘画、雕塑、建筑及造园艺术的私人赞助，以及对长期被忽略的古典设计的重新发现。艺术家们用丰富的想象力来构思完美无缺的城市空间透视图，并加以大众熟识的英雄人物作为点缀（图1.23）；甚至连城镇领导人也开始美化市集广场。这些广场或在市政厅前，或在大教堂前，或在被长期遗忘的古罗马竞技场遗址上。它们不仅仅是为人们提供买卖的场所，也是发布通告和参与庆典活动的平台。似乎是为了追忆罗马奥古斯都时代的光辉岁月，早期的意大利城镇运用开放式广场（piazza）的概念，对大众开放使用。事实上，这些意式广场没有几座是方形的，大多甚至并不完全符合欧几里得对长方形的定义，而且其空间都经历了几个世纪的调整演变。

威尼斯的圣马可广场在世界城市开放空间榜单上总是名列前茅。拿破仑把这1.3公顷的空间称为欧洲最美丽的会客厅。在一幅1496年的油画里，我们可以看到人们在这个空间里怡然自得（图1.24）。这个空间是"L"形的，两部分的汇聚点是一座*钟楼*，稍长的那部分看似是矩形的，肉眼很难觉察其空间的不规则性，而是被两侧的拱廊所吸引，更多地注意到圣马可大教堂和道奇宫表面的强烈反差（图1.25）。较短的那部分是座小广场，占地0.4公顷。在小广场上放眼望去，水面风光无限，海岛上的圣乔治大教堂被石柱廊加上了边框，而这些石柱其实是城市徽章的基座。威尼斯的其他广场跟圣马可广场的设计语汇相同。我们不能忽视的一点是水城威尼斯的大运河和小河道纵横交错，共同组成了别样的通达性高的公共开放空间。大小船只穿梭其间，每逢喜庆节日，就有大批市民驻足观赏。

我们印象中伟大的意大利广场都是城墙之间的石砖铺地，这种经典印象成了永恒，似乎很难让人接受在意大利广场上搭配软质乔木。意大利锡耶纳坎波广场占地1.2公顷，呈贝壳状，古朴优雅却张力十足，依偎在建城的三山环抱之中。一共有11条街道汇聚到广场，从其中任意一条进入都会立刻被广场空间所吸

What is now known as Taoranting (Joyous Pavilion) Park in Beijing was a place of public retreat as early as the Yuan Dynasty (1271-1368) known as the Kuanyin (Goddess of Mercy) Nunnery. Visitors became so numerous by 1695 that Municipal Council Officer Jiang Zao took the unusual step of building the evocatively named Joyous Pavilion, the name taken from a famous poem expressive of the delight of relaxing in Nature. Poetic inscriptions on the Joyous Pavilion commemorate numerous eminent poets, writers, celebrities and famous calligraphers. In 1952 the setting evolved into a full-fledged public park with a willow-lined lake and many pavilions. The park remains a heavily visited urban public open space to this day.

Open space ambition in Europe, like science and invention, had to wait until the light at the end of the Dark Ages, the dawn of the Renaissance, truly a rebirth of human impulses long dormant in the West that now for the first time gave rise to the middle class. As conditions for trade improved, the private accumulation of wealth fostered the rise of private patronage, the secularization of the arts—from painting to sculpture to architecture to garden making—and the rediscovery of long-neglected classical canons of design. Artists imagined immaculate urban spaces in perfect perspective populated with heroes from the people's shared narrative tradition, (1.23) even as the city fathers began to beautify the utilitarian market squares in front of town halls or cathedrals or on the site of long forgotten Roman arenas as places for people not only to buy and sell goods but to gather for proclamations and participate in celebrations. As if recalling the grandeur of its Roman Augustan age, Italy early adopted the concept of the *piazza* (open square) as a place open to the people. Few were in fact square; most were not even purely rectangular in a strictly Euclidean sense, and all evolved with adjustments and additions over the centuries.

图 1.24
金泰尔•贝利尼（1429-1507 年），《圣马可广场的处决十字架》，威尼斯，1496 年
Gentile Bellini (1429-1507), *Procession of the True Cross at Piazza San Marco, Venice*, 1496

在这个最早描述供人使用的城市公共开放空间的西方作品中，我们可以看到广场这种我们在现今熟知又热爱的形式早在五个世纪前就形成了它大型公共广场的雏形。
In one of the earliest Western depictions of peopled urban public open space, we can see that the Piazza which we know and love today already reached its essential form as a grand civic plaza five centuries ago. (Credit: Gallerie dell'Accademia Collection, Venice)

图 1.25
乔瓦尼•安东尼奥运河，圣马可广场，威尼斯，约 1730 年
Giovanni Antonio Canal ("Canaletto") (1697-1768), *Piazza San Marco, Venice*, ca. 1730

在贝利尼之后的两个世纪中，广场的边缘增加了与之匹配的柱廊，也重新进行了铺装，无论是否有宗教的庆典，广场都吸引着人们前去漫步和休息。
Over two centuries after Bellini, the edges of the Piazza now have matching arcades and the pavement has been refined, attracting people to stroll and relax whether there is a religious festival or not. (Credit: Fogg Art Museum Collection, Harvard University)

Piazza San Marco in Venice stands near the top of any list of iconic urban public open spaces for people. Napoleon called the 1.3-hectare space the most beautiful drawing room in Europe. In a painting from 1496, we see it lovingly depicted as a people space.(1.24) The space is L-shaped, the two legs meeting at the pivotal *campanile* or bell tower. The longer leg of the L-shaped space is pseudo-rectangular: the eye hardly notices the irregularity, and dwells on the bilateral arcades leading the eye to the richly contrasting façades of the Cathedral of San Marco and the Doges Palace. (1.25) The shorter leg (Piazzetta or *little* piazza, 0.4 hectares) reveals a stunning water view of the island church of San Giorgio framed by columns that serve as pedestals for city insignia. Many of the other squares in Venice share the same vocabulary as Piazza San Marco. Nor should we overlook the role of canals as accessible open space in a city like Venice with its Grand Canal and many intersecting canals, always teeming with boats and often hosting festivals attracting crowds of onlookers on the periphery.

We are so used to the timeless forms of the great Italian squares paved from wall to wall that to imagine them with trees seems like sacrilege. The austere yet theatrical Piazza del Campo (paved in 1413) in Siena is a 1.2-hectare shell-shaped square nestled between the three hills on which the city is built. (1.26) Entering from any of its eleven converging streets, one grasps the space instantly. Marble stripes radiate out from the focal city government building (Palazzo Pubblico) like rays from the sun or a scallop shell.

The exuberant Piazza della Signoria in Florence, Siena's rival city, with its collection of oversized sculpture and the arcade-framed colonnade vista to the riverfront is too complex to grasp at first glance. (1.27) In these spaces in both Siena and Florence, fountains for public water supply became heroic works of public art, but Florence wins for provision of public art and amenities like the *loggia* (recalling a Greek *stoa*) and the Uffizi (government offices), a grand gesture of urban design (1560-74) by architect Giorgio Vasari for Cosimo de Medici, a giant among civic art patrons.

Beginning in the Renaissance, the liberation of the individual visionary from medieval canons of anonymity henceforth allowed a grateful world to give credit where credit was due. Let their example inspire us as well. Often a successful space is the result of a long wait after the original design, or the collective work of a succession of talented designers working over generations. The then thousand-year-old equestrian statue of Marcus Aurelius mentioned earlier was so revered in the Renaissance that when Pope Paul III had it moved to the Capitoline Hill, he commissioned Michelangelo to redesign that space as the heart of Rome. Let no designer today complain of a site too difficult to create a masterpiece. After Michelangelo's makeover of the hilltop hodgepodge of buildings, the ugly ducking emerged as a swan for the ages. The Piazza del Campidoglio (begun 1536) is widely admired as perhaps the greatest masterpiece of urban design ever created. (1.28-1.30) It is only 0.4 hectares in area. Michelangelo was the first architect to organize exterior space with the same order as buildings,

图 1.26
坎波广场，锡耶纳
Campo, Siena

从市政厅的塔楼的顶部看去，（塔楼的阴影落在广场上），坎波广场炫耀着它的标志性的贝壳的形状，它略微的不规则就是它的魅力之一。广场上禁止车辆通行，这给户外用餐者提供了一个相当开敞的空间。喷泉是广场上唯一的公共艺术小品。
Seen from the top of the tower of the Town Hall whose form casts a shadow over the piazza, the Campo reveals its iconic shell-like form, whose slight irregularity is one of its charms. Car-free, the space provides a majestic setting for quiet open air dining. The only public art is the fountain.　(Credit: anonymous)

引（图 1.26）。大理石条纹以市政厅大楼为焦点，向四周发散，犹如四射的阳光，又如贝壳褶皱的纹理。

充满生机与活力的贵族广场位于锡耶纳的竞争城市佛罗伦萨。广场里有大型雕塑，拱廊环绕四周，从乌菲齐到河岸的框景难以一眼尽收（图 1.27）。两个广场空间里的喷泉都是公共艺术的典范，但是佛罗伦萨在公共艺术和设施上更胜一筹，它有能让人想起*古希腊柱廊*的凉廊，有建筑师乔治•瓦萨里的杰作乌菲齐政府办公室。后者是乔治应当时城市艺术赞助巨头科西莫•德•美第奇之邀于 1560 年至 1574 年创造的城市设计杰作之一。

从文艺复兴时期开始，个性与个人远见从中世纪默默无闻的教条中解放出来，社会也更开明地承认个体的贡献。这个过程至今还激励着我们。一个成功的空间从最初的设计到最终被公众认可通常需要很长一段时间，或者要经历几代才华横溢的设计师跨时代的倾力合作。前面提到过的古罗马皇帝马可•奥略流斯骑马铜像在文艺复兴时代备受推崇，当教皇保罗三世将它移到卡庇多神殿山时，他委托米开朗琪罗重新设计新址，使其成为罗马城的中心。后世的设计师若是要抱怨场地制约条件太多，想到这个案例就该三缄其口。在米开朗琪罗对山顶大杂烩式的建筑群进行清理之后，丑小鸭华丽转变成了白天鹅，并成为几个世纪的经典。始建于 1536 年的这座坎皮多里奥广场被公认为城市设计的瑰宝，（图 1.28 - 图 1.30）这个占地面积只有 0.4 公顷的广场空间赢得了世人的景仰和礼赞。米开朗琪罗也是第一个按照建筑秩序组织外部空间的建筑师，可惜他在有生之年都未能亲眼看见这一设计的完成，这一伟大壮举直到 20 世纪才得以实现。骑士"坐镇"广场中央，四周是精心排列的雕像群，地面精巧复杂的几何椭圆图案铺装以骑士雕像为中心呈螺旋状向外发散，跟周围梯形的建筑表面协调吻合。广场几何铺装直到 1940 年才按照原设计得以

图 1.27
贵族广场，佛罗伦萨
Piazza della Signoria, Florence

丰富的公共艺术装饰着这个位于城市中心的公共开放空间，这里作为意大利文艺复兴的诞生地，人们前来漫步闲坐、交谈或者观看表演。
Profuse public art adorns this public open space in the heart of the city synonymous with the birth of the Italian Renaissance, as people relax by strolling, sitting, conversing, or watching a performance. (Credit: private collection)

图 1.28 - 图 1.30
坎皮多里奥广场，罗马
Campidoglio, Rome

这个城市广场设计的杰作调节了随意建造的山顶场地的混乱，吸引了全世界慕名前来的崇拜者以及模仿者。很少的游客知道这个广场宏大的铺装纹理是由米开朗琪罗设计的，并且在他设计的 400 年后终于被建造出来。
This masterpiece of civic plaza design reconciling the chaos of a haphazardly built-up hilltop site has attracted admirers and imitators worldwide. Few visitors realize that its magnificent paving pattern designed by Michelangelo was only finally installed 400 years after he designed it. (Credits: alainlm, Prasenberg at en.wikipedia, and Jensens)

图 1.31
纳沃纳广场，罗马，1699 年
Caspar van Wittel (1656-1736), *Piazza Navona, Rome*, 1699

这个杰出的绘画庆祝着它的时代，在那个时代经过几个世纪演化的设计终于达到了生动与统一的形态，这种形态沿用至今。在今天，人们热爱它的壮丽，它富有戏剧性的喷泉，它外围的零售活动，以及游客们的漫步。
This excellent painting celebrates the era in which the design which had evolved over centuries essentially reached the lively and unified form which it still maintains today. Then as now, people love its grandeur, its theatrical fountains, its peripheral retail activity, and its parade of visitors. (Credit: Thyssen-Bornemisza Museum collection, Madrid)

yet he never lived to see this design completed, which took until the 20th century. The equestrian statue at the space's center presides over a carefully orchestrated array of sculpture punctuating the perimeter of the space. The oval form of the geometrically sophisticated paving pattern that spirals out from the equestrian statue brilliantly makes peace with the necessarily trapezoidal alignment of the facades. Only added in 1940 according to the great artist's original design, the spiral radial paving pattern is much imitated in China, though usually without the subtlety of the original's grading of radial slopes.

Perhaps the second most important urban public open space in Rome, Piazza Navona, 1.4 hectares, continues to be cited as iconic because of its lively yet unified architectural enclosure and its three fountains made into high art, the total effect so powerful that if it had never been created, we would truly miss it. (1.31) The over-the-top theatricality of Giovanni Lorenzo Bernini's Four Rivers Fountain tells a transcontinental story about water with water and marble with which no abstract design from our era could possibly compete. Occupying the elongated footprint of a Roman sports arena, the timeless space is rarely empty of grateful visitors of all ages sharing one multi-faceted space, with sidewalk restaurants, performers, and local artwork on display.

Wild Nature outside the garden wall and city wall was also starting to capture the popular Western imagination. Mountainous landscapes like the Alps, unlike those in China, had long been feared and not worshipped, but now they began to inspire artists like Leonardo and Titian, as the distant background far from the city. While royal palaces

实施。这个放射状的铺装样式在中国广受模仿，却少了原作在竖向上对斜坡的精妙处理。

罗马第二大开放空间或许要属纳沃纳广场，它占地 1.4 公顷。屡被援引为广场设计的经典之作，概因其有着俏皮但不失统一性的建筑轮廓，还有三个具有极高艺术价值的喷泉，令人叹为观止。其整体效果是如此震撼，倘若这个空间未被建成，那真是人类的一大损失。乔瓦尼·洛伦佐·贝尔尼尼的四河喷泉极具戏剧风格，用流水与云石讲述横贯大陆与汪洋的故事，是我们这个时代任何抽象设计难以媲美的（图 1.31）。四河喷泉选址于某古罗马竞技场的旧址延伸区域，在这个经典的空间里，游客络绎不绝，老老少少共同分享一处多面空间，路边餐厅、表演者及本地艺术展览为其带来了活力。

花园及城墙外的原始大自然开始令西方大众心驰神往。诸如阿尔卑斯山的山脉景观常常让人感到畏惧，而不是崇敬，这与中国不太一样；但文艺复兴时期的达·芬奇和提香等艺术家却从中得到了灵感，将其作为远离城市的背景。虽然市中心华丽的皇宫都会置有美轮美奂的花园或园林，但这些空间并不对大众开放。大自然是以一棵树木、一个花园等元素形式逐渐融入人们的城市生活当中的；历史上最早的行道树可追溯至 1590 年的巴黎林荫小径，此小径主要是类似槌球运动的场地，伦敦的蓓尔美尔街从法语而来，也是以这种运动得名的。

在欧洲人到达新大陆之前，美洲的土著居民住在水牛皮小木屋、部落的长形木屋及普韦布洛式住宅中，他们都有公共开放空间以供跳舞、庆祝、货物交易和辩论等活动之用。值得一提的是，在玛雅和阿兹特克文化中，城市是仪式活动的中心，城市规

图 1.32
球场广场，科潘，洪都拉斯
Ballcourt Plaza, Copan, Honduras

从墨西哥到秘鲁的原始美洲文化在神圣的金字塔边孕育出有着综合城市广场的城市。玛雅文明中有一种在一个狭长的球场上举行的神圣球类活动，广场的遗址坐落于科潘的金字塔旁边，大概可追溯到公元前 738 年。
Native American cultures from Mexico to Peru evolved cities with complex urban plazas adjacent to sacred pyramids. Mayan culture featured a sacred ball game performed on a narrow court, such as survives in a plaza near the base of several pyramids at Copan, dating from 738 AD. (Credit: HJPD)

图 1.33
普韦布洛的广场，新墨西哥州，美国，约 1911 年
Plaza at Isleta Pueblo, New Mexico, USA, ca. 1911

在新墨西哥最大的印第安村庄的中心，这个围合广场被一棵孤植树所装点着，它是几个世纪来丰收舞蹈的传统基地。人们在这个干旱气候的地区跳舞祈祷雨水和大丰收。印第安居民和游客可以在四周矮墙之外来观赏这种舞蹈表演。
In the heart of the largest pueblo in New Mexico, the enclosed plaza adorned by a single tree is the traditional setting for the centuries-old harvest dance, offered as a prayer for rain and bountiful harvests in this arid climate. Pueblo residents and visitors observe the performance from outside the low perimeter walls. (Credit: Fred Arvey)

划布局复杂而精密，围绕神殿或金字塔并面向城市空间。其城市空间可与古希腊和古罗马的城市设计相媲美，有时还像公元 600 年的科潘市一样，设有带雕刻的石柱群和石头铺砌的球场，用以开展神圣球类运动（图 1.32）。

美国西南部的印第安土著所居住的普韦布洛住宅构造可追溯至 1000 多年前，它促使人们形成了聚集在广场（西班牙语里露天广场叫 "piazza"，是 "场所" 的意思）上举行庆祝或娱乐活动的传统（图 1.33）。建筑历史学家文森特·斯库利曾提到，"人们在普韦布洛前的露天广场跳舞，此举是美洲最优美的建筑艺术表现形式。"² 相似地，16 世纪后期及 17 世纪初期的西班牙殖民者和基督传教士来到北美西南部、墨西哥、南美洲及菲律宾开拓新殖民地的时候，都根据西印度法去监管殖民地，并形成了公共开放空间。这些空间又名 "playas"，面积一般都是 130m × 200m，空间的边缘会有教堂、修道院、守备部队或总统府等重要公共机构建筑物。当葡萄牙及西班牙分别在 1550 年及 1572 年进驻和统治澳门及马尼拉时，也同样在城市规划上作了相应调整。

同样的，欧洲城市在 17 世纪见证了中轴式大城市广场的出现。这些大广场威仪庄严，往往坐落在主要的十字路口或目的地，以其宏伟气势给人以强有力的震撼感。人们在其间举行公共庆典，常见的有阅兵和宗教聚会。贝尔尼尼的椭圆形双柱廊，如双手合捧，环托梵蒂冈的圣彼得广场，这样的设计能尽量让更多的人汇聚于此瞻仰圣地。美中不足的是，这个大面积铺装的倾斜着的公共广场缺乏宜人尺度，也没有公共设施。1816 年，建筑师基赛匹·瓦拉迪埃延续了同样的传统，重建了椭圆形的波波洛广场，建筑周界井然有序、铺装一丝不苟，中心的方尖碑巍然矗立于罗马城内。整个空间秩序感无可挑剔，可惜并不适合人们在其间驻足休憩（图 1.34）。

相似尺度的公共开放广场空间随后在亚洲出现。天安门广场

in the center of cities boasted extensive gardens and even parks, these were not generally open to the public. Nature and public recreation entered the people's city slowly, one tree and one garden and one game at a time: the first street-tree planting seems to have been an *allée* planted in Paris in the 1590s for playing a croquet-like game called *palmai*, an association that lives on in the name of London's Pall Mall.

Before the arrival of Europeans in what they called the New World, Native American communities, whether made up of buffalo hide lodges, long houses, or pueblos for the most part had a public open space for dancing, celebration, trade and debate. Notably, in Mayan and Aztec culture, cities were architecturally sophisticated ceremonial centers, organized around temple pyramids facing urban spaces almost as sophisticated as those of ancient Greece and Rome and sometimes, as in Copan (600 AD), included sculptured stelae and a stone paved court for playing the sacred ball game. (1.32)

The Native American pueblos of the American Southwest, some dating back millennia, offer a rich tradition of *plazas* (the Spanish equivalent of *piazza*, place) bringing the people together in celebration and recreation. (1.33) "The People dancing on the *plaza* in front of their Pueblo is the finest act of architecture in America", noted architectural historian Vincent Scully. ² Likewise, as they colonized the North American Southwest, Mexico and South America and the Philippines, the Spanish Conquistadors and Holy Fathers in the late 16th and early 17th centuries in obedience to the Laws of the Indies regulating colonization formed urban public open spaces or *playas*, typically 130 by 200 meters, edged by a church/monastery, garrison (*presidio*) and governor's palace. In Asia the Portuguese and Spanish provided much the same in Macau (1550s) and Manila (1572).

Likewise in European cities, the 17th century also witnessed

图 1.34
波波洛广场，罗马
Piazza del Popolo, Rome

由于受到米开朗琪罗天才般坎皮多里奥广场设计的启发，在两个世纪之后，巴黎以及其他地方都建造了人民广场，在 1816 年的瓦拉迪耶在混乱的建筑布局中创造了新的秩序，但是为车行驶用，并不完全符合它的名字。
Inspired by Michelangelo's design genius at the Campidoglio, and after two centuries of plaza building in Paris and elsewhere, in 1816 Valadier created order out of chaos, but not a place for people, despite its name. (Credit: WolfgangM)

图 1.35
红场，莫斯科
Red Square, Moscow

这个大型的集市广场见证了很多政治场面。1804 年铺设的铺装相对于土地是一个很大的进步，但是极简主义的设计使这个公共场地仍然让踱步和放松的人们略感不自然。
For this large market square that has witnessed many political spectacles, the paving that arrived in 1804 was an improvement over dirt, but the minimalism of the design of this public space remains hostile to lingering and relaxing. (Credit: Raul P)

the advent of big urban plazas on axis with major vistas at major intersections or destinations, designed to impress the multitudes with the power of the state, and accommodate public ceremony, often military or religious. Bernini's magnificent oval double colonnade like cupped hands embracing Saint Peter's Square (Piazza San Pietro, 1656-67) in the Vatican was designed to allow the greatest number of people to see the Papal blessing. Nonetheless for all its beauty the massive paved and obliquely tilted public plaza lacks human scale and public amenity. In the same tradition, architect Giuseppe Valadier's rebuilding of the oval Piazza del Popolo (1816) with orderly architectural perimeter, paving, and central focal obelisk inside the city walls of Rome is a masterful ordering of space, but not one for people to relax in. (1.34)

Similarly scaled public open space in Asia came later. It is a mistake to assume that the scale of Tian'anmen Square in Beijing is as old as the Forbidden City itself; enlarged fourfold in 1958, it attained a scale that may actually owe something to the large squares of Europe by way of Red Square in Moscow, an elongated rectangular market square outside the Kremlin (the word that now means "red" originally meant "beautiful") first paved in 1804, but without other public amenity. (1.35)

The 17th century ushered in the Parisian counterpart of the Italian *piazza*, the *place*. With its formal garden and five fountains, the 2-hectare square Place des Vosges (1612), surrounded by a uniform arcade, though not a palace was initially more a private garden retreat than a public square, primarily intended for the shared use of the abutters and until 1850 constituted half of the public park open space in Paris (1.36). Signaling the priority of design unity over private aspiration, the absolutely uniform architectural facades enclosing the 1.7-hectare Place Vendome (1702) in Paris were constructed even

图 1.36
孚日广场，巴黎
Place des Vosges, Paris

追溯到 4 个世纪前，这个巴黎最古老的设计广场，全欧洲之后的广场原型，为了至高无上的建筑统一性而牺牲了自己的独有风格，在中世纪的广场上，建筑的统一性高于一切。很久以后公众才允许进入这个公园。
Dating back four centuries, the oldest planned square in Paris and prototype of all subsequent European squares subordinated individuality to shared architectural unity surpassing anything seen in medieval squares. Public access to this garden retreat came much later. (Credit: Poulpy)

的现状尺度并非跟紫禁城一样历史悠久。1958 年，天安门广场的面积被扩大了 4 倍，这种尺度大概是受欧洲大广场的影响，比如莫斯科的红场。红场是一个细长的长方形广场，坐落在克里姆林宫（克里姆林在俄语里原意为"美丽"，后来引申为"红"）外面，于 1804 年第一次铺装，却没有其他公共设施（图 1.35）。

17 世纪巴黎出现了意大利露天广场（piazza）的对等物——法式广场（the place）。建于 1612 年的孚日广场占地 2 公顷，尽管不是宫殿，其规整的花园和 5 个喷泉却被统一的拱廊围绕；它最初的用途并非公共广场，而是一个私家花园，为周围地主们的休息之处；1850 年前，它一直是巴黎一半的公园开放空间（图 1.36）。

Chapter One　Origins of the People's Parks　**031**

建于1702年的旺多姆广场占地1.7公顷，其绝对统一的建筑表面围合格局甚至在内部空间设计完成前就已建好，这体现出设计的统一优先于个体的愿望。中心柱是后来增加的，其周围的空间全部被铺装起来（图1.37）。巴黎的发展趋势是开辟大面积的公共空间，便于交通穿行，却不鼓励人们停留。骚乱时有发生，安全问题是一大隐患。

在17、18世纪，巴黎和伦敦的市中心都是皇家花园，平民不得入内，尽管多年以后都衍变成了为民众服务的公园，在当时却警戒森严。彼时的罗马却是大方地将别墅古堡对民众开放，而这与民共享的传统要追溯到尤利西斯·凯撒大帝时期，他将罗马花园慷慨地遗赠给罗马市民，这也是史料记载的最早类似事件。17世纪，伦敦的一些以往的皇家园林已经逐渐向市民开放了，这些园林通常有大片大冠幅的主林木和草地。查尔斯·布里奇蒙设计的142公顷的海德公园于1637年对民众开放（图1.38），它证明了在闹市中心设置一片开放公园空间的重要性，这远早于19世纪的公园运动；之后，亨利·怀斯和查尔斯·布里奇蒙于1730年共同设计了111公顷的肯辛顿公园、格林公园（图1.39-图1.40）、圣詹姆斯公园（图1.41）以及其他皇家园林也陆续对外开放。人们聚集起来不仅仅是为了游行、表演和庆典，也是为了消遣，哪怕只是沿着圣詹姆斯公园的林荫大道走走。沿着由碎石铺就、两旁浓荫覆盖的林荫大道游行的方式迅速传到了其他城市的开放空间，比如波士顿的特雷蒙特大道（1650年建）、葡萄牙里斯本的自由大道（1755年建）和维也纳的普拉特公园（1766年建）。英国也是最早推广需要利用公共空间运动和进行团队活动的国家之一。人们可以到公园玩保龄球、曲棍球、骑马，冬天在冰上滑冰、夏天在草地上打盹。当初匀称端庄的皇家水景经过改造和植栽的重新设计后，呈现出一种更为自然的风貌。法国著名的景观建筑师勒诺特以其对凡尔赛花园典雅的中轴式设计闻名于世（凡尔赛花园的奢华程度只有同时期稍晚的北京颐和园能相与之媲美），

before the individual house interiors were designed. (1.37) The space surrounding the central focal column added later is entirely paved. Paris was set on a course toward grand open spaces through which traffic was to move smoothly but people were not encouraged to linger. Security from riots was an issue.

In the 17th and 18th centuries, the royal gardens that today serve as public parks in the heart of the city, in Paris and in London, remained inaccessible to the people, unlike many ancient Roman villas which had a tradition of being generously opened to the public going all the way back to Julius Caesar's bequest of his Roman gardens to the citizens of Rome, perhaps the earliest such instance recorded. The 17th century also saw the opening up to the public of formerly private royal parks in the middle of London, with vast tracts of canopy trees on grass. Opened to the people in 1637, 142-hectare Hyde Park, first among equals, laid out by Charles Bridgeman (1.38) followed by 111-hectare Kensington Gardens (designed 1730 by Henry Wise and Charles Bridgeman), Green Park (1.39-1.40), St. James Park (1.41) and the other royal parks, demonstrated the value of a large public park in the heart of a teeming city long before the parks movement of the 19th century. People gathered not just for parades, performances and celebrations but for recreation as simple as a good brisk walk down The Mall (1650) along St. James Park. The mall or promenade idea—wide gravel walks shaded by double rows of trees—spread to other city open spaces like Boston's Tremont Mall (1728), Lisbon's Passeio Público (1755), Vienna's Prater (1766). Britain was also an early adopter of organized sports of all kinds, and many group and team activities required shared open space. People came to the parks to bowl, play croquet, ride horseback, skate on the ice in winter, nap on the grass in summer. Redesign of once formal water features and

图1.37
旺多姆广场，巴黎
Place Vendome, Paris

更多的早期巴黎式的广场都有铺装，几乎没有任何配套设施来吸引人们前来逗留，它更像是一个优雅的让人陷入沉思的空间，它的设计统一性和壮丽超过了个人的设计偏向。
More typically the early Parisian squares were paved and offered hardly any amenities to attract people to linger other than the experience of immersion in an elegant space where design unity and grandeur prevailed over individual design preference. (Credit: http://flickr.com/photos/http2007/ http2007)

图1.38
海德公园，伦敦，航拍图
Hyde Park, London, Aerial View

在小尺度的孚日广场震惊巴黎的同时，这个巨大的、前身是私人皇家公园的海德公园对公众开放了，并且之后一直是主要活动的庆典地点，同样也是进行各种休闲活动的绝佳地点。
This vast, formerly private Royal Park opened to the public about the same time that Place des Vosges on a much smaller scale transformed Paris and continues to host major events as well as accommodate leisure of all kinds. (Credit: Andreas Praefcke)

图 1.39 - 图 1.40
格林公园，伦敦
Green Park, London

18 世纪开始，人们在各个季节都会去远离城市一段距离的公园散步。人们在这里放松，与朋友会面，做做白日梦，清理清理想法，或者做运动。种植在草坪上的树阵和树丛相结合，回到了最开始的样子。
In all seasons, since the 1700s people have gone for a walk in the London parks to put the built up city at a distance for a while, to relax, connect with friends, daydream, clear their thoughts, and get some exercise. The mix of allées and informal planting of trees on turf goes back to the beginning. (Credit: "Photo by DAVID ILIFF. License: CC-BY-SA 3.0" and Benkid 77)

图 1.41
约瑟夫·尼克尔斯购物广场，圣詹姆斯公园，伦敦，约 1771-1772 年
Joseph Nickolls (attributed), Mall, St James Park, London, ca. 1771-1772

伦敦人愿意外出来到一个枝繁叶茂的公园度过一段美好的时光。这肯定了在这个没有那么多城市化和疯狂喧嚣的世界中城市开放空间给我们带来的好处。
The throngs of Londoners out to have a good time in the leafy park in the heart of London affirm the benefits of urban open space in a world much less urbanized and frenetic than our own. (Credit: Royal Collection, London)

图 1.42
邱园，伦敦
Kew Gardens, London

中国塔是 1761 年由威廉·钱伯斯爵士建造的，这个塔意味着在西方设计中不断增加的中式影响。这些影响源于贸易与去中国的旅行。邱园作为一个植物园和树林园在 1840 年对外开放。1814 年在伦敦，第二个中国塔在圣詹姆斯公园建起。
The Chinese pagoda built by Sir William Chambers in 1761 suggests the increasing Chinese influence on Western design triggered by trade and travel to China. Kew Gardens opened to the public as a botanical garden and arboretum in 1840. In 1814 a second pagoda was built in St. James Park in London. (Credit: Targeman)

replanting restored a more naturalistic character. The great French landscape architect André Le Nôtre, known for his formal radial axial design at Versailles (its only rival for extravagance of private open space is its near contemporary, the Summer Palace in Beijing), actually turned down a commission to redesign St. James Park, suggesting that it should be left alone. London now led the world in the provision of public parkland.

The Age of Enlightenment in the 18th century West ushered in not only the notion of individual liberty but also a new appreciation of Nature at the same time that Europeans who visited China such as Sir William Temple, Sir William Chambers, Father Mattheo Ripa

他在接到重新设计圣詹姆斯公园的任务时辞而不受，并建议保留其原状。伦敦如今在世界公园用地储备上独占鳌头。

18 世纪启蒙时期的西方不仅引进了个人自由的理念，也同时引进了新的自然观。当时造访了中国的威廉·坦普尔爵士、威廉·钱伯斯爵士、马国贤神父，以及乾隆皇帝御用画师正王致诚等欧洲人都被中式的风景园林深深吸引，这是继马可·波罗之后的第一批欧洲来访者。相对西方双边对称的花园设计而言，中国道法自然的山水园林给西方带来了新的启迪，并带来了革命性的突破。对中国事物的品位和追求形成了一股*中国风*，刮遍欧美，这比日本风的兴起要早 100 年（图 1.42）。1792 年，弗里德里希·路德维希·冯·斯凯尔和出生于美国的本杰明·汤普森爵士

图 1.43
英式花园中的塔，慕尼黑，德国
Pagoda in English Garden, Munich, Germany.

约 1972 年，中国塔在非正式英国种植园里的出现，意味着结合了东西方设计和欧式汇集风格的景观第一次在英格兰兴起。
The ca. 1792 Chinese pagoda in the informally planted English Garden suggests the European convergence of Eastern and Western design forces that first sprang up in England. (Credit: "Photo by DAVID ILIFF. License: CC-BY-SA 3.0")

图 1.44
徐扬，职业辉煌期 1764-1770 年，《姑苏繁华图》
Xu Yang (fl. 1764-1770), *Suzhou's Golden Age*

我们也许可以把这个卷轴画卷的细节称之为沿着苏州大运河的绿道。它预言了我们现如今的绿色基础设施和都市化景观。
This detail from the scroll depicts what we might call a greenway along the Grand Canal in Suzhou anticipating today's green infrastructure and landscape urbanism. (Credit: Palace Museum, Beijing)

将德国慕尼黑一个私家狩猎苑设计改造成了 373 公顷的英式公园时就特地修建了一座中式塔（图 1.43）。在西方出现公园之前，中国城市的不规整水岸给人们的休闲生活带来了无限乐趣。现藏辽宁博物馆的乾隆时期长卷《姑苏繁华图》生动地展现了*民间*的生活场景（图 1.44）。苏州大运河水岸就是一个绿色长廊。水乡里通常有纵横交错的水路。从某种意义上来讲，今天景观都市主义的绿色基础设施也是想要重新获得以往的那种平衡。

在同时期的英国，威廉·肯特和"全能布朗"等景观设计师正在努力寻求突破，他们将规整对称的旧式花园向不对称的*自然桃源*发展：铺草地、种果树、造岩穴、借溪谷、现海角等造景手法让人居风景环境进入了一个黄金时期，仿佛古希腊罗马神话里众神仙子的自然居所。这在克劳德·罗兰和尼古拉·普桑的风景画里均有体现。天人合一的和谐共存与中国道家所描绘的仙人神峰之境相呼应。中国园林对英式园林的影响是深远的。有时候，

图 1.45
罗素广场，伦敦
Russell Square, London.

这个公园位于一个 19 世纪联排别墅区的附属地带。它于 2002 年被重修。这次重修使公园恢复了它 1800 年的布局，此布局是由 18 世纪著名的景观师汉弗莱·雷普顿设计的。
This park in a district of attached 19th C. townhouses was restored in 2002 to its 1800 layout by noted 18th C. landscape gardener Humphrey Repton. (Credit: Chris Nyborg)

图 1.46 - 图 1.47
塔维斯托克广场花园，卡姆登，伦敦
Tavistock Square Gardens, Camden, London

这个城市公共开放空间可追溯到 19 世纪 20 年代，它吸引了一定数量的和平纪念事件，包括圣雄甘地的纪念雕像、纪念广岛原子弹爆炸受害者的和平之树以及反战人士的纪念石碑。
This urban public open space dating from the 1820s has attracted a number of peace memorials, including a statue of Mahatma Gandhi, a tree planted to commemorate the victims of the Hiroshima bombing, and a Conscientious Objectors memorial stone. (Credit: Stephen McKay and John Winfield)

图 1.48
塞勒姆公园，塞勒姆，马萨诸塞州，美国
Salem Common, Salem, Massachusetts, USA

从17世纪初开始，新英格兰地区（北美的6个英属殖民地）的新社区布局包括了可共享的或"共有的"开放空间。这个开放空间从使用主义（放牧以及军队训练之用）开始，到19世纪搭建典型的室外音乐演奏台，从而演变成文化和娱乐的休闲之地。当时塞勒姆是和中国最早有贸易往来的城市之一。
From the beginning in the 17th C., the layout of new communities in New England, six British colonies in North America, included shared or "common" open space whose use evolved from utilitarian (cattle grazing and military drilling) to cultural and recreational by the 19th C. when a bandstand pavilion was typically added, as here in one of the first cities to trade with China. (Credit: Fletcher 6)

图 1.49
拉德洛公园，拉德洛，马萨诸塞州，美国
Ludlow Common, Ludlow, Massachusetts, USA

到了19世纪，即使一个小型美国社区也增添了自己的演奏舞台，在夏季那里经常会有免费的音乐会。表演通常在有草地和树荫的软质景观上进行。
By the 19th C. even a small American community added its own bandstand pavilion where free concerts were performed in the summertime, in a soft landscape typified by turf and shade trees. (Credit: John Phelan)

and Qianlong Emperor's court painter Jean-Denis Attiret were falling in love with the Chinese garden and landscape design for the first time since Marco Polo. To a Western world of bilaterally symmetrical garden design, the imitation of Nature in Chinese garden design was a revelation, and revolutionary. The taste for things Chinese (*chinoiserie*) spread across Europe and America, predating fascination for things Japanese by a century. (1.42) In Munich a Chinese pagoda was included in the transformation of private hunting grounds into the 373-hectare public English Garden (1792) designed by Friedrich Ludwig von Sckell and the American-born Sir Benjamin Thompson. (1.43) In this age before public parks in the West, urban waterfronts in China now provided informal recreational enjoyment. The *Suzhou's Golden Age* (gu1 su1 fan2 hua2 tu2) in the Liao Ling Museum vividly depicts a world not of officials in ceremony, but among the people (min2 jian1). (1.44) Suzhou's Grand Canal waterfront was a greenway corridor. Water towns often included covered walkways along canals. In a sense, the green infrastructure of today's landscape urbanism seeks to recapture that balance.

Concurrently in England, landscape designers like William Kent and Lancelot "Capability" Brown were reinventing landscape design from a mere matter of formal bilateral symmetry into asymmetrical *arcadia*: turf and lawn, groves and grottoes, vales and promontories evoking a "golden age" of living in unspoiled Nature like the gods, naiads and nymphs of classical Western mythology as shown in the highly influential landscape paintings of Claude Lorrain and Nicholas Poussin. This vision of living in harmony with Nature had its counterpart in the sacred mountain landscape of the Immortals of Chinese Taoism.

"自然"风格承袭了大尺度的自然风景改造、围湖、迁移建筑，在一定尺度内激发了未来景观建筑师的创作灵感。现在惯用的借景法——将远景整合到近景画面中——大约就是从中国传来的。

在开放空间向民主化平稳迈进的过程中，意大利、法国以及伦敦的广场启迪了许多西方其他城市。建于1631年的科文特加登广场拱廊尺度较小，入口有守卫，当时只向领主贵族开放，现在是首个已消失的广场，它的名字也是从其灵感源地意大利而来的。紧随其后的是建于1635年的莱切斯特广场、建于1775年的贝德福德广场、建于1800年的罗素广场、建于1825年的贝尔格雷夫广场以及其他半公共景观。它们共同向世人昭示了一个理念：都市生活需要有草地和花圃等自然元素，不管大小如何，至少对担负得起的人来说是不可或缺的（图1.45 - 图1.47）。很多广场都有栏杆围绕，在没有汽车之前，这些有的是用来停马车的。1666年伦敦大火之后，重建方案都不约而同地提出需要在伦敦建造更多的广场，而且在最优秀的设计方案中还沿用了意大利的广场说法"piazza"。这引来了大批模仿者，譬如18世纪的爱丁堡、费城、萨凡纳，以及19世纪的波士顿和巴黎。

与此同时，受到英国宗教迫害的清教徒殖民主义者辗转来到美洲新大陆，并从1620年起把他们的定居地叫做新英格兰，以纪念家乡。他们从新英格兰向南发展，按照其家乡地域风格修建了很多建筑和聚居地，这些聚居地的特点在于预留土地由各户共享，特别是外围牧场、农田、草地以及可以用以放牧或军队演习的中心公共绿地。到后来，新的定居者利用这些公共用地建立起了教堂、公立学校、市政厅和政府。很多公共用地被称为绿色空间，因为它们本来就是城镇中心的一小片自然绿地，汇聚了所有的道路。它们形态各异，从不规则的、有机的，到矩形、三角形、椭圆形或方形（图1.48 - 图1.49）。从英国独立出来后，美国民众每年聚集在公园绿地庆祝国家独立日。经典的新英格兰村庄风格大受欢迎，这不仅仅是因为其外表赏心悦目，还因为它倡导了当地民众聚集起来参与公共活动，最重要的是参与"村镇会议"以亲自决定关系到当地生活质量的事务。19世纪，很多村镇形成了农民公共设施委员会，一起美化村镇中心并种植行道树。新英格兰的村落建筑风格不断给世界各地包括中国的规划和设计提供灵感，在19世纪90年代的中国广州，一个庄园别墅的广告牌上甚至展示着新英格兰的村庄建筑。³

在公共开放空间方面，苏丹统治下的爪哇与新英格兰平行发展，爪哇的所有村落和城镇的中心都是由大面积的开放性方形草坪构成的，其上种有遮阴树；这些草坪广场被称为阿伦阿伦，它们延续了其作为公开表演、官方庆典以及公共娱乐平台的功能，这其中包括表演传统的罗摩亚娜舞蹈。万隆的阿伦阿伦广场建于1810年，是由摄政首领万然纳塔库苏马二世（1794-1829年）部署的。他是万隆城邦的建立者，其建城的指导者是当时治理东印度群岛的荷兰将军丹德尔斯，其建城的哲理是依照爪哇文化的禅宗哲理（图1.50 - 图1.51）。

图 1.50 - 图 1.51
巴杜珊卡尔的公共广场和弟索尔潘的公共广场，爪哇岛，约 1910-1940 年
Alun-alun in Batusangkar, Java. and Alun-alun, Tjisoeroepan, Java, ca. 1910-1940

印尼的穆斯林公共广场毗邻穆斯林统治者的宫殿群，而且为了提供庆典活动的空间，它持续举办大型公共活动、宫殿庆典、节日和娱乐消遣，这点和新英格兰的公园很相似。对于世世代代在那里的印尼人们来说，他们不分昼夜地聚在这里，有时会在晚上观看罗摩传的皮影戏。
In Muslim Indonesia the *alun-alun* squares adjoin the sultan's palace compound and in addition to providing ceremonial space, continue to accommodate public spectacles, court celebrations, festivals, and entertainment, a lot like a New England common. For many generations, Indonesian people have come together day and night, sometimes to see night performances of the Ramayana by shadow puppets. (Credits: Tropenmuseum of the Royal Tropical Institute)

The influence of Chinese garden design on the English style is profound. Sometimes the "natural" style entailed large-scale redesign of a natural landscape, damming ponds, removing buildings on a scale that would inspire future landscape architects. Likewise the notion of borrowed landscape—integrating distant features into the foreground design—may have been borrowed from China.

In the steady march toward democratization of open space, like Italian *piazzas* and French *places*, London squares inspired many other cities in the West. Small in scale, gated and accessible only to the abutters, the now lost first square, Covent Garden Piazza (1631) bowing in its use of "piazza" in its name to its Italian inspiration, followed by Leicester Square (1635), Bedford Square (1775), Russell Square (1800), Belgrave Square (1825) and other shared semipublic landscapes made the point that urban life, at least for those who could afford it, required access to Nature in the form of lawn and flower beds, however limited in scale. (1.45-1.47) Most squares were enclosed by a fence. Some were used for the parking of carriages, long before cars. After the London Fire of 1666, competing plans for its reconstruction seemed to agree that London needed more squares, actually called *piazzas* on the preferred design. These inspired a legion of imitators like Edinburgh, Philadelphia and Savannah in the 18th century and Boston and Paris in the 19th.

In the meantime, when colonists fled religious persecution in England to settle in what they loyally called New England beginning in 1620, and colonies further south soon thereafter, they brought with them a vernacular tradition of building and settlement pattern that featured the setting aside of land to be shared in common by all landowners, typically peripheral pastureland, cropland and meadows as well as a central "common" where citizens could graze their cattle and gather for military practice. Over time in newer settlements the common also became the location for the church, public school, and the town hall or seat of local government. Many commons came to be called greens, for that is what they were, a snippet of Nature in the heart of the town upon which all roads converged. The form varied from irregular or organic to rectangular, triangular, oval or square. (1.48-1.49) After they won their independence from Britain, people gathered on their commons and greens each year to celebrate the National Holiday, Independence Day. The classic New England village type is much admired not only for its visual quality but for its expression of local citizens coming together to participate in their civic institutions, most importantly "town meeting," in which they decide for themselves matters affecting local quality of life. In the 19th century many towns formed private village improvement societies to beautify the town center and plant street trees. The archetype of the New England village continues to inspire development master plans and design, even in China, like a Majesty Manor Villas billboard showcasing a New England village seen outside Guangzhou in the 1990s. ³

In a parallel development, in Java under the Islamic sultanates, the alun-alun or large central open lawn squares with shade trees formed the heart of all villages, towns and cities, and continues to function as the center for public spectacles, court celebrations and public entertainment, including performances of the Ramayana. The Alun-alun (1810) in Bandung was first laid out by regent Wiranatakusumah II (1794-1829), the founder of Bandung city, at the instruction of the Dutch Indies governor General Daendels in the Javanese cultural philosophy of Catur Gatra (1.50-1.51).

The so-called New World offered an unprecedented opportunity for French and English colonists to design without the constraint of prior human intervention. Designers were free to borrow the best ideas from any era. If the topography was not too irregular, they were free to impose geometrical order and often chose the simple grid, like New Haven (1639) with its central square, New Haven Green. Thomas Holme's plan for Philadelphia (1685) and James Oglethorpe's plan for Savannah (1733) illustrate a much more ambitious hierarchy of streets and a regular and equitable distribution of public squares, in a tradition dating back to the classical Roman architect Vitruvius' *De Architectura*, with its parallel tradition in many ancient Asian grid cities. Even in Roman camps and colonial towns of the classical world (Paris and London among them), the regular orthogonal grid in the West revolved around an intersection of main streets aligned to true north-south and east-west, with a central square, an idea continued in the 19th century settlement of the Midwest and Western U.S. Spanish and French colonial cities did little more. In Philadelphia, the five original squares, each serving a quadrant of the city but not visible one from the other, have each evolved independently. In Savannah, one can feel a more powerful visual connection between the squares because they are closer together. Each square and its surrounding 12 blocks became

所谓的新世界的发现，为英法殖民者创造了前所未有的机遇，他们可以去重新设计，不受窠臼的限制。设计师可以从任何时代汲取灵感。如果地形并非很不规则，设计师可以大胆运用几何图形，他们通常选择简单的方格，比如建于 1639 年的纽黑文绿地及其中心的纽黑文广场。托马斯·赫尔姆斯的费城规划（1685 年）以及詹姆斯·奥格尔索普的萨凡纳规划（1733 年）里的街道等级更为严格，公共广场分配更为有序均等。这一传统要追溯到古罗马建筑师维特鲁威的《建筑十书》，同时亚洲其他城市也保留了类似传统。在古典年代的罗马营地和殖民小镇（也包括巴黎和伦敦），规整的直角网格图样围绕主要街道的十字路口，呈南北和东西方向，中央有广场，这一设计理念一直延续到 19 世纪美国中西部。西班牙和法国殖民城市也大抵如此。在费城，5 个最初的广场各据城市一角却互不干扰，各自演化发展。在萨凡纳，广场彼此紧邻，相互呼应，给人的视觉联系感更加强烈。每个广场和周遭的 12 个街区在 150 年的发展期间形成一个整体区块。带中央喷泉或雕像的绿地广场逐渐发展，使街区景致生机盎然，而生机勃勃的街景让整个城市成为有机统一体，并为大家提供了亲民的社区公园。直到 21 世纪的今天，人们还在讨论环境的公平权益如何在公园的分布中体现出来。（图 1.52）。对于一个连续的公园系统，使其仅仅从视觉联系发展到空间实体联系，还得等到下个世纪景观建筑学的发展。

为了突显美国这个民治、民权、民享的新民族，来自法国的外聘建筑师皮埃尔·查尔斯·朗方于 1792 年为华盛顿做了城市规划，他将政府建筑沿着中心绿地分布，这片中心绿地向市民开放，并向波托马克河这一主要自然景观敞开（图 1.53）。首都华盛顿采用古典建筑风格，寓意经典世界里的民主理想。巴洛克式的对角线林荫大道结合城市广场的规划设计最初被安纳波利斯（1718 年）所采用。这样的规划设计模式可追溯到罗马教皇希

图 1.52
福塞斯公园，萨凡纳，约 1870 年
Forsyth Park, Savannah, ca. 1870

这个照片捕获了这个美国城市非凡公共空间的三维空间体验，具体来说，是在一排排行道树的缝隙中见到的阴影下的城市广场以及它的精致喷泉。每个广场都曾经是社区公园。
This view captures the three-dimensional spatial experience of this American city's ambitious open space plan of shaded urban squares with elaborate fountains visible from one another along tree lined streets. Each square was a neighborhood park. (Credit: Thomas M. Paine Stereoview collection)

图 1.53
国家广场，华盛顿
The Mall, Washington

最初的设计是，把国会大厦建立在最高的山上，国家广场被构想成一个在首都中心的开放空间，没有周边的篱笆或者围墙，直接通向波托马克河，随时向人们展开怀抱。政府建筑、纪念碑、博物馆很快包围了国家广场，广场一直在举行活动或者节日庆典。
From its initial design in 1792, the Capitol crowned the highest hill and the Mall was conceived as an open space, without a surrounding fence or wall, leading to the Potomac River and always open to the people, in the heart of the U.S. capital. The Mall was soon bordered by government buildings, monuments and museums, and continues to host events and festivals. (Credit: Carol M. Highsmith)

a module for the orderly development of the city over 150 years. The vistas down alternating streets animated by a progression of green squares with central fountains or statues unify the entire city and democratize it by providing a neighborhood park for all, anticipating 21st century arguments for environmental equity in how parks are distributed. (1.52) To make the leap from a merely visual connection to a physical connection in a continuous park system had to wait for the advent of the profession of landscape architecture in the next century.

Signifying the arrival of a new nation governed *of* the people, *by* the people and *for* the people, imported French architect Pierre Charles L'Enfant's brilliant plan for Washington D.C. (1792), organized the national government around a central green space open to the people it served, and open in another sense to the main landscape feature of the unoccupied site, the Potomac River. (1.53) In its use of classical architecture the national capital alluded to democratic ideals invented in the classical world. In its use of baroque diagonal boulevards combined with squares the plan was preceded in the New World only by Annapolis (1718), an urban design tradition going back to the diagonal axes and squares of Baroque Rome under Pope Sixtus V (1585-90) and the unrealized plans for rebuilding London after the Fire of 1666, and one to be continued in cities like Paris and Philadelphia, and Tokyo and New Delhi in the 20th century.

After the French Revolution toppled the monarchy, ordinary citizens of Paris flocked to the formerly off-limits royal gardens such as the Tuileries and the Jardin du Luxembourg, at last transforming works designed by the great landscape architect André Le Nôtre in his signature axial style into public parks. A truly iconic urban public open space, the Jardin du Luxembourg balances *bosque* (woods) and *parterre* (flowers and lawn) and today combines its traditional manicured gardens, lawn, clipped trees, gravel, and central water feature with movable chairs, café and play areas that invite lingering and frequent revisiting. (1.54-1.56) Bigger and more people-friendly than squares, such public gardens first functioned as full-fledged public parks in a time when the term "park" still usually applied to private estate lands. In Florence, Grand Duchess Elisa Bonaparte opened the 118-hectare riverfront Medici estate to the public as a promenade park, Le Cascine (1814), still the largest park in that city, where it remains a favorite of joggers, horseback riders and picnicking families. In London's Regents Park (1812) the visionary early 19th century architect John Nash redesigned a royal park into a 166-hectare public park in the heart of a high-end private residential development. (1.57)

The perimeter of Regents Park was defined by the uniform architecture of private townhouses. Nash's boldly conceived and geometrically interesting Regent Street used uniform architecture and scale, along with dramatic changes of vista, to connect the outlying Regents Park to St. James Park, another royal park in the heart of the great city. The street soon became synonymous with fashion and wealth. If this enclave attracted mostly the well-to-do, it continued the

图 1.54 - 图 1.56
卢森堡公园，巴黎
Jardin du Luxembourg, Paris

这个真正标志性的公共花园保存了原有皇室花园的特点，它毗邻以前的宫殿（现在是法国参议院的地点）。它的宁静吸引了每天不计其数的游客慕名前来。这个公园包括了草坪和树丛、一个行驶玩具船的水池、旋转木马、咖啡店、游戏场和100个雕塑。在公共空间里流行的可移动座椅可以说是从这里起源的。
This truly iconic public garden preserves the character of the original royal garden, adjoining the former palace that now houses the French Senate. The calmness of this retreat attracts thousands of visitors daily. The park includes lawn and bosque, a basin for sailing toy boats, carousel, café, playground, and 100 statues. The popularity of the movable chair in public space could be said to have begun here. (Credit: Serged, Benh Lieu Song, and Dinkum)

克斯塔斯五世（1585-1590年）统治时期巴洛克罗马的交叉轴线和广场设计，以及1666年伦敦大火事件后制定的伦敦重建方案（未实现）。这一模式还在20世纪的巴黎、费城、东京和新德里等城市中得到延续。

当法国大革命推翻了君主制后，巴黎的平民百姓成群结队地来到了以前禁止对外开放的皇家园林，比如杜伊勒里宫及卢森堡公园，他们最终将景观建筑师勒诺特的这些标志性的中轴风格杰作转变成了公园。卢森堡公园的设计平衡了灌木丛和花坛的分

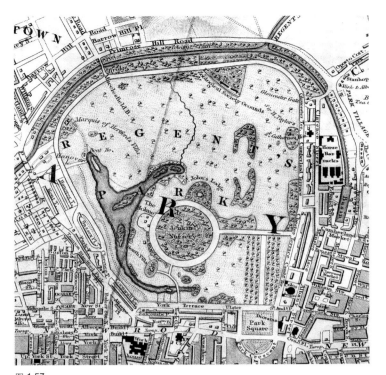

图 1.57
伦敦，丽晶公园，1830 年
Regent's Park, London, 1830

这个早期的地图反映了 1811 年英式园林的随意性，在某种程度上反映了东方思想的随意性。中心的高端别墅区在 1835 年首次对外开放。室外音乐演奏台俯看着湖景。
This early map captures the informality of the English style of landscape design of 1811 that to some degree owes its informality to Eastern thought. The centerpiece of a high-end villa development first opened to the public in 1835. A bandstand overlooks the lake. (Credit: Pointillist at en.wikipedia)

trend toward truly democratic open space, but it would take the new profession of landscape architecture to complete the task.

Frederick Law Olmsted would later come to found the profession of landscape architecture only after key formative experiences in China and England. China was opening up to trade with the West. Among the many young sailors who shipped out to China from the east coast of the United States was young Olmsted, who survived a rough voyage to China in 1843. The experience of mingling freely among all walks of life in Canton (Guangzhou) made a strong impression on him. As a landscape architect in the late 1850s Olmsted advocated public parks as places where people from all walks of life could come together and experience what we might today call *communitas*, or feeling good together in a public place. Such thinking had in part been hatched in his visit to China in 1843.

In the Industrial Revolution of the 19th century, rapid urbanization gave rise to an interest in Transcendentalism on the part of Western intellectuals who like Taoists and Buddhists before them longed for a more spiritual connection to Nature, and sought ways of living more simply and in harmony with it. Mount Auburn Cemetery (1831), Cambridge, Massachusetts, was the first large-scale designed landscape open to the public in the U.S. (1.58), followed by Philadelphia's Laurel Hill Cemetery (1836). To this day the harmony

布，在现代还融合了传统的园艺技术和现代配套设施，是公共开放空间的典范。修剪整齐的花园、草坪、树木、鹅卵石及有特色的喷泉，辅以活动的椅子、露天咖啡厅及游乐场，令人流连忘返（图 1.54 - 图 1.56）。当时"公园"一词一般都还只是应用在私有土地上，这些比一般的广场更大更亲民的公共花园已经是具有成熟功能的公共空间。在佛罗伦萨，埃莉萨·波拿巴大公爵夫人于 1814 年以散步公园的形式对公众开放了占地面积 118 公顷的原美第奇河边庄园。这座公园被称为卡施奈公园，它现今仍是市内最大的公园散步长廊，并且依旧是大众慢跑、骑马、野餐郊游的好去处。在伦敦，开放于 1812 年的丽晶公园占地 166 公顷，是由建筑家约翰·纳什重新设计的，此公园原是皇室御用花园，后成为高尚住宅区中心的公共花园（图 1.57）。

丽晶公园周边的私家联排别墅建筑风格统一。约翰·纳什大胆地以摄政街的景观、尺度和街道形状配合邻近建筑物的风格，将市郊的摄政公园和市中心的另一座皇家公园——圣詹姆士公园连接起来。自此，摄政街演变成为时尚和财富的代名词。如果这样的区域吸引的主要是小康人士，那么开放空间的大众化趋势便将会得到延续，但这一目标的实现还有待景观建筑学的进一步发展。

奥姆斯特德在游历了英国和中国后，开创了景观建筑学这个专业。当时中国正对西方开放贸易。1843 年，年轻的奥姆斯特德和一群水手从美国东岸出发，历经万水千山才来到中国。在广州的日子里，奥姆斯特德结交了来自五湖四海的朋友，这给他留下了深刻的印象。作为一名 19 世纪 50 年代晚期的景观设计师，奥姆斯特德提倡的理念是公园是供社区各阶层人士共享的，而这想法刚好是从 1843 年的中国之旅中萌芽的。

19 世纪的工业革命加速了城市化进程，部分西方智者原本醉心于东方的道家释家思想，此时则追求与自然建立精神层面的联系，这催生了超经验主义。超经验主义的精髓在于天人合一，与大自然产生心灵上的共鸣。1831 年，美国马萨诸塞州剑桥镇的奥本山公墓是首个对公众开放的大型景观设计（图 1.58）；1836 年费城的月桂山公墓紧随其后。值得强调的是，这些花园式墓园并不是公园，但时至今日，在很多城市中，这些和大自然融合的花园墓地吸引着成千上万的游客来此散步、观鸟或赏花。

到了 1830 年，美国的思想家开始衡量城市生活中康乐设施的价值。虽然英国的德比植物园（1840 年建）、默西塞德郡公园（1945 年建）及索尔福得的皮尔公园（1846 年建）都宣称一开始就是公有公园而不是限制开放的私有物业，但真正意义上的第一个大众公园，应该是占地 20 公顷的波士顿公园（1634 年建）（图 1.59），其间的蒙特林荫大道（1728 年建）模仿英国圣詹姆士公园中的人行长廊。1830 年以前，波士顿公园中的草地是供牧牛之用，但后来公园便加建了围栏以保护草皮和榆树丛。虽然设计的原意并不是一座公园，但经过数代的演变，便发展形成了位于市中心供群众闲暇娱乐的场所。1830-1840 年间的新英格兰地区，公园或休憩场所中一般都种植当地有树冠的树木并设有围栏以保护草地和珍贵树木。美国内战过后，为纪念壮烈牺牲的勇士，公园内便设有各类纪念碑。

19 世纪，城市的居住环境肮脏恶劣，在英国情况尤为严重，这一现象加速了社会改革精神的发展。这些精神中包括 19 世纪

图 1.58
奥本山公墓，剑桥市，美国马萨诸塞州
Mount Auburn Cemetery, Cambridge, Massachusetts, USA

这是世界上第一个花园公墓也是首次以地形和植物的多样化为主的大型景观设计。西方对于公墓的前卫想法把一个拥挤的并且不卫生的墓地转变成为一个适于生活和游览的地方，就像是一个公共花园。
The world's first garden cemetery was also the first large designed landscape in the U.S. and featured topographical and botanical variety. The West pioneered the idea of the cemetery replacing the crowded and unsanitary graveyard as a place for the living to enjoy visiting, just like a public park. (Credit: Daderot)

40 年代广为流行的公园运动，它甚至传播到了亚洲；公园运动旨在赋予民众享有大尺度的、精心设计的公共开放空间的权利。当时，火车站前的广场在其最初产生时很难作为人民驻足休憩的开放空间。皇家园林也并非长期开放，而民众却再也不满足于看皇室贵族的脸色和心情行事，只能偶尔进去游览。即便是可以出入，彬彬有礼地拜访私家宅邸、参观墓园总敌不过自由自在徜徉于公共开放空间，后者带来的舒畅喜悦之情莫可言状。城市居民强烈要求享有自己的公园空间。

初期的公园除了有一般城市广场空间拥有的喷泉及雕像以外，还设有为大众带来舒适和方便的公共配套设施。基本设施包括长凳、整齐而等级分明的道路、可达的草坪及洗手间；高等设施包括凉亭、小卖部、咖啡屋、游乐场、锻炼设施、有标识的植物和历史遗迹。史上第一座城市游乐场，即建于 1843 年的哥本哈根蒂沃利公园，证明了私有企业亦可满足公众康乐设施尚未满足的需求（图 1.60）。园内设有东方风格的建筑物、规整的花园、旋转木马、过山车、烟花会演等项目。后来的月亮公园及迪士尼乐园都是以蒂沃利公园为蓝本，但不同的是这些都是商业机构而不是城市的公共开放空间。

19 世纪中叶，西方城市陆续投资绿色基建项目。1852-1870 年的巴黎，乔治·欧仁·奥斯曼首次以绿色通道作为城市规划的配套设施，为巴黎市规划出完善的景观林荫大道，贯穿市内各重点区域。这个构思启发了其他的欧洲城市，如维也纳的环城大道，是由弗朗茨·约瑟夫一世大帝于 1857 年下令建成的。这些林荫大道显示出国家的实力并能防止暴乱，而且大规模种植行道树能够防止火灾蔓延（图 1.61 - 图 1.62）。这些巴黎和维也纳的笔直林荫大道发展形成了大型公园，为后来美国的绿化道形成做了铺垫，其中最早的范例是在 1849 年的芝加哥。

of Nature and memorialization of such garden cemeteries in many cities attracts thousands of visitors to walk, bird-watch, and admire the botanically rich plant collections. But they were not parks.

By 1830 leading thinkers in the U.S. began to argue for the value of recreation in urban life. Although Derby Arboretum (1840), Birkenhead Park (1845) and Peel Park, Salford (1846), all in England, are often cited as the first parks that were public from the moment they were first created, and not simply an opened-up private estate, a leading contender is Boston Common, 20 hectares set aside by and for the people in 1634. (1.59) Its first pedestrian promenade, Tremont Mall (1728) imitated the Mall at St. James Park; the grazing of cows was ended in 1830; and its turf and grove of elms were soon enclosed in an ornamental fence. Although it was not designed but had evolved over generations, it was still a park—a place where people could spend leisure time in a relaxing natural setting in the heart of the city. In the 1830s-40s, New England greens and commons were being enclosed with ornamental fences and planted with groves of the most beautiful indigenous canopy trees scattered over turf. They attracted historic monuments, and after the Civil War statues typically honoring the sacrifice of the common man in war, and still later bandstand pavilions.

More broadly, squalid urban conditions especially in England gave rise to a spirit of social reform that was to accelerate through the 19th century. By the 1840s that spirit included the Parks Movement, which spread city by city, country by country as far as Asia and focused on granting people the right of public access to more ambitiously scaled and designed open space designated as parks. The advent of plazas in front of the railway station hardly counted as open space for people to linger in and relax. Nor were the people willing any longer to settle for access to royal gardens only when royalty was in the mood to allow the public onto their lands. And even when access was forthcoming, politely visiting an estate or garden cemetery was not the same thing as lingering and lounging in a public park, the delights of which were just now being imagined. City dwellers were demanding parks of their own.

Apart from the ornaments they shared with other urban space types like fountains and statues, parks would sooner or later provide public amenities for the comfort and convenience of the public, both basic ones like benches, graded and groomed paths, accessible lawns and restrooms, and high amenities like pavilions, refreshment stands and cafes, playgrounds and exercise equipment, labeled plantings and historic markers. The world's first urban amusement park, Tivoli Gardens in Copenhagen (1843) made the case that private enterprise could fill an unmet public recreational need. (1.60) Buildings with oriental character, formal gardens, and open water were combined with merry-go-round and roller-coaster rides and fireworks. The many Luna Parks, Disneylands, and theme parks to follow drew inspiration from Tivoli, but they are commercial enterprises, not urban public open space.

From the mid 19th century, Western cities underwent massive investment in green infrastructure. In Paris from 1852 to 1870 Baron Haussmann laid out a system of landscaped boulevards through the fabric of Paris, pioneering the use of greenways as a city planning device. The idea inspired many other European cities like Vienna's celebrated Ringstrasse (1857), ordered by Emperor Franz Joseph I; the boulevards showcased state power, warded off mob rule, and their large scale and generous tree planting prevented the spread of fires. (1.61-1.62) Boulevards in Paris and Vienna led to large parks, foreshadowing later parkways in the United States that had been advocated as early as 1849 in Chicago.

It was a chance visit to Birkenhead Park in Liverpool in 1850 that inspired an excited twenty-eight year old farmer-journalist named Frederick Law Olmsted to marvel at the novelty of a People's Park with lakes, plantings, curving roads, bridges and pavilions. (1.63-1.64) In this moment the die was cast. Olmsted's words quoted at the beginning of this chapter appeared in 1852, the same year that the media-savvy American promoter of public parks and picturesque landscape gardening, Andrew Jackson Downing, died in a steamboat accident at age thirty-seven, leaving a void that Olmsted now filled. In Olmsted's subsequent career, the profession of Landscape Architecture emerged to face the growing challenge of reconciling the rapidly urbanizing Western city with Nature. Superseding landscape gardening, the first environmental design profession, landscape architecture began in the largest and fastest growing city in North America, New York City. Teamed with Downing's former partner, British architect Calvert Vaux, Olmsted was the first design professional to designate himself a landscape architect. (1.65) Olmsted and Vaux went on to win the competition for a new central park, originally advocated by Downing,

图 1.59
水之庆典，波士顿公园，1848 年 10 月 25 日
View of the Water Celebration, Boston Common, October 25th, 1848

一个长期市政供水系统的引入的确是带来了庆祝活动，波士顿把这场庆祝的地块选在了世界上第一个不是由私家庄园演变而来的市中心公共公园中。用来展示水压能量的是一个喷泉，它位于青蛙池塘，这是一个天然池塘，至今还是一处最受人们喜爱的景点，请注意图中中心偏左的中国游客。
The introduction of a long-needed municipal water supply system was indeed cause for celebration, which Boston chose to hold in what is the first downtown public park in the world that had not originated from a private estate. Demonstrating the power of water pressure itself is the fountain, placed in a natural pond, the Frog Pond, still a favorite feature today. Note the Chinese visitor, left of center. (Credit: U.S. National Archives and Record Administration)

图 1.60
蒂沃利公园，哥本哈根
Tivoli Gardens, Copenhagen

这个世界上最老的主题公园依旧保持着它原有的特色，其中包括了一个派生的中国建筑形式的戏台。这种私有的场所帮助启发了公共开放活动空间的设计和活动的举行。
The world's oldest theme park still retains much of its original character, including a derivative Chinese architectural style for a theater. Such privately owned destinations have helped inspire public open space design and activities. (Credit: Sissew)

1850 年，28 岁的农民兼记者奥姆斯特德偶然造访了利物浦的伯肯黑德公园，他流连惊异于园中湖泊、林木、曲径、桥梁及凉亭等景致，这为其日后的职业发展做好了铺垫（图 1.63 - 图 1.64）。本章开段的引言来自 1852 年奥姆斯特德的著作。同年，推崇公共公园主义及艺术造园的美国杰出景观园林先驱唐宁死于水上事故，享年 37 岁，唐宁去世所留下的空缺最终由奥姆斯特德填补。在奥姆斯特德其后的职业生涯中，景观设计这个行

图 1.61 - 图 1.62
弗朗茨·约瑟夫路,内环路,维也纳,约 1890 年
Franz Joseph's Quai, Ringstrasse, Vienna, ca 1890s. and Ringstrasse and Natural History Museum, Vienna, ca. 1900

以 19 世纪中期的巴黎为样本精心设计过的维也纳,树木列植的林荫道和公园开始取代城墙以及贫民区。北京在一个世纪之后也这么做了。
Following the example of Paris in the mid 19th C., tree-lined boulevards and parks began to replace city walls and slums in the beautifully designed city of Vienna, as Beijing was to do a century later. (Credit: LC-DIG-ppmsc-09230 and 09234, U. S. Library of Congress, Prints and Photographs Division, Photochrom Prints Collection)

with an entry they called "Greensward." Central Park was designed in the naturalistic English style but owed much to the latest example of it, Bois de Boulogne (1853-58) in Paris. In fact Olmsted went to Paris and accompanied "engineer" Adolphe Alphand to his major park project then under construction, and returned seven times. Besides this park of 845 hectares, Alphand designed Parc des Buttes-Chaumont (1867), important because it transformed a crime-infested rock quarry into picturesque wildness safe for people to walk over, and incorporated Chinese and English garden design. (1.66) Likewise, in Central Park, a rough and abused site was to be cleaned up, regraded, smoothed over in some areas, and made more wild and irregular in others, planted naturalistically and provided with irregular water features for boating and skating, all to look as if Nature itself had provided them, leaving some in later generations to assume, wrongly, that the designers had simply preserved pastoral scenery. (1.67-1.69) There was only one symmetrical feature, a promenade, but all else—turf, separate paths and roadways for different uses—conformed to the irregular topography. The interference of cross traffic was minimized by being

业逐渐成形，以应对西方加速城市化与大自然的矛盾和统一的挑战。继风景园林设计后，首个环境设计专业——景观建筑在北美最大、发展最快的城市纽约开始发展。奥姆斯特德与唐宁的前合作者，英籍建筑师沃克斯共同合作，并且奥姆斯特德是第一个称自己为专业景观建筑师的人（图 1.65）。他与沃克斯共同设计的中央公园的"绿地"方案从众多参赛方案中脱颖而出。这一方案也是当初唐宁所极力拥护的。中央公园受自然风格的英式景观设计的影响，但是很大程度上归功于近期的巴黎布洛涅森林公园（1853-1858 年）的模式。实际上，奥姆斯特德曾经七次到访巴黎，陪同"工程师"吉恩·查尔斯·阿尔道夫·阿尔方视察兴建中的主要公园项目。除了占地 845 公顷的布洛涅森林公园以外，阿尔方还设计了 1867 年的比特·休蒙特公园。这个公园在景观建筑界里地位甚高，原因是阿尔方不但把罪案率高的荒废采石矿场改建成一个景致优美、充满野趣又安全的散步游玩场所，而且还融入了中式和英式造园要素（图 1.66）。相似地，中央公园的旧址亦曾是一处粗糙混乱的场地，需要大费周折去清理、改变地形，一些区域需要平整，而另一些区域则需要以不规整的形式增添野趣，还要搭配自然种植，并设置不规则的水体以供夏日泛舟或冬季滑冰之用。所有这些特征看似自然本身所赋予的，这使后人想当然地误以为设计者只是单纯地保留了天然的田园风光而已（图 1.67 - 图 1.69）。公园内唯一一个对称的景观节点是一条步行道，但其他元素，如草地、分离的小径、不同用途的道路设计各异，这是为了适应不规整的地形。马车道或汽车道布置于

图 1.63 - 图 1.64
伯肯黑德公园，利物浦，英国
Birkenhead Park, Liverpool, England

这是第一个由公共资金建成的自然景观公园，位于英国港口城市利物浦的市中心。奥姆斯特德受它启发成为了一名风景园林师并且在那时他逐渐开始构思纽约的中央公园。作为园林师和建筑师的约瑟夫·帕克斯顿在 1847 年设计了这个公园。
The natural landscape of the first publicly funded park in Britain in the heart of the port city of Liverpool inspired Olmsted to become a landscape architect and shaped his thinking about how to design Central Park in New York. Landscape gardener and architect Sir Joseph Paxton designed the park in 1847. (Credit: Sue Adair and Eric The Fish)

图 1.65
约翰·辛格·萨金特，弗雷德里克·劳·奥姆斯特德画像，1895 年
John Singer Sargent, Frederick Law Olmsted, 1895

几乎在赢得纽约中心公园设计竞赛的 40 年之后，奥姆斯特德在他最大的私人项目——巴尔的摩庄园中挂出了他的画像。从此他没有停下他前进的脚步。
Almost four decades after winning the competition to design Central Park, Olmsted was hardly slowing down when he posed for his portrait at Biltmore estate, his largest private project. (Credit: The Biltmore Co., Asheville NC.)

图 1.66
比特·休蒙特公园，巴黎
Parc des Buttes-Chaumont, Paris

它引入了中式和英式的花园设计，这个法国园林将一个犯罪热点采石场变成一个吸引游客、风景如画的场地。
Incorporating principles of Chinese and English garden design, the French park transformed a crime-infested rock quarry into a picturesque retreat that continues to delight visitors. (Credit: Jean-Louis Vandevivière)

Chapter One Origins of the People's Parks 043

图 1.67
中央公园，纽约
Central Park, New York City

继 1873 年在几乎没有城市化的土地上建完之后的一个世纪里，公园的周边被城市建筑所包围，正如奥姆斯特德和沃克斯明智的预言，到某天这里会被人造的"墙"所包围，它比中国的长城还要高 2 倍，将由城市建筑组成。
A century after its 1873 completion on barely urbanized land, the wisdom of Olmsted and Vaux, who predicted that some day it would be "surrounded by an artificial wall, twice as high as the Great Wall of China, composed of urban buildings," was abundantly evident. (Credit: "Image by Alfred Hutter")

图 1.68
中央公园，纽约
Central Park, New York City

从最开始，中央公园的中心湖就吸引着人们前来划船，逃离城市生活的烦嚣。
From the beginning, the Lake in the heart of Central Park has attracted city dwellers to go rowing and escape the stress of city life. (Credit: Daniel Case)

图 1.69
中央公园，纽约
Central Park, New York City

在池塘（桥旁边的）附近的滑冰场比池塘本身更多地提供给不同年龄的人们可信赖和安全的条件。
An ice skating rink near the Pond (beyond the bridge) offers more reliable and safer conditions for people of all ages than the Pond itself can. (Credit: Tomás Fano)

sunken below the level of the surrounding terrain. Olmsted and Vaux defended the scale of Central Park by describing the future park as "surrounded by an artificial wall, twice as high as the Great Wall of China, composed of urban buildings." [4]

Olmsted passionately advocated the necessity of the natural scenery of country parks to provide overstressed city-dwellers a quiet setting in the middle of a crowded city to relax or get exercise, restore their spirits, and commune with Nature, an idea popularized in the West by the Transcendentalists, whose spiritual connection to Nature had its much earlier counterparts in Asian thought. (1.70)

Parks as large as Central Park (341 hectares) have long been affectionately called "the green lungs" of the city, anticipating the concept of environmental sustainability. Nowadays, such large urban parks, greenways or riverfront corridors of open space provide environmentally sustainable design. Parks on this scale both protect precious wetland and groundwater resources, wildlife habitat and indigenous plant communities, and still remain accessible for public enjoyment.

After Central Park was established, the English-style parks movement spread across America. Every major city sought to emulate Central Park. Even the then separate city of Brooklyn across the East River from Manhattan wanted its own. Prospect Park is a true

地势较低的位置，使人车分流，避免了不同交通流线的混乱。奥姆斯特德就中央公园的尺度和未来景象作了如下描述："将高楼大厦作为公园的人工屏障和围墙，足有中国万里长城的2倍高。"[4]

奥姆斯特德强烈提倡公园自然布景的必要性，并认为其可为饱受压力的城市居民提供休憩和放松的空间，在闹市中辟出一块净土，让人们远离尘嚣，通过锻炼身体、重塑心灵，与大自然亲密接触。这个理念在超验主义者的推动下盛行于西方，而这种人与大自然的精神纽带在亚洲早已被人们发掘并推崇（图1.70）。

中央公园占地341公顷，这类大型公园因其环境的可持续性而被喻为城市的"绿肺"。时至今日，大型城市公园、绿道、滨河廊道的开放空间提供了环境的可持续性设计。这类规模的公园可保护珍贵的湿地、地下水资源、野生动物栖息地及本土生长的植物群落，同时供人们娱乐休闲之用。

中央公园建成后，英式风格的园林设计思想传遍美国。每一个主要城市都争相模仿中央公园。当时距离曼哈顿只有一河之隔的布鲁克林也想拥有自己的中央公园。奥姆斯特德的代表作"展望公园"的景观如田园诗画，一反轴线空间的常规，长草坪的弧度空间将人的视野拉长，一眼望不到头。这是城市公园中最长的空间，奥姆斯特德也视其为自己的经典杰作（图1.71 - 图1.73）。随着这种设计风格的流行，其他杰出设计师相继出现，比如旧金山的威廉·哈蒙德·霍尔和约翰·迈凯伦，把沙丘转变成郁郁葱葱的金门公园。金门公园于1874年开始兴建，占地412公顷，比

图1.70
中央公园，纽约
Central Park, New York City.

日光崇拜者聚集在最大的草坪区域——绵羊草坪，在城市中心寻找自己的特定空间。
Sun worshippers on the largest lawn area, the Sheep Meadow, find plenty of space to be left alone in the heart of the city. (Credit: Ed Yourdon)

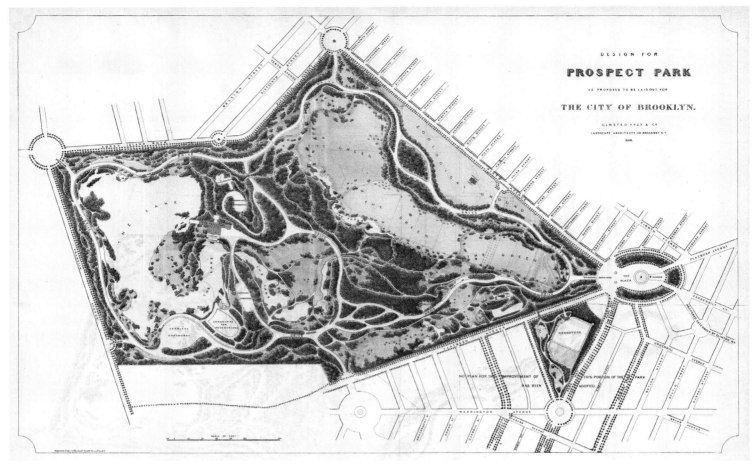

图 1.71
展望公园,奥姆斯特德和沃克斯事务所,布鲁克林,纽约,1868 年
Olmsted, Vaux & Co., Design for Prospect Park, Brooklyn, New York, 1868.

奥姆斯特德把公园设计提高到了一个更高的层次。他认为这个公园是他的杰作之一。它的伟大之处可以从他的设计原图中看出,精心设计的交通系统仿佛与自然本身融为一体。
Olmsted took park design to an even higher level in the park which he considered his masterpiece, whose greatness is revealed in his design proposal drawing and its sophisticated circulation system that seems like a part of nature itself. (Credit: Bishop, William, Manual of the Common Council of the City of Brooklyn, 1864)

中央公园面积大 20%,而且维护水平一样良好(图 1.74 - 图 1.75)。

中央公园建成后的数十年间,林肯和格兰特两任总统分别于 1864 年和 1872 年对优胜美地和黄石这两个国家公园进行了维护,就此引发了全新的国家公园运动。与此同时,衍生出的州立公园运动亦于 1885 年在尼亚加拉保护区的州立公园建立后兴起(图 1.76)这个州立公园虽然是热门的蜜月胜地,但可惜过度的商业开发导致了环境的破坏。国家公园这个构想最近被纪录片制片人肯·伯恩斯喻为美国最美妙的主意。虽然国家公园的选址都位于较遍远地带,但随着人口流动,偏僻的选址反而成为优点。这些雄伟壮丽、引人入胜的景观在任何时候都应该得到所有人的精心保护。人们可选择组团游览,也可以独立探险,给自己一个机会来进行精神上的反思,此举正契合亨利·大卫·梭罗的超验主义精神。梭罗是散文集《瓦尔登湖》(又称为《丛林里的生命》)(1854 年)的作者,也是环境运动的先驱。国家公园的思想迅速地传遍全美及其他国家。在一个世纪内,美国国家公园管理局把城市附近的土地,包括古迹、林荫大道甚至自然奇观都纳于其管理之下。1982 年,中国产生了自己的国家公园系统,当中采纳了多项美国的管理思想,但直到 2007 年才建立了真正

evocation of pastoral scenery: transcending the too obvious solution of axial vista space, the curving space of Long Meadow irresistibly pulls people on toward an end that cannot be seen; it is the longest such space in any urban park. Olmsted considered it his masterpiece. (1.71-1.73) As the movement spread, other skilled designers emerged. Notably, in San Francisco William Hammond Hall and John McLaren transformed sand dunes into the lush landscape of Golden Gate Park (begun 1874). At 412 hectares, it is twenty percent bigger than Central Park and as well maintained. (1.74-1.75)

Less than a decade after the creation of Central Park, President Lincoln's preservation of Yosemite in 1864 and President Grant's preservation of Yellowstone in 1872 launched a new idea in the world, the National Park movement, alongside a State Park movement, which gained momentum after 1885 with the creation of Niagara Reservation State Park, a popular honeymoon destination already degraded by excessive commercial development. (1.76) The National Park has recently been called America's Best Idea by documentary film maker Ken Burns, and surely by any reckoning it is one of the very best.

图 1.72
大草坪，展望公园，布鲁克林，纽约，1873 年
Long Meadow, Prospect Park, Brooklyn, New York, 1873.

为了避免这个轴线景观中太过明显的设计痕迹，长草地的曲线型空间和温和波浪线不自觉地吸引人们往前走向看不到的终点。这个标志性照片中富有诗意的雾喷场景反映了东方思想上的"天地之气"。
Avoiding the too obvious design choice of the axial vista, the curving space and gentle undulations of Long Meadow irresistibly pull people on toward an end that cannot be seen. This mist in this iconic photograph in a poetic sense is suggestive of the "cosmic breath" of Eastern thought. (Credit: reproduced with permission Frederick Law Olmsted NHS/NPS)

图 1.73
低草坪，展望公园，布鲁克林，纽约
Nethermead, Prospect Park, Brooklyn, New York

和谐的水面、林地和草坪就是奥姆斯特德的设计风格，甚至沿着小溪旁的小路漫步都会让人感到那么惬意，这就是自然的感觉。
The harmonious balance of water, forest, and meadow defines Olmsted's design style, to be felt even along secondary paths beside small brooks, all capturing the feeling of nature. (Credit: Garry R Osgood)

图 1.74
金门公园，旧金山
Golden Gate Park, San Francisco.

1873年纽约中央公园接近完工，同时也把公园运动的风潮带到了太平洋海岸。金门公园比中央公园面积大了20%，它的设计饱含了奥姆斯特德的心血。
Bringing the parks movement to the Pacific Coast at about the time that Central Park was being completed in 1873, this beautiful park, 20% bigger than Central Park, owes much to Olmsted. (Credit: Hispalois)

图 1.75
金门公园，旧金山，约1900年
Golden Gate Park, San Francisco, ca. 1900

在公园最初建造的几十年后，植物逐渐成熟，人们现在可以享受草坪或者温室里的异域植物。大多数人都感到意外，如此繁茂的景象取代了以前荒凉、被风横扫的沙丘。
A few decades after its initial construction, plantings have matured, and people then as now enjoy a visit to the lawns or the exotic plant collection inside the conservatory and are mostly unaware that this lush scenery has replaced an inhospitable landscape of bare windswept sand dunes. (Credit: Thomas M. Paine stereoview collection)

图 1.76
尼亚加拉大瀑布，约1880年
Niagara Falls, ca. 1880

在公园运动进行的同时另一个运动也在进行着，远离城市的自然风景保护运动是为了让城市的居民可以来到这种地方度蜜月等等，甚至在冬季景色也很壮观。
In a movement paralleling the parks movement, the preservation of spectacular scenery far from the city for the benefit of urban dwellers who could travel there began in such places as this honeymoon destination, magnificent even in the middle of winter. (Credit: Thomas M. Paine stereoview collection)

Though located far from city-dwellers, national parks existed for their benefit as people became more mobile. These spectacular, inspiring, spiritually powerful landscapes were to be preserved for all time for all the people. People could choose to visit them in groups, or seek solitude, in the Transcendentalist spirit of Henry David Thoreau, author of *Walden; or, Life in the Woods* (1854), which spearheaded the environmental movement that is still with us. The idea spread across the country and the world. Within a century, the U.S. National Park System included lands near the cities, historic sites, and parkways as well as natural wonders. China's own system launched in 1982 adopted many of the characteristics of the U.S. invention, although meeting truly international standards awaited the creation of Pudacuo National Park in 2007.

The world's first continuous system of urban parks is Boston's Emerald Necklace which began even before Olmsted's initial involvement in 1876. (1.77) Linking parks rich with water features and culminating in a great country park, Olmsted's plan built on two earlier extensions of the no longer sufficient Boston Common, first to the Boston Public Garden (1837), then Commonwealth Avenue (begun 1865), a tree-shaded promenade bisecting a fashionable Parisian-styled neighborhood and often compared to Baron Haussmann's Parisian boulevards. Commonwealth Avenue anticipated the parkways that Olmsted was to design for Boston and elsewhere, such as in Buffalo, New York (1868) and Riverside, Illinois (1869), and became an important subtype of open space for people on the move. Commonly misunderstood is the subtle fact that the "parkway" is not just a road, it is a greenway corridor with a road in it. (1.78) An early example is Horace Cleveland's master plan for a park and parkway system for Minneapolis (1883). After the Emerald Necklace, greenway corridors quickly became an important form of urban public open space in many other cities, notably Chicago (1.79), and many later examples have reclaimed waterfront brownfields and abandoned railroad rights of way. Boston itself enlarged the idea even further by inaugurating the first Metropolitan Park System in 1893, the vision of landscape architect Charles Eliot to protect and open up uplands, river greenways and beaches in a greenbelt circling Boston long before the term greenbelt had been invented.

By the time of the Chicago World's Fair of 1893, expositions had been a recurrent event for a century, dominated by France, showcasing the benefits of free trade among nations in a spirit of friendly competition. The technological breakthrough of iron structures supporting glass walls and roofs enlarged the orangery/conservatory/palm house world of the rich into the indoor "winter gardens" and exposition halls for the people. The first such palace for the people was landscape gardener Joseph Paxton's Crystal Palace in Hyde Park for the Great Exhibition in London, built in 1851 and symbolically 1851 feet (570 meters) long. (1.80-1.81) This indoor world set a high standard for landscape design. In the Universal Exposition of 1889 in Paris, open space design became a matter of vast paving to move

图 1.77
人们在牙买加池塘上面溜冰，波士顿
Skating on Jamaica Pond, Boston

1876 年以后，奥姆斯特德的翡翠项链把波士顿市区的开放空间和未完全开发地区的重要自然景色区域连接在了一起。尤其包括了这个城市里最大的水池，它长时间地吸引了大量的休闲活动，甚至在冬天也是一样。就像这张 1876 年的印刷品所表现出的一样。
After 1876, Olmsted's Emerald Necklace linked Boston's downtown open spaces with the most significant landscape features in the less developed parts of the city, notably including the largest pond in the City that had long been attracting intense recreation even in winter, as shown in this charming 1859 lithograph. (Credit: J. H. Bufford, lithographer, ca. 1859)

图 1.78
洛克小溪公园大道，华盛顿特区
Rock Creek Parkway, Washington DC.

1870 年左右，始于美国的公园大道——沿着自然排水系统的绿色廊道，并包含景色秀丽的车行道和游览步道，通常有丰富的植被和自然地形——已经预见了当今的绿色基础设施。车行道的排列道通常是缓的弧独线段，避免了直线和突然的转弯。洛克小溪公园大道就是其中一个例子。它建于 1923-1936 年。
Beginning in the U.S. around 1870, parkways—greenway corridors following the natural drainage system and containing scenic roadways and trails with intense planting and natural shaping of the ground—anticipate today's green infrastructure. Roadway alignment was typically curved in gentle arcs, avoiding straight lines and abrupt turns, as in this example constructed between 1923 and 1936. (Credit: Bachrach44)

图 1.79
湖岸公路，芝加哥
Lakeshore Drive, Chicago

建于 1937 年。芝加哥这条沿密歇根湖的著名公园大道延伸了高标准的公园系统。今天不仅仅是给机动车，还提供给行人、慢跑者、骑自行车者和赛格威驾驶者使用。
Built in 1937, Chicago's famous Parkway along Lake Michigan extended the ambitious park system and today accommodates not only vehicles but walkers, joggers, cyclists, and Segway riders. (Credit:Dori, License: Dual GFDL, CC)

符合国际标准的普达措国家公园。

世界上首个连贯的城市公园群，就是波士顿的"翡翠项链"，奥姆斯特德在 1876 年首次参与到这个项目中来（图 1.77）。翡翠项链在规划上将市内各个公园以水景和绿地连接起来，并以一个大型郊野公园作为终点。奥姆斯特德的设计基于波士顿公园的两个扩展部分——波士顿公共公园（建于 1837 年）和联邦大道（始建于 1865 年）。这条绿树成荫的人行道从一个时尚的巴黎式街区中央贯穿而过，并常常用来与奥斯曼男爵的巴黎大街作比较。联邦大道影响了奥姆斯特德在波士顿和其他城市的园林大道设计，比如在纽约州的布法罗市（1868 年）和伊利诺伊州的里弗塞得市（1869 年），这种设计也成为开放空间的一个重要子类型，服务于快节奏的都市繁忙人群。绿化大道常被误以为是一条普通的道路，其实它是条绿化通廊，而道路只是其间的一部分（图 1.78）。类似的先例是霍拉斯·克里弗为明尼阿波利斯城的公园及园林大道设计的总体规划图（1883 年）。翡翠项链竣工后，绿化道路很快便成为城市开放空间的重要形式（图 1.79）。以芝加哥及其他大城市为例，当地政府成功翻新滨海废弃工业区及废置铁路等用地。波士顿进一步于 1893 年为首个城市公园系统举行了开幕仪式。这个项目是由景观建筑师查尔斯·艾略特发起的，目的是要保护开放的高地、绿色通道及海滩。受保护的地区形成了一条绿化带包围着波士顿，而当时还没有绿化带这个专用名词。

当芝加哥在 1893 年举办世界博览会时，世博会已经历时百年了，当时法国的展览占据了主导地位，他们展示了国际自由贸易及良性竞争的益处。铁质结构在支撑玻璃幕墙和天花板方面的技术突破大大提升了建筑物内的跨度，将昔日有钱人世界里的橘园、温室、棕榈室转换成对民众开放的大型室内"冬季花园"和展示厅。第一座这样的宫殿是由景观园林家约瑟夫·帕克斯顿在海德公园内为伦敦博览会建造的水晶宫。水晶宫建于 1851 年，长 1851 英尺（570 米）（图 1.80 - 图 1.81），用于象征这个特殊的年份。这个室内天地为景观设计树立了一个高标准。在 1889 年的巴黎环球博览会上，开放空间设计利用大面积铺装将人群从树池疏散出来，分流至著名的埃菲尔铁塔，铁塔是根据其具有远见的设计者而命名的。芝加哥是世界摩天大楼的诞生地，在这里，建筑师丹尼尔·伯翰召集了包括奥姆斯特德在内的一群才俊，着手设计 1893 年的芝加哥世界博览会（图 1.82 - 图 1.83）。这个团队被雕塑家圣·高登称为自 15 世纪以来最杰出的西方艺术家组合。在奥姆斯特德的坚持下，整个项目取址于他于 1871 年规划的一个潟湖，自然形态的湖泊内有岛屿，岛屿上树木成林。整个项目如同一座最新改造的古罗马广场，心怀自然。这是首个由景观建筑师规划的世博会。此次会展中的摩天轮是对埃菲尔铁塔的回应，近年来，被伦敦和中国等世界其他地方广为模仿。丹尼尔·伯翰以他的名言"只做大手笔"而闻名，他在博览会"大白城"中的经典风格的怀旧倾向，成了城市美化运动的催化剂，其影响波及远在亚洲的马尼拉。虽然博览会结束以后，展馆必须像电影布景一样被拆除，但博览会至今仍是富有想象力的设计作品的沃土。位于三栋世界级的摩天大厦旁边，2010 年的上海世博会承继了前人的"大手笔"，成为史上最大规模的世博会，也是第一个在发展中国家举办的世博会。

图 1.80 - 图 1.81
水晶宫，大博览会，伦敦，19 世纪 60 年代
Crystal Palace, Great Exhibition, London, 1860s

水晶宫前所未有地使用了铁结构和大幅玻璃板。而且它室内和室外的景观设计也为以后的世界级博览会定了新的高标准。它同样深刻影响到城市公园的设计，正如我们看到的金门公园。
The use of iron structure and large sheets of plate glass pioneered here, and the interior and exterior landscape design set a high standard for all subsequent world's fairs and greatly influenced urban park design, as we saw in Golden Gate Park. (Credits: Thomas M. Paine stereoview collection)

图 1.82 - 图 1.83
H.D 尼克尔斯（1849-1939 年），芝加哥世界博览会，彩色平版印刷，1893 年
H.D. Nichols (1859-1939), Chicago World's Fair, chromolithograph, 1893. and Chicago Worlds Fair, 1893

奥姆斯特德参与了世博会的平面规划，确保了人们在需要大型城市广场的同时，自然仍是设计的核心。这张历史照片摄于树林岛上。
Olmsted's involvement in the master plan of the world's fair assured that while the crowds required grand civic plazas, nature was at the heart of the design, in Wood Island, seen in the historic photograph. (Credits: Boston Public Library Print Department Collection and "Unidentified photographer")

the crowds between planters on their way to the Eiffel Tower, named for all time for its visionary designer. It was in Chicago, birthplace of the skyscraper, that architect Daniel Burnham corralled what sculptor Saint-Gaudens called the greatest assemblage of Western artists since the 15th century, a group that included Olmsted as the master planner, to design the 1893 Chicago Worlds Fair as a full-immersion experience of classical buildings surrounding, at Olmsted's insistence, a naturalistic lake with a Wooded Island on the site of a proposed lagoon from his own plan of 1871. (1.82-1.83) This updated Roman forum had Nature in its heart. This was the first exposition master-planned by a landscape architect. Its answer to the Eiffel Tower, the Ferris Wheel has been much imitated, lately in London and China. Daniel Burnham is famous for saying, "Make no small plans" and his thinking and nostalgic preference for the formal classical style in the "Great White City" of the Fair formed a catalyst for the City Beautiful movement that guided urban design in permanent cities as far afield as Manila, even if expositions themselves had to be dismantled like movie sets. Expositions continue to be fertile ground for visionary design. Next door to three of the world's tallest skyscrapers, Shanghai Expo 2010 is heir to that Big Idea, too: the largest world's fair ever and the first in a developing country.

Meanwhile large downtown urban plazas continued to provide settings for parades and celebrations. In the United States there was nothing as perfect as Place de la Concorde despite many City Beautiful efforts aspiring that high. Instead, an awkward intersection like Times Square combined with its newspaper-media ambience evolved into the must-experience crowd-pleaser place that persists to this day. Times Square was celebrating the New Years Eve countdown to midnight as early as 1904.

The 19th century was also the high point of worldwide colonization and empire by Britain, France, Spain, the Netherlands and other Western nations. In the latter half of the 19th century the parks movement spread to China and the rest of East Asia. Western designers flocked to the opportunities of exporting European civic design ideas to the parks, central squares and governmental buildings in the capitals of colonies across Asia.

China spent most of the 19th century dealing with social unrest highlighted in the Taiping rebellion and foreign domination forced on a weak central government by the unequal treaties after the Opium War in the 1840s. Into the 20th century expatriates from England, France, Germany, the Soviet Union and elsewhere were living in Guangzhou, Hong Kong, Xiamen, Shanghai and other treaty cities in enclaves including public gardens in the English or French style from which the Chinese were excluded.

Shanghai was in the forefront. Roughly contemporaneous with Central Park in New York City, before 1863 the British community in Shanghai established a racetrack several kilometers in circumference on a site that later became the site of Peoples Square. Originally called Public Garden, Huangpu Park (1868) in Shanghai was the first

图 1.84
上海，黄埔公园，约 1900 年
Huangpu Park, Shanghai, ca. 1900

在中央公园建成的五年前，在上海的英国外籍团体于 1868 年在亚洲建立了第一个西方风格的城市公园，公园建在黄浦江的西岸。不幸的是，公园不对中国人开放。
Five years before the completion of Central Park, in 1868 the English foreign community in Shanghai built the first Western style urban park in Asia, on the west bank of the Huangpu River. Unfortunately, it was closed to native Chinese people. (Credit: postcard by unidentified photographer, Xiaowei Ma Collection)

图 1.85
上海公众公园，上海
Shanghai Public Garden

19 世纪 80 年代，上海在沿着苏州河并且临近黄浦公园的地方开放了一个公园，在那里当地的居民可以休闲放松，就像世代的杭州人在西湖边上那样。这个图片的拍摄早于 1911 年废止排队的措施。
In the 1880s the City provided a park near Huangpu Park along the Suzhou River where local Shanghainese could relax, as generations had done along the edge of West Lake in Hangzhou. The image predates the abolition of queues in 1911. (Credit: postcard by unidentified photographer, Xiaowei Ma Collection)

图 1.86
法国花园，现复兴公园，上海
French Park, now Fuxing Park, Shanghai

在上海的法租界区，旅居的人民在 1906 年建立了一个法国样式的花园。公园 1928 年后终于对中国人开放。如今孩子们依旧在池塘里划船，就像照片中所示。
In the French Concession area of Shanghai, the expatriate population created a French style garden in 1906 that was finally open to Chinese people after 1928. Children still play with boats in the pond shown here. (Credit: postcard by unidentified photographer, Xiaowei Ma Collection)

与此同时，市中心的大型广场继续为公众提供游行和庆祝的场所。虽然美国各地积极进行城市美化活动，目标高远，但成效都没有达到如协和广场般的完美。相反地，像时代广场这种拙劣的交叉路口，再加上大型广告牌的霓虹灯景观，竟演变成游人必须体验的人群聚集地，而且一直延续至今。时代广场的新年倒计时于 1904 年便开始了。

19 世纪亦是全球性殖民统治的高峰期，西方大国如英国、法国、西班牙、荷兰等都纷纷开拓殖民地。在 19 世纪后半叶，公园运动传至中国及东亚其他地方。众多西方设计师不失时机地将欧洲式的公园、中央广场及政府建筑设计运用到亚洲各个殖民地的首都市政空间中。

中国在 19 世纪后半叶面临着巨大的内忧外患，对内有太平天国起义，对外，自 1840 年鸦片战争后，软弱无能的清政府便在西方列强船坚炮利的威胁下签订了一系列不平等条约。进入 20 世纪，来自英国、法国、德国及俄国的侨民住在广州、香港、厦门、上海及其他通商口岸城市的租界中，这些华人不得入内的租界里有英式或法式的公共花园。

上海可算得上是城市公园的先驱。早在 1863 年，上海的英国人社区建造成了周长数公里的赛马场，这基本上与纽约的中央公园同时期，后来赛马场成了人民广场。上海的黄浦公园建于 1868 年，原名公共花园，由一位苏格兰人设计，风格为欧式，内设音乐演奏台、草坪及网球场。很不幸的是，1928 年以前，黄浦公园都只限于外籍人士入场。然而，这里和其他通商口岸的租界区内的公园为中国引进了公园设计的新思维，并传播到了租界以外的社区（图 1.84 - 图 1.86）。

自此之后，其他东亚国家亦开始建造公园。曼谷的隆比尼公园（建于 1868 年）是泰国首座公园。深受西方创新观念的影响，拉玛四世国王（1804-1868 年）用 32 公顷的皇家用地建立了公共公园，并以佛陀出生地命名。上野恩赐公园（建于 1873 年）建于一座毁于内战的庙宇旧址上，是日本第一座公园，今天的上野公园有动物园、博物馆、湖泊、寺院，每年 4 月吸引了众多游客前来赏樱（图 1.87）。霹雳州的太平湖公园（建于 1880 年）是马来半岛（现在的马来西亚）的首座城市公园，这里曾经是一个锡矿，把废矿湖改建成湖滨公园是罗伯特·桑迪兰·弗洛得·沃克上校提出的理念（图 1.88）。公园占地 64 公顷，共由 10 个风景秀丽的湖泊组成，至今仍是全马来西亚保养得最好及最美丽的公园。吉隆坡的湖滨公园（建于 19 世纪 80 年代）位于市中心，是前侨民住宅区，占地 92 公顷，郁郁葱葱的热带植物景观围绕着两个湖。湖泊公园内的周边种植有丰富的热带植物，亦有大型热带树围合的大片草地、花圃、兰花园、池塘、喷泉、长满绿草的山丘、雀鸟公园、蝴蝶公园、鹿苑及山谷中的表演场。在英国殖民时代之前，还不能称为公园的休闲花园，早已见之于传统马来语的文献中，例如马来诗歌班顿、传奇小说、印尼传统皮影戏及玛蓉舞。

韩国的城市开放空间和中国大致相同，传统上这些空间都是供户外市场或社区活动之用的。自由公园是韩国首个西式现代公园，建园目的是纪念美国将军麦克阿瑟在第二次世界大战中解放了韩国。其历史可追溯至 1888 年，公园建立于仁川中的一处欧洲领土上，设计师是俄罗斯建筑师兼土木工程师瑟拉丁·萨玛汀，

图 1.87
上野公园，东京，日本
Ueno Park, Tokyo, Japan

如今每个春季人们在这个第一个日本公共公园里享受樱花的盛开。1873年这个公园取代了一个在内战中被摧毁的寺庙。
Today people enjoy the cherry blossoms each spring at the first public park established in Japan, which in 1873 replaced a temple destroyed in civil war. (Credit: Bernard Gagnon)

图 1.88
太平湖公园，霹雳州，马来西亚
Taiping Lake Gardens, Perak, Malaysia

在马来西亚的文学中，愉乐公园要早于英国人在马来西亚的出现，但是这个国家的第一个公共公园是当时一位英国居民的愿景，那时马来西亚还是英属马来西亚殖民地。公园的前身是一个矿场，现如今一直都是当地引以为傲的资源。
In Malaysian literature, pleasure gardens predate the British presence there, but the first public park in the country was the vision of a British resident when Malaysia was the British colony of Malaya. It healed a former mining site and remains a source of local pride. (Credit: Johnwxh30 at en.wikipedia)

public park in China, designed by a Scotsman in European style, with a bandstand, lawn, and a tennis court. Unfortunately, it was restricted to the foreign public until 1928. However, here and in other treaty ports the parks in the foreign districts had imported thinking about park design that was already spreading into the wider community. (1.84-1.86)

Soon after, other East Asian nations were creating parks. Lumpini Park (1868), Bangkok, became the first public park in Thailand. King Rama IV (1804-1868), who embraced Western innovations, created the public park on 32 hectares of royal land and named it after the birthplace of the Buddha. Japan's first public park, Ueno Park (1873) replaced a temple destroyed in civil war. (1.87) Today its features include a zoo, museums, a lake, temples and Sakura cherry trees that attract throngs every April. The Taiping Lake Garden (1880) in Perak was the first urban public park in Malaya (now Malaysia), interestingly on a former mining site, and the vision of Colonel Robert Sandilands Frowd Walker. (1.88) Its 64 hectares with ten scenic lakes and ponds remain the most beautiful and well maintained park in Malaysia. Lake Gardens (1880s) is a 92-hectare tropical botanically rich designed landscape surrounding two lakes on the site of former expatriate estates in the heart of Kuala Lumpur. The lake garden itself consists of large lawns lined with big tropical trees, flower beds, orchid gardens, ponds, fountains, grassy hills, bird park, butterfly park, deer park and a valley venue for performances. Pleasure gardens, if not public parks, from well before the British colonization era were alluded to in traditional Malay literature such as the Pantun, Hikayat, Syair, Wayang Kulit and Mak Yong.

As in China, Korean urban public space traditionally accommodated an outdoor market and community events. The first modern, Western-style public park in Korea dates from 1888 and has been called Jayu Gongwon (Freedom Park) since Gen. McArthur liberated Korea. Established in a European enclave in Incheon, the park was designed by a Russian architect and civil engineer, Seredin Samatine, and was originally called Gakgook Gongwon (International Park).

In the Philippines, Luneta (Rizal) Park (1901) transformed an execution ground in the cleared area outside the original Spanish walled city into an axial central park of 22 hectares in the governmental heart of Manila, honoring the poet Jose Rizal, whose execution in 1896 sparked the first successful uprising in Asia against a Western colonial power. (1.89) In 1946 the Park became the setting for the Declaration of Independence from the United States. Today this iconic space offers outdoor concerts and includes Chinese and Japanese gardens, the national library and national museum. A grandstand hosts political events, and a pond features an island shaped like the Philippines.

In 1903 the Japanese created the first European-style urban park in Taiwan, Taichung Park, featuring a relocated city gate and a pond, followed in 1908 by Taihoku Park, in Taipei, later renamed the 228 Peace Memorial. (1.90)

图 1.89
黎刹公园，马尼拉，菲律宾，1899 年
Luneta (Rizal) Park, Manila, Philippines, 1899

为了纪念在 1896 年被处死、第一个成功在亚洲发起的抵抗西方殖民运动的诗人约瑟•黎刹，1946 年独立宣言在黎刹公园宣布，从此菲律宾从美国独立。如今这个标志性的公园比最初的设计多了更多有趣的活动。它提供户外音乐会，包含中式和日式花园、国家图书馆和国家博物馆也在此。
Honoring the poet Jose Rizal whose execution in 1896 sparked the first successful uprising in Asia against a Western colonial power, in 1946 the Park became the setting for the Declaration of Independence from the United States. Today this iconic space with a much more interesting design than it had originally offers outdoor concerts and includes Chinese and Japanese gardens, the national library and national museum. (Credit: Alden Marsh, The history and conquest of the Philippines, 1899)

公园原名为国际公园。

菲律宾的黎刹公园（建于 1901 年）位于首都马尼拉的政府核心地段，这是原先位于西班牙式城墙外的空地上的一个刑场改造成的中轴式公园，占地 22 公顷，是为了纪念 1896 年在此地被处决的诗人荷西•黎刹。黎刹的遇难导致了亚洲第一次成功地反抗西方殖民统治起义（图 1.89）。1946 年菲律宾从美国获取独立，黎刹公园成为独立庆典的重要会场。今日的黎刹公园为户外演唱会提供了场地，而园内另设有中式及日式花园、国家图书馆和国家博物馆、政治活动看台以及一座放置着菲律宾群岛模型的人工水池。

1903 年，日本人在其强占的中国领土台湾建造了第一座欧洲风格的城市公园——台中公园，园内设有日月湖，并保留了台湾府的古城门。台北大天后宫于 1908 年落成，后来改名为二二八和平纪念公园（图 1.90）。

在中国，1904-1905 年间，两座大概最早开放给所有市民的公园同时开园，这两个公园相距 2500 公里，一样是免费入场。在位于太湖的古城无锡,当地的市绅筹集资金建造了锡金公园。占地 3.6 公顷的锡金公园内设有一条湖滨小径、人行桥、归云坞、龙岗、天绘亭、兰花亭、松崖白塔并放置镌刻怀素《四十二章经》（图 1.91 - 图 1.92）。另外，通过月洞门及石子长廊，可到达别有洞天的紫园（寓意紫气东来），园内有一个八卦形状的水池及兰茶室，与此不同，厦门的一个临时西式公园忽略

图 1.90
台中公园，台北，中国台湾
Taihoku Park, Taipei, Taiwan

公园建于 1908 年，是台湾第一个公共公园。这张图片显示了早年间虽然被日本占领，但是公园依旧给许多台湾人带去快乐，如今这里伫立着一个和平纪念碑。
Created in 1908, the first public park in Taiwan, here shown in its early years pleasing many happy Taiwanese, though living under Japanese occupation, today hosts a peace memorial. (Credit: postcard by unidentified photographer, Xiaowei Ma Collection)

图 1.91 - 图 1.92
锡金公共花园，无锡，约 1905 年
Xijin Public Garden, Wuxi, ca. 1905

这个公园是中国最早所知的由私人捐资建造的公共公园。它向所有人开放，包括了沿湖步道、人行桥、花架、亭台、月洞门、假山和在中国传统私家园林中通常能看到的著名书法家的碑文。在设计上缺少的是视线的遮挡。
The earliest known privately funded public park in China open to all the people included a lakeside path, footbridges, trellis, pavilions, moon gate, rockery, and inscriptions from famous calligraphers as one might find in a traditional Chinese private garden. What it lacks in design confidence it more than makes up for in its vision. (Credit: postcard by unidentified photographer, Xiaowei Ma Collection)

了中国的文化符号（图 1.93）。所有这些让人想到私家花园的精心设计和文化。与此同时，在东北北部的齐齐哈尔，占地 64 公顷的龙沙公园成了黑龙江省的第一个公园（图 1.94）。唐代诗人李白有"将军分虎竹，战士卧龙沙"的诗句，龙沙公园由此而得名。园内的寿山祠是为纪念清末爱国名将寿山将军所建。建于 1907 年的望江楼是园内最早的建筑，位于劳动湖东畔的石山上。面积 20 公顷的劳动湖，湖中有岛，而岛上有跨水而建的曲桥、彩虹式拱桥与玉带桥。自 1963 年起举办的一年一度的冬季"龙沙冰景游览会"举世闻名。园内另设有图书馆和动物园，饲养有东北虎和棕熊等珍兽。上述两座举足轻重的公园的建园日子

In China in 1904-05, perhaps the first two parks for all the people opened almost simultaneously 2500 kilometers apart, free of charge. Leading citizens in Wuxi, an ancient city on Lake Tai, raised funds to create Xijin Public Garden. The 3.6 hectare park featured a lakeside path, footbridges, trellis, Dragon Knoll, octagon pavilion, orchid pavilion, white pagoda, and stone inscriptions from famous calligraphers. (1.91-1.92) In addition, through a moon gate and beyond rockery lay the hidden dreamy world of the Purple Garden (alluding to purple air coming from the east, an auspicious symbol) with a pool shaped like a *bagua* (the equally auspicious eight diagrams symbol)

图 1.93
市政花园，广州，厦门，约 1905 年
Municipal Gardens, Xiamen, Guangzhou, ca, 1905

这个早期明信片上广州公共开放空间的景色展示了当时在这个通商口岸很受欢迎的独特的西式风格。
This early postcard view of a public open space in Guangzhou reveals a distinctly Western style that one might well expect in the treaty port formerly known as Canton. (Credit: postcard by unidentified photographer, Xiaowei Ma Collection)

and orchid tea room. All this recalls the elaboration and culture of private gardens.Conversely a contemporary Park in Xiamen largely ignored Chinese culture (1.93) Meanwhile in Qiqihar in northern Manchuria, Longsha Park (64 hectares) became the first park in Heilongjiang Province. (1.94) Its name ("dragon sand") alludes to a Tang dynasty poem about the fought-over lands north of the Great Wall. In that spirit, Shoushan Temple honors a recent military hero. Among the pavilions that crown the hills, the oldest building, on a rocky promontory on the shore of Labor Lake, dates from 1907. The 20-hectare lake includes an island with ponds, zigzag bridge, rainbow bridge, and jade belt bridge. The ice carving festival held each winter since 1963 has become famous. The park includes both a library and a zoo with Manchurian tigers and bears. These two immensely important parks predate the 1911 Revolution by seven years. (1.95-1.102)

In the West a profound technological change was underway—the assembly-line mass production of the automobile. After 1910, open space design had to mitigate the automobile's destructive effect on the quality of urban life. The first limited access parkway appeared in 1907 in the Bronx River Parkway in New York, ushering in perhaps the most ambitious metropolitan parkway system anywhere, largely created by "master builder" Robert Moses, sometimes compared to Baron Haussmann of Paris.

比 1911 年的辛亥革命还要早 7 年之久。（图 1.95 - 图 1.102）

当时的西方正在进行一场彻底的技术革新，就是汽车装配线的大规模生产。1910 年以后，开放空间的设计须缓和汽车对城市生活质量造成的破坏性影响。首个限制出入的绿化道路——朗克斯河公园大道于 1907 年在纽约出现，引入了可能是工程量最为浩大的大都会绿化道路系统。此系统主要由"建筑大师"罗伯特·摩西担纲主创，人们常把他跟负责巴黎城市规划的奥斯曼男爵相提并论。

将汽车停泊在原本以行人为主导的市集广场等开放空间中，或更糟的是，清除传统的城市肌理来适应汽车，这些行为是对城市生活质量发出的全球性挑战，连伊斯兰圣城麦加和麦地那亦不能幸免。

花园城市运动由英国传到斯堪的纳维亚、荷兰、美国等因汽车而令郊区城市化加速的国家和地区。新市镇规划提倡在一片中央开放空间，比如村镇绿地的周围设置禁止车辆通行的住宅或商业区。从规划上来看，这些车辆禁行区很容易被误认为是当今中国正在兴建的住宅项目，但有一个很大的差别，就是中国的项目更高更稠密，以致将开放空间建在地下停车场的顶部。

1911 年清帝退位以后，孙中山先生成为中华民国大总统，曾是清朝皇族御花园的北京颐和园开始对外售票开放，作用和纽约的中央公园相似。1925 年，陆续有更多的御花园对外开放，北京的北海公园就是其中之一，当时在西方，北京被喻为有史以来最美丽的城市之一。1927 年，苏州建立了大公园，从开园

图 1.94
龙沙公园，齐齐哈尔，约 1930 年
Lungsha Park, Qiqihar, ca. 1930

这个公园几乎和无锡的锡金公园同时开放，这个是在中国东北地区第一个对公众开放的公园，自 1906 年对外开放，1/4 的世纪过去了，公园有让人印象深刻的入口大门以及茂密的种植。
Almost as old as Xijin Public Garden in Wuxi, this is the first park to be publically funded, the first in northeast China, a quarter century after it opened in 1906, with an impressive entrance gate and thick plantings. (Credit: postcard by unidentified photographer, Xiaowei Ma Collection)

图 1.95 - 图 1.96
龙沙公园，齐齐哈尔，约 1920 年
Ryusha Park, Qiqihar, ca. 1920

这个公园可以追溯到日本占领中国东北地区的时候，它还提供了一个大型鸟类养殖场和足够的荫凉邀请人们前去休息。
This public park also dating back to the Japanese occupation of northeast China features an aviary and ample shade to invite people to sit and relax. (Credit: postcard by unidentified photographer, Xiaowei Ma Collection)

图 1.97
夏家河子公园，大连，1930 年
Hsia-Ho-Tzu, Dalian, ca. 1930

游泳的人穿过绿树苍苍的公园，许多家庭在花园中休息。这种健康的娱乐活动在大连要比中国的其他城市开展得早。
Swimmers walk though the leafy park passing families relaxing among flower gardens. This healthy range of recreational activities was made available in this attractive city earlier than in most other cities in China. (Credit: postcard by unidentified photographer, Xiaowei Ma Collection)

图 1.98
人们在电气公园中垂钓，大连，约 1920 年
Fishing at Electric Park, Dalian, ca. 1920

在这个开敞的游乐园中，精心设计过的有鱼的水景旁的亭子提供了荫凉。
Within a spacious amusement park, pavilions provide shade next to an elaborately designed water feature stocked with fish. (Credit: postcard by unidentified photographer, Xiaowei Ma Collection)

图 1.99
中华公园，沈阳，约 1930 年
Chinese Park, Shenyang, ca. 1930

当地的中国居民在这个远离城市其他日本公园的自己的公园中享受着团体感，他们聚集在遮阴亭下，或带着小孩散步。
Local Chinese residents enjoy a sense of community in their own park, separate from Japanese parks elsewhere in the city as they gather in the pavilion and take their children for a walk. (Credit: postcard by unidentified photographer, Xiaowei Ma Collection)

图 1.100
千代田公园，沈阳，约 1930
Chiyoda Park, Shenyang, ca. 1930

追溯到沈阳被称作奉天府的时期，这个公园在水景上有着新文化运动的特点，照片中家庭漫步在很宽的步道上。缺少成熟的大树也许体现出这个公园是当时新建的。
Dating from the era when the city was known as Mukden, this park reveals an art-deco design in the water feature, as families stroll on the wide paths. The absence of mature trees may indicate the newness of construction. (Credit: postcard by unidentified photographer, Xiaowei Ma Collection)

图 1.101
奉天满铁公园，沈阳，约 1930 年
SMR Park, Shenyang, ca. 1930

座椅旁参天的大树和临近的鸟舍给空间带来了生机。我们从紧实的土地可以看出公园的高使用性。
Shady seating is in abundance in this space animated by mature trees and featuring an aviary enclosure. The compaction of the ground indicates heavy use. (Credit: postcard by unidentified photographer, Xiaowei Ma Collection)

Chapter One Origins of the People's Parks

图 1.102
长春西公园，长春，约 1955 年
West Park, Changchun, ca. 1955

公园中充满吸引力的水景被种有遮阴树丛的公园式小地形所环绕，而水景中的溜冰池塘则会让人误以为是西方公园中的冬季活动。
Pond skating in this attractive water feature surrounded by park-like terrain with clumps of shade trees could be mistaken for the same wintertime activity in a Western Park. (Credit: postcard by unidentified photographer, Xiaowei Ma Collection)

图 1.103
江滩，以前的汉口，武汉，1930 年
Bund, Wuhan formerly Hankou, ca. 1930

在 1950 年汉口合并武昌和汉阳很久之前，汉口和上海一样也拥有一个迷人的滨水绿道。
Long before it was combined with Wuchang and Hanyang to form Wuhan in 1950, Hankou had an impressive waterfront greenway corridor like Shanghai. (Credit: postcard by unidentified photographer, Xiaowei Ma Collection)

图 1.104
汉口中山公园，约 1930 年
Zhongshan Park, Hankou, ca. 1930

这个精美的明信片中集合了在城市开放空间里的人们各种活动的可能性，在那个时代中国有很多公园为了纪念孙中山先生而命名为中山公园。
This wonderfully evocative postcard filled with human activity sums up the potential of urban open space that was unfolding in the age when so many parks across China were named in honor of Dr. Sun Zhongshan. (Credit: postcard by unidentified photographer, Xiaowei Ma Collection)

The parking of cars in what had once been pedestrian-dominated open spaces like market squares, and even worse, the clearance of the traditional urban fabric to accommodate them, posed a global challenge to the quality of urban life, not least in Islamic holy cities like Mecca and Medina.

The Garden City movement spread from England to Scandinavia, the Netherlands, and the United States, where the automobile accelerated suburbanization. New town plans called for car-free superblocks organized around a central public open space like a village green. In plan, some of these superblocks might be mistaken for residential projects being built in China now, with one big difference. In China the projects are taller and denser, and open space is forced to be built on deck over underground parking garages.

After the fall of the Qing Dynasty in 1911, when Dr. Sun Zhongshan became President of the Republic of China, the formerly private imperial gardens of the Summer Palace in Beijing were opened to the people and thereafter played a role similar to Central Park in New York. In 1925 more imperial parks like Beihai Park in Beijing were also opened to the people, and Beijing was being considered in the West to be one of the most beautiful cities ever created. In 1927 Suzhou created Big Park (Da Gongyuan), open to all people from the outset. Big Park featured Min De (People's Virtue) Pavilion, overlooking two gourd-shaped ponds to the south from an artificial hill, in correct *fengshui* orientation, and ornamented with imported green and red glass, rosewood furniture and famous Chinese calligraphy and paintings.

Today there are more than one hundred Zhongshan Parks across mainland China and Taiwan named in honor of Dr. Sun, who died in 1925. One of them is next to the Forbidden City. Another is the 11-hectare Zhongshan Park in Wuhan (formerly Hankou), created in 1927-9 and incorporating a private garden dating from 1910. (1.103-1.104) The cultural and recreational facilities including a 6-hectare lake, people's meeting place, Chinese-Western style gardens, playground, library, swimming pool, skating rink, tennis court, and pavilions today attract over 5 million visitors a year. Nanjing also has a parkway honoring Dr. Sun. (1.105)

Many Western style parks and city plans designed by Western consultants spread across China, particularly after 1934, when Nationalist leader Chiang Kai-shek's wife Soong Meiling launched the New Life Movement advocating a modern healthy lifestyle, counteracting with athletic fields the unhealthy way of life of opium addiction. Civil war halted this progress, and many parks were destroyed. (1.106-1.107)

To commemorate the historic milestone of the People's Liberation in 1949, many examples of sculptural memorialization adorned plazas and bridges. Likewise, the urban parks in the imported English or French style survived, now open not just to the foreign public but to all people. The racetrack created by and for the expatriate British in Shanghai before 1863 became People's Square and People's Park when it opened to the people in 1951. From 1949 through the

起就对所有人开放。大公园的重点建筑民德亭坐落于人工山丘上，俯视南面的两个葫芦形池塘，符合好的风水朝向，亭身饰以进口红绿色玻璃，亭内陈设红木桌椅及著名的中国书法和国画作品。

时至今日，中国大陆及台湾地区共有超过100座中山公园，全部都是纪念1925年逝世的孙中山先生。其中一座中山公园的位置，就在紫禁城附近。另外一座是占地11公顷的武汉汉口的中山公园。此公园建于1927-1929年，前身是建于1910年的一座私家花园（图1.103 - 图1.104），园内的文化休闲设施包括面积6公顷的湖泊、聚会场所、中西合璧的园林、儿童游乐场、图书馆、游泳池、溜冰场、网球场及凉亭，每年吸引500多万人游览观光。南京也有一条纪念孙先生的公园大道（图1.105）。

很多由西方顾问设计的西式公园和城市规划在中国大量出现，尤其是在1934年之后，蒋介石及其夫人宋美龄推行的"新生活运动"，提倡以现代健康的生活方式来抵御鸦片烟瘾的恶习。连年内战使这场运动被迫中断，同时也有很多公园毁于战火（图1.106 - 图1.107）。

1949年，为庆祝新中国成立，大批纪念性的雕刻装饰了广场和桥梁。另外，大量的英式或法式风格的城市公园得以保留，不再仅仅只对外国人开放，而是对所有人民。1863年以前由上海英国侨民兴建的跑马场已演变成人民广场和人民公园，并于

图 1.105
中山陵园路，南京
Zhongshan Lingyuan Road, Nanking

同样是为了纪念孙中山先生，1929年建了这条美丽的公园大道。
Also honoring Dr. Sun is this beautiful example of a parkway, dating from 1929. (Credit: Weiming Ran)

Chapter One Origins of the People's Parks 059

图 1.106
旅顺游泳场，旅顺
Park Lake Swimming Facility, Lüshunkou

在 1934 年新生活运动之后，人们开始倡导锻炼和健康的生活方式，游泳是其中之一，越来越多的中国人开始游泳。
After 1934 the New Life Movement ushered in a focus on exercise and a healthy life style, in which swimming played a role for increasing numbers of Chinese. (Credit: postcard by unidentified photographer, Xiaowei Ma Collection)

图 1.107
工人海滨疗养院，大连，约 1956 年
Workers Seaside Sanatorium, Dalian ca. 1956

工人们可以在沿海设施中休息，这个设计考虑到了滨水的公共空间运用。
Workers could rest at seaside facilities, the design anticipates how to apply the shorefront public space. (Credit: postcard by unidentified photographer, Xiaowei Ma Collection)

图 1.108
梅隆广场，匹兹堡，约 1960 年
Mellon Square, Pittsburgh, ca. 1960

这个最早的现代景观广场建在一个地下车库之上，它是全球无数的实例的灵感来源。
The first modern landscape plaza built over an underground garage has inspired countless examples globally. (Credit: John L. Alexandrowicz)

1951 年对公众开放。从 1949 年直到 20 世纪 80 年代，中国很多城市都兴建了自己的人民公园，这些城市包括保定、成都、大连、广州、喀什、南昌、南宁、深圳、太原、西宁及乌鲁木齐。这与一个世纪以前西方的公园运动相似，然而中国在新中国成立后的 30 年内仍受位于其西方和北方的邻居苏联的影响，改造方式欠柔和，并使城市空间退化到了早期的思想：用以展示政权的力量，而不是供人民休闲娱乐的大型铺砌广场。

自 1980 年邓小平提出改革开放以来，以上这些都改变了，新思想层出不穷，各个城市开始提供不同类型的公共开放空间，例如绿化比铺装多的公园或铺装比绿化多的广场，大大提升了人

1980s many cities across China introduced People's Parks, including Baoding, Chengdu, Dalian, Guangzhou, Kashgar, Nanchang, Nanning, Shenzhen, Taiyuan, Urumqi, and Xining. This movement was similar to the parks movement in the West a century earlier, except that the dominant influence for the first three decades after Liberation was the unsubtle approach of the neighbors to the north and west, the Communist Soviets, in which urban space reverted to earlier thinking: large paved plazas for the display of power but not for the recreation of the people.

All of that changed after 1980 with Deng Xiaoping and the "opening up" of China to the free flow of ideas and improvement of the quality of life in cities by providing urban public open space of all types, parks with more greenery than paving, plazas with more paving than greenery (广场 or *guang3chang3*). As leisure time became available, people now had places to congregate not only during traditional festivals but on weekends throughout the year. Nowadays, on weekends Chinese citizens show a love for their parks that dwarfs weekend usage in the West. People gather in large groups to sing songs from the recent glorious past, dance the fox trot, show off water calligraphy or perform *taiji* on pavement rather than jog or sit on a lawn as one might in the West, truly the convergence of *communitas* and 热闹 *re4nao4*.

In the West, the pace of new park projects accelerated in the latter twentieth century. A summary of all the worthy accomplishments in urban public open space design from the last seventy years must include but not be limited to these milestones:

• Integrated urban design in Scandinavian design-conscious cities like Stockholm and Copenhagen emphasized quality of life such as parks and streetscape amenities for pedestrians and bicyclists.

• Historic preservation and restoration of landscapes and townscapes for education and recreation with obvious tourism economy benefits has grown dramatically since the pioneering Colonial Williamsburg, Virginia restoration in the 1930s funded by John D. Rockefeller. (Arthur A. Shurcliff, landscape architect).

• Roof Gardens capped underground parking garages. Mellon Square, Pittsburgh (John O. Simonds, 1955), is the first modern landscape plaza built over a parking garage and forerunner of green roofs today. (1.108) Union Square, San Francisco (M. D. Fotheringham Landscape Architects, 2002), is the site of many public concerts, events, art shows, and impromptu protests.

• Private developers now provided public amenity. Rockefeller Center (Raymond Hood, principal architect, 1930-39) integrates lavish public art, Channel Garden and Lower Plaza with a seasonal ice rink in the art deco complex. (1.109) Urban public open space has the power to help gain project approval from an often reluctant city council, but requires regulation to hold the developer to his promises.

• "Pocket parks" transformed vacant lots. Paley Park, New York City (Zion and Breen, 1967), in a space of only 13 x 30 meters, boasts a robust water wall whose soothing sound masks traffic noise. (1.110) Keller Fountain Park, Portland, Oregon (Lawrence Halprin 1966) creates a "new type of people's park" and uses abstract concrete forms and waterfalls to evoke the High Sierras. (1.111)

• Intensely designed and immaculately maintained small urban parks restore luster to faded downtown areas. Norman B. Leventhal Park, Boston (Halvorson Design Partnership, 1991) (1.112) and redesigned Bryant Park, New York City (Hanna and Olin, 1992) feature a cafe, accessible lawns and movable chairs. (1.113)

• Strong-design parks laden with cultural facilities replaced industrial or transportation uses. Parc de la Villette (Bernard Tschumi, 1982) replaces abattoirs with Paris' largest park (55 hectares) with

图 1.109
洛克菲勒中心，纽约
Rockefeller Center, New York City

这个新艺术运动时期的杰作现在依旧是纽约最受欢迎的名胜，尽管它是被包围在庄严市区和商业建筑之中的一块私人场地，它仍然极大地影响了公共开放空间的设计。
This masterpiece from the art-deco era remains one of the most popular destinations in New York, hugely influential in public open space design, though a private property, in its civic grandeur and commercial restraint. (Credit: Gabriel Rodriguez, Seville Spain)

民的生活质量。由于大家的休闲时间增多了，群众对聚集场所的需求不再局限于传统节日，而是一年中的所有周末。如今，百姓会在周末表现出他们对公园的喜爱，而公园在周末的使用率也远远高于西方。群众聚在一起唱怀旧金曲、跳狐步舞曲、打太极，或沾水作墨、以地为纸，秀传统书法，而不是像西方人那样在公园缓步跑或坐在草坪上。中国的公园将"市民共同体"和"热闹"很好地融合在一起。

在西方，兴建公园项目的进度在20世纪后期急速加快。以下是过去70年来城市公共开放空间设计有里程碑意义的成就，但这个领域的成就并不仅限于此：

• 以斯德哥尔摩和哥本哈根为代表的斯堪的纳维亚城市在综合性城市设计中强调生活质量：如供步行者和骑行者使用的公园和街道景观设施。

• 20世纪30年代约翰·洛克菲勒赞助完成的弗吉尼亚州威廉堡殖民复兴花园修复计划（由景观建筑师亚瑟·舒克利夫主创），对历史保护、景观及城镇风光修复、教育、文化休闲以及由旅游业带来的经济效益的提升具有显著的成效。

• 屋顶花园覆盖于地下停车场之上。匹兹堡的梅隆广场（由景观建筑师约翰·欧·西蒙兹设计，建于1955年）是第一个建于停车场之上的现代景观广场，亦是今日屋顶花园的先驱（图1.108）。旧金山的联合广场（由当地佛斯林曼景观建筑事务所设计，建于2002年）是很多公共演奏会、公共活动、艺术展览及临时发起抗议示威的场地。

• 现在由私人开发商提供公共设施。洛克菲勒中心（雷蒙德·胡德为主创建筑师，建于1930-1939年）是装饰艺术的集合群，整合了大量的公共艺术、海峡花园、下层广场及季节性的溜冰场（图1.109）。城市开放空间能帮助说服市议会批准项目，同时也需要利用规范来约束发展商坚守承诺。

• "口袋公园"转变了空置场地。纽约的佩利公园（由锡安

图 1.110
佩利公园，纽约
Paley Park, New York City

无论多小的空间都可以提供人们愉悦和沉醉的情感。水幕墙的声音掩盖住了城市的声音。座椅也可以移动。极简的设计在这个小尺度空间里创造了无限空间的感觉。这个私有公共空间的例子补充了公共开放空间。
No space is too small to provide urban amenity and sensory immersion. The sound of the water wall masks the sounds of the city. The furniture is movable. Minimalist design at this intimate scale creates a feeling of spaciousness. This example of privately owned public space complements publicly owned public open space. (Credit: Jim.henderson at en.wikipedia)

图 1.111
伊拉·凯勒喷泉广场，波特兰，俄勒冈州，美国
Ira Keller Fountain, Portland, OR, USA

风景园林师劳伦斯·哈普林率先使用了由山中瀑布启发而来的大体积叠水瀑布水景，并且允许人们近距离接触它。他因此创造了一种新的公园类型。
Landscape architect Lawrence Halprin pioneered in the use of fountains inspired by mountain waterfalls and open for people to enjoy close up. He thereby created a new kind of people's park. (Credit: Hagar66)

图 1.112
诺曼·利文萨尔公园，波士顿
Norman B. Leventhal Park, Boston

这是一个在小场地中拥有丰富空间的安静公园，它就像繁华市中心地下车库上的屋顶花园一样。公园的停车费用于公园的维护管理。
A serene park fits a variety of spaces into a small site that is like a green roof over underground parking in a busy downtown area. Parking fees pay for park maintenance. (Credit: Thomas M. Paine)

图 1.113
布莱恩特公园，纽约
Bryant Park, New York City

这个公园曾被忽视，在重新设计之后变得很受欢迎。它紧邻纽约公共图书馆，以咖啡店、可进入的草坪、喷泉、可移动的座椅以及悬铃木的树荫闻名。悬铃木在中国也很受欢迎。
This once neglected and now redesigned and popular park adjoining the New York Public Library features a café, accessible lawn, fountain, movable chairs, and the shade of plane trees, also a favorite tree in China. (Credit: Kamel15)

图 1.114
拉维莱特公园，巴黎
Parc de la Villette, Paris

在这个地标性景观中的红色构筑物仿佛打扰了"绿色公园"的简单宁静。也许受它的影响，近期中国的景观设计中增多了对红色的使用。公园的前身是一个屠宰场。
The bold red structures in this instantly iconic landscape snub the easy serenity of a "green" park and have perhaps been influential in encouraging the use of red in recent Chinese landscape design. The park replaced a slaughterhouse. (Credit: Pline)

图 1.115
雪铁龙公园，巴黎
Parc André Citroën, Paris

一处中心草坪，两个被喷泉广场分开的温室亭子，"变形花园"，"移动花园"，以及6个主题依次为一种金属、一颗行星、一周中的某一天、一种色彩、一种水的状态和一种鼓励游客进行无预期联系的感觉的主题花园，所有这些取代了之前的汽车制造厂。
Replacing a car factory are a central lawn, two greenhouse pavilions separated by a fountain plaza, a "Garden of Metamorphosis," a "Garden in Movement" and six "Serial Gardens," each themed with a metal, a planet, a day of the week, a state of water, and a sense encouraging visitors to make unexpected connections. (Credit: Gilles Clément)

图 1.116
千禧公园，芝加哥
Millennium Park, Chicago

作为20世纪最后十年中美国最创新和最成功的公园之一，千禧公园取代了一个铁路停车场。人们可以在大草坪上观看表演。右侧云门的反光表面吸引着参观者，在前面是它的皇冠喷泉，喷泉的两个双子塔的LED显示屏播放着日常人们的脸部，并且从他们的嘴中有水喷出。
One of the most innovative and successful parks in the U.S. of the last decade, the park replaces a railroad yard. The Great Lawn accommodates audiences for performances. To the right is Cloud Gate whose reflective surface enchants visitors, and above that is Crown Fountain, whose twin towers display LED images of everyday people who playfully spout water from their mouths. (Credit: J. Crocker)

themed gardens showcasing a deconstructivist aesthetic among its signature red "follies" that, despite their name harking back to an 18th century fascination with fake ruins, appeal to a Chinese modernist sensibility. (1.114) The designer deliberately uses provocative red to snub the easy serenity of the green park. Parc André Citroen, Paris (Clément and Provost, 1992), a 14-hectare park, replaces a car factory with a central lawn, two greenhouse pavilions separated by a fountain plaza, a "Garden of Metamorphosis," a "Garden in Movement" and six "Serial Gardens" themed with a metal, a planet, a day of the week, a state of water, and a sense. (1.115) Millennium Park (Frank Gehry, Kathryn Gustafson, Jaume Plensa, Anish Kapoor and others, 2004), Chicago, is a popular 10-hectare Park replacing railroad yards. (1.116) The High Line (James Corner Field Operations, Piet Oudolf, and Buro Happold, 2009), New York City, a 2.3 km linear park replaces an elevated rail line on Manhattan with rail inspired detailing. (1.117)

与布林建筑事务所设计，建于1967年）只有13米×30米的空间，公园以一道水幕墙瀑布著称。瀑布的潺潺流水声音掩盖了交通的嘈杂声（图1.110）。俄勒冈州波特兰的凯勒喷泉公园（由景观设计师劳伦斯·哈普林设计，建于1966年）创造了一种"新类型的人民公园"，设计师运用了抽象的混凝土形状及叠水瀑布，令人们联想到内华达山脉（图1.111）。

• 紧凑设计及完美维护的小型城市公园可令褪色的市中心恢复光彩。波士顿的诺曼·利文萨尔公园（由霍尔沃森设计联合公司设计，建于1991年）（图1.112）及纽约市翻新的布莱恩特公园（由汉纳/奥林景观建筑事务所设计，建于1992年）（图1.113），以小茶座、可通达的草坪及可移动的椅子为特征。

• 拥有强烈设计感的公园背负着把工业或运输用途的设施更新成文化设施的重任。拉维莱特公司（由伯纳德·屈米设计，建于1982年）取代了昔日的屠场，是巴黎最大的公园（55公顷），内有主题花园，其间陈列着解构主义的审美观及标志性的红色构造物"荒诞"（图 1.114）。虽然这些装饰性构筑物的命名源于18 世纪对假遗址废墟的着迷，但其实是中式现代主义的感觉。设计师故意采用富有煽动性的红色以使宁静的绿色公园成为陪衬。巴黎的安德烈·雪铁龙公园（由克莱蒙和普沃斯景观建筑事务所设计，建于 1992 年）是一座 14 公顷的公园，以一个中央草坪、两个被喷泉广场分隔的温室展馆、一座"变形园"、一座"运动园"及以金属、行星、一周的一天、水的形态及感官为主题的六个序列花园取代了先前的汽车制造厂（图 1.115）。芝加哥的千禧公园（由弗兰克·盖里、凯瑟琳·古斯塔夫森、普连萨、安尼诗·卡普尔以及其他优秀设计师联袂设计，建于 2004 年）是一座广受欢迎的公园，占地 10 公顷，原来是火车站（图 1.116）。纽约的高线公园（由詹姆斯·科纳·菲尔德设计事务所、设计师皮特·多夫和工程师布罗·哈帕德合作完成，建于 2009 年）取代了曼哈顿的一条高架铁轨，成为以铁路细节为灵感、总长 2.3 公里的线状公园（图 1.117）。

- 大规模的城市公园如面积 1000 公顷的阿姆斯特丹博斯公园（由设计师梵·伊斯特仁、穆德和生物学家泰杰斯博士合作完成，建于 1931 年）（图 1.118）或古阿兹特克遗址上占地 850 公顷的墨西哥城查普特佩克公园（由迭戈·里维拉和其他艺术家合作完成，建于 1940 年）（图 1.119），抑制了废弃的工业用地的有毒排放物，并且提供了公众休闲设施。有时这些公园会向其昔日的工业历史致敬，如占地 7.7 公顷的西雅图煤气厂公园（由景观建筑师理查德·哈格设计，建于 1975 年）（图 1.120）。

- Large urban parks on the scale of 1000-hectare Bos Park, Amsterdam (C. van Esteren and J. H. Mulder and biologist Dr. J. P. Thijsse, 1931) (1.118) or 850-hectare Chapultepec Park, Mexico City (Diego Rivera and others, 1940, on an ancient Aztec site) (1.119) may henceforth be restricted to brownfields sites that seal off toxins and accommodate public recreation. Sometimes they celebrate former

图 1.117
高线公园，20 大街，纽约
High Line, at 20th Street, New York City

取代了一个高架铁路的高线公园是另外一个在美国近 5 年来很创新和成功的例子，这个线形公园提供了一个令人愉快的体验和一个不被打断的连续交通空间。
Replacing an elevated railroad is another of the most innovative and successful parks in the U.S. of the last five years, a linear park providing an exhilarating experience of uninterrupted movement above the traffic. (Credit: Beyond My Ken)

图 1.118
博斯公园，阿姆斯特丹
Bos Park, Amsterdam

在荷兰紧临海平面的一个平坦场地上，大型公园可以屏蔽都市环境，并且提供沉浸其中的自然环境。未来的大型公园将主要会是恢复生态健康的废弃地。
In a flat landscape closely tied to the sea level as in the Netherlands, large parks can screen out urbanization and offer immersion in Nature. Large parks of the future will be mainly brownfields sites restored to ecological health. (Credit: Benutzer:AlterVista)

industrial history, like 7.7-hectare Gasworks Park, Seattle (Richard Haag, 1975). (1.120)

Over the last four decades sustainability has emerged as a theme that is not going away. The Design with Nature movement championed by Ian McHarg in his book of that name took the greenbelt idea to the regional scale, preserving the natural resources that needed to be sustained, and allowing development only in the rest. In *To Heal the Earth*, for example, McHarg's writings in turn inspired landscape architect John T. Lyle to envision regenerative design. Open space has always played a pivotal role, and that truth may now gain wider recognition. Parks once called "green lungs" are now called green infrastructure, but that only hints at the levels of meaning they must continue to impart in the world we are entrusted to create—sustaining not just the environment but community as well. Landscape architect Lawrence Halprin regarded urban public open space as the essence and heart of what cities are all about: a creative environment allowing for freedom of choice, and a maximum of interaction between people and their surroundings.

As we have seen in this chapter, over the centuries the recognition of the need for public open space in urban life has steadily grown. Good ideas indeed are freely borrowed and adapted from other cultures, from East to West, West to East, and we are all better off for it. The advent of the public park in the 19th century broke new ground.

可持续发展这一主题持续了40年，且没有消退的迹象。麦克哈格通过他的著作《设计结合自然》一书掀起了一场"设计结合自然"的运动，并把绿化带这一想法运用到区域规模层面上来，保护可持续的自然资源，仅在其余地方做有限开发。以《拯救世界》为例，麦克哈格的文章启发了景观建筑师约翰·莱尔对再生设计的设想。一直以来，开放空间都演绎着一个关键的角色，而这个事实到今天可能会得到更多的认同。一向被视为"绿色之肺"的公园现已被称为"绿色基础建设"，但这都只是字面的意思，我们要揭示出这些公园在我们所创造的世界里被赋予的深层内涵——不但维护环境，而且使我们的社区得到可持续发展。景观建筑师劳伦斯·哈普林认为城市公共开放空间是一个城市的灵魂精髓所在：一个具有创造性的环境会给予选择的自由，并实现人类与周围环境最大程度的相互融合。

由这一章可见，在过去几个世纪里，城市生活对公共开放空间需求的认知依然持续增强。好的构思会被其他文化借用和融会贯通，从东方传到西方，或从西方传到东方，而我们也会因此更加受益。19世纪公共公园的出现开辟了新天地。20世纪见证了这个构思如何被广泛地传播到世界各地，亦变成度量城市生活质量的领先指标，但可惜西方公园的开支很容易成为众矢之的，以至被迫缩减预算。在过去的20年里，对公园以及新式公园类型的投资都有复苏的趋势，这在其后的各章将会有描述及讨论。与此同时，在中国，无论是在杭州的西湖、上海的鲁迅公园或中山公园、大连的人民广场、成都的活水公园、香港的九龙公园，或

图 1.119
查普特佩克公园，墨西哥城
Chapultepec Park, Mexico City

公园在某种程度上补偿了几个世纪以前被西班牙摧毁的阿兹台克古城，这个公园保留了一个阿兹台克统治者祈神的地点，它提供着重要的生态服务功能。它是世界上最大的城市公园之一。
In some measure compensating for the loss of the ancient Aztec city design destroyed by the Spanish in centuries past, this park preserves a site held sacred by Aztec rulers that performs important ecological services. It is one of the largest urban parks in the world. (Credit: Thelmadatter)

是其他更多的公园中，总有大批百姓聚集其中，此情此景令所有西方公园都羡慕不已。

在 21 世纪，我们为大众提供有生命探寻意义的、天人合一的开放空间体验，我们的想象力有多丰富、共同慷慨度有多高，机会就有多大。有远见的城市领导者会继续做出改变。鉴于中国以往已有许多杰出的成就，中国将发挥其 2500 年的城市公共开放空间历史的优势，并将其应用于已历时 5000 年的文明发展历程中，寻求新的突破，鼓舞并启发世界其他地方。[4]

The 20th century saw the idea spread globally, to become a leading metric of urban quality of life, and yet in the West parks expenditures became easy targets for budget cutting. In the last twenty years, there have been signs of resurgence of investment in parks, and of new park types, which will be described in later chapters. Meanwhile in China the people congregate in numbers that would do any Western park proud, whether in West Lake in Hangzhou, Luxun Park or Zhongshan Park in Shanghai, Peoples Square in Dalian, in Living Water Garden in Chengdu, Kowloon Park in Hong Kong, or many others.

In the 21st century our opportunity to provide life-affirming and earth-affirming open space experiences to all is limited only by our imagination and our collective generosity. Civic leaders with vision can continue to make a difference. With many outstanding achievements already behind it, China is poised to take advantage of this two-thousand-five-hundred year history of urban public open space and adapt it to the ends of its own five-thousand year civilization, break new ground, and inspire the rest of the world.

图 1.120
煤气厂公园,西雅图
Gasworks Park, Seattle

纪念旧时代的遗迹已经变成了景观设计的一种主题,这里将旧的发电厂转换成一个新的公园。中国的工业建筑再利用有朝一日也许会和时髦商业建筑一样为公园增色。
The celebration of ruins from former eras has been a theme in landscape design for centuries, here updated to include a former power plant in a new park. China's recycled industrial buildings may someday enhance parks as readily as trendy retail developments. (Credit: USGS)

注释
1. 马可·波罗.《马可·波罗的游记》(伦敦:人人出版社,1908)
2. 文森特·斯库利.《普韦布:山,村,舞蹈》(芝加哥:芝加哥大学,1989)
3. 吴凯堂 (Thomas J. Campanella).《混凝土文化——中国的城市化革命及其对世界的影响》(纽约:普林斯顿建筑出版社,2008),208
4. 奥姆斯特德和沃克斯.纽约市公共工程部备忘录.

NOTES
1. Marco Polo. *The Travels of Marco Polo* (London: Everyman, 1908)
2. Vincent J. Scully. *Pueblo, Mountain, Village, Dance* (Chicago: University of Chicago Press, 1989)
3. Thomas J. Campanella. *The Concrete Dragon, China's Urban Revolution and What it Means for the World* (New York: Princeton Architectural Press, 2008), 208
4. Olmsted and Vaux. Memo to New York City Department of Public Works,1872

出处
埃德蒙·N·培根.《城市设计》(纽约:维京出版公司,1967)
斯蒂法诺·比安卡.《阿拉伯世界的城市结构:过去与现在》(纽约:泰晤士·哈德逊公司,2000)
希格弗莱德·吉迪恩.《空间·时间·建筑》(剑桥:哈佛大学出版社,1967)
科达恩.《庭院艺术的历史》(纽约:达顿出版社,1928)
劳伦斯·哈普林.《城市》(剑桥:麻省理工出版社,1972)
朱莉娅·卡齐米,乔治·哈格雷夫斯.大型公园(纽约:普林斯顿建筑出版社,2007)
杰弗里,苏珊·杰里柯.《环境塑造史论》(伦敦:泰晤士·哈德逊公司,1995)
芒福德.《城市发展史》(纽约:哈考特世界图书公司(Harcourt Brace & World),1961)
诺曼·牛顿.《土地上的设计史》(剑桥:哈佛大学出版社,1971)
瑞溥思.《美国城市的历史——美国的城市规划历史》(纽约:普林斯顿大学出版社,1965)
保罗·朱克.《城镇与广场》(纽约:哥伦比亚大学出版社,1959)

Sources:
Edmund N. Bacon. *Design of Cities* (New York: VikingPress, 1967)
Stefano Bianca. *Urban Form in the Arab World Past and Present* (New York: Thames & Hudson, 2000)
Siegfried Giedion. *Space, Time and Architecture* (Cambridge: Harvard University Press, 1967)
Marie LuiseGothein. *A History of Garden Art* (New York: Dutton, 1928)
Lawrence Halprin. *Cities* (Cambridge: MIT Press, 1972)
Julia Czerniak and George Hargreaves eds. *Large Parks* (New York: Princeton Architectural Press, 2007)
Geoffrey and Susan Jellicoe. *Landscape of Man* (London: Thames & Hudson, 1995)
Lewis Mumford. *The City in History*, (New York: Harcourt Brace & World, 1961)
Norman T. Newton. *Design on the Land* (Cambridge: Harvard University Press, 1971)
John W. Reps. *The Making of Urban America* (Princeton, NJ: Princeton University Press, 1965)
Paul Zucker. *Town and Square* (New York: Columbia University Press, 1959)
for Asian parks online references sourced May 26, 2010 include:
① http://pioneer.chula.ac.th/~sbussako/Articles/articles%20for%20publication/lui%20chang%20paper.pdf
② http://www.lifeinkorea.com/Travel2/339
③ http://www.artisandevelopers.com/web/tokyo/ueno1.htm
④ www.timeoutkl.com/aroundtown/venues/Lake-Gardens
⑤ http://en.wikipedia.org/wiki/Huangpu_Park, en.wikipedia.org/wiki/Taiping_Lake_Gardens

为什么城市开放空间比以往更重要
Why Urban Public Open Space Matters More Than Ever

Chapter **Two**
第二章

前人种树，后人乘凉。
One generation plants the trees, the next enjoys the shade.
——中国谚语 (Chinese Proverb)

采菊东篱下，悠然见南山。久在樊笼里，复得返自然。
Picking mums by the eastern fence, I lose myself in the southern hills. Too long a prisoner, captive in a cage, Now I can get back again to Nature.
——陶渊明（365-427年）Tao Yuanming (365-427)

我们需要这么一个地方，人们下班后会自然而然来到这里，散一小时的步，在这里，看不到、听不到、也感觉不到街上的嘈杂和喧嚣，他们可以暂时逃离凡尘俗世的纷扰。
We want a ground to which people may easily go after their day's work is done, and where they may stroll for an hour, seeing, hearing, and feeling nothing of the bustle and jar of the streets, where they shall, in effect, find the city put far away from them...
——奥姆斯特德，1870年[1] Frederick Law Olmsted, 1870[1]

花点时间在城市的开放空间中徜徉，是健康城市生活的保证——因为它在这个日益多元和跨文化的世界里，为我们构筑了共享和宽容，并最终成为了我们繁荣城市生活的支柱。
Spending time in urban open space is necessary to healthy urban life—in that it builds a sense of community and tolerance that in turn provides the underpinnings for a thriving urban life in an increasingly diverse, multicultural world.
——克莱尔·库珀·马库斯，1998年[2] Clare Cooper Marcus, 1998[2]

可持续性的景观设计有三点原则——生态健康、社会公平和经济繁荣。我提倡再移入美学的观点到可持续性发展的概念中。身临其境的美学体验可以引导我们对环境的认知、共鸣、爱、尊重和关怀。
Sustainable landscape design is generally understood in relation to three principles—ecological health, social justice and economic prosperity. I call for reinserting the aesthetic into discussions of sustainability. An immersive, aesthetic experience can lead to recognition, empathy, love, respect and care for the environment.
——伊丽莎白·梅耶尔，2008年[3] Elizabeth K. Meyer, 2008[3]

日常生活中，任何人都无法失去交流与沟通。相反，我们都需要快乐、放松、自然的生活。而每一个开放空间和公共公园都应该成为大众交流、沟通和游乐的佳所。
One cannot let people lose communication and connectedness. Instead, we all need happy, relaxed, and natural lives. Every open space and public park should be full of people chatting and enjoying themselves.
——王石，万科总裁，2011年 Wang Shi, Chairman, Vanke, 2011

从上一章可见，城市公共开放空间一直以来都有着重大的影响。本章将会通过具有说服力的事例来证明现在的城市公共开放空间比以往更加重要。城市公共开放空间的专业人士可通过令人信服、合乎逻辑的论据去证明提供更多更优良的空间和提升我们已有空间质量的必要性，这也是本章的主旨。

许多专家曾以各种不同的方式来证明城市公共开放空间的必要性。本书无法对其进行一一探讨，下面只举其中一例。

一个来自澳大利亚的观点

研究资料表明开放空间是由三个相互关联的元素来界定的：物理特性（古德，1980）；现场活动，如休闲娱乐活动（戈尔，1987）；使用者的感知（伍雷，2003）。开放空间会带来一系列的环境、社会、经济、心理及物质利益，包括为儿童及青少年提供精神层面的体验（源恩，1996），以及供儿童消耗过剩精力的安全而有价值的活动（博格斯，1988 第462页）。鉴于建筑密度的增加，设计和提供高质量的开放空间的需求也增强了。

（资料出处：http://www.urbandesignaustralia.com.au/images/Papers09/Alan%20March%20final%20paper.pdf，2010年1月8日）

我们需要思考这些需求，并在最广泛的层面上考虑它们的意义。前人说过：人不能只依靠面包来生存。为了生存，人类需要很多东西，有一些更基本，有一些被广泛分享，还有一些更私密，尤其是对生活在拥挤的城市里的人们来说。对此最为清晰的阐释或许要属人文社会学家马斯洛的人类需求层次架构了。马斯洛将构成人类体验的需求分成多个层次，从自然生存到个人成就感和修养，或是人生价值的实现，也就是他所说的"自我实现"（图2.1）。当一个层次得到满足，人类的心灵便可自由地去追求下一个更高的层次，直至达到自我实现的需求。马斯洛的思想体系影响了几代思想家，而且经常以金字塔的图示表现出来。他的理论架构可解释大部分的个体经验，当然也会有例外，但对大部分人来说，马斯洛的理论是获得广泛认可的。这一架构通常会被商业界用作

In the previous chapter we have seen that urban public open space has mattered for a long time. This chapter will make the compelling case that urban public open space matters now more than ever. Urban public open space professionals can justify the need to provide more and better space and to improve the space we already have by considering the compelling logical arguments which are the subject of this chapter.

Many experts have made the case for the necessity of urban public open space, in a variety of different ways. A complete survey of them all is beyond the scope of this book. We include one example.

An Australian Perspective

The literature suggests that three interrelated elements constitute a definition of open space: physical properties (Gold, 1980); the activities onsite, such as recreation (Gehl, 1987); and user perception (Woolley, 2003). Open space delivers a host of environmental, social, economic, psychological and physical benefits, including providing children and youth groups with a set of non-materialistic experiences (Yuen, 1996), functioning as a "safety valve where children can burn off energy" (Burgess, 1988 pg 462). As housing densities increase the need to plan for and provide quality open space intensifies.

(Source:http://www.urbandesignaustralia.com.au/images/Papers09/Alan%20March%20final%20paper.pdf, January 8, 2010)

Our approach is to think about needs and to consider them in the broadest sense. As has been said, man lives not by bread alone. For their very survival people need many things, some more basic than others, some shared widely, others more personal, especially when living in crowded cities. Perhaps the clearest framework for thinking about just what that means is what humanistic sociologist Abraham Maslow called the Hierarchy of Needs. Maslow organizes the range of

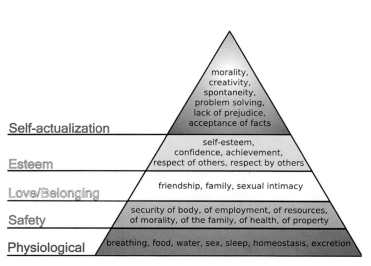

图 2.1
马斯洛需求层次论
Maslow's Hierarchy of Needs (Source: J. Finkelstein)

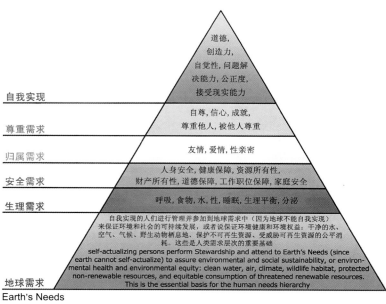

图 2.2
土地和人类需求层次论
Hierarchy of Earth's Needs and Human Needs

needs that comprise human experience as a series of levels ranging from basic physical survival all the way up to personal fulfillment and refinement, or what he memorably terms "self-actualization," living a fully realized life.(2.1) As each level is satisfied, the human spirit is freed to aspire to the next higher level, all the way up to self-actualization. His framework has influenced generations of thinkers, and is frequently diagrammed as a pyramid. While the framework generally explains the experience of most individuals, of course there are exceptions, and they can be poignant, but for people in general Maslow's framework is widely accepted. Often used to explain what motivates consumers in the marketplace, the framework can also help us understand the benefits which consumers derive from urban public open space.

For each level of need, urban public open space matters in essential ways in enabling us to aspire to live a better life. Admirable though this framework is, the reader may have noted that something is missing: the needs of the Earth. Maslow may have taken it for granted when he published his framework in 1954 but would not take it for granted now. Following Maslow's Hierarchy of Human Needs updated to include the Earth (2.2), this chapter examines the many ways in which urban public open space matters.

■ Earth's Needs

As we saw in the last chapter, large urban parks have been called "the lungs of the city" since the beginning of the parks movement in the nineteenth century, an era of rapid urbanization, industrialization, and massive air pollution. In addition to centrally located parks, perimeter open space, much it left in its natural state, was considered essential for what is now called "environmental protection," or more elegantly, "stewardship". We increasingly realize our responsibility to pass on to future generations an environment left in at least as healthy a condition as we found it, rather than to assert a right of development and profit at any cost, no matter what the environmental consequences may be. In addition, as good stewards we must concern ourselves with living in harmony with other forms of life, and preserve wildlife habitat even close to the city. In considering future generations and their healthy regeneration, for people and all other living things, we must indeed provide regenerative systems.

In recent decades, most experts have come to agree that despite all the progress in environmental protection, not enough is being done. In our time a number of disruptive forces are at work. Industrialization and urbanization have reached a tipping point: climate change, the dwindling supply of non-renewable natural resources, and the decline of air and water quality cannot continue much longer without irreversible consequences for life on earth. If we fail to act boldly enough, it is not hard to imagine that a sea level rising a meter or more over the rest of the century will threaten indiscriminately both low-lying islands and heavily populated low-lying coastal areas, dwarfing the problems that the dyke-protected "Netherlands" have faced for millennia. More than 70 percent of the world's population lives on coastal plains, and 11 of

解释消费者的动机,也可以帮助我们去理解城市开放空间对消费者的益处。

对于任何一个层次的需求,城市公共开放空间发挥的本质作用都是鼓励我们去追求更好的生活。这个架构虽然堪称绝佳,但读者可能觉察到其中缺乏的一些东西:地球的需求。1954年马斯洛发表其理论之初,由于年代和意识的局限性,地球上一切条件的提供被认为是理所当然的,但这并不适用于当今。下面是经过调整的马斯洛人类需求层次理论,包含了地球一项(图2.2),本章亦会从多方面探讨城市公共开放空间的重要性。

■ 地球的需求

如上一章所述,自19世纪公园运动开始,大型城市公园被喻为是"城市之肺",当时正值急速都市化、工业化及大面积空气污染的年代。位于市中心的公园和周边的许多开放空间都保持着自然状态,这在现在看来是很重要的,就是所谓的"环境保护",或更优雅的说法是"护养自然"。我们逐渐意识到应该传承一个健康的环境给我们的子孙后代,而不是不计后果地妄自宣称有权开发环境和从中牟利。另外,秉承优良的自然护养传统,我们需要与其他生物和谐共处,保护野生生物栖息地,甚至是靠近城市的野生物族群。考虑到人类及其他生物的下一代的健康再生能力,我们急需建立再生系统。

最近几十年,大部分专家都承认环保取得了一些进展,但还是远远不够。在我们这个时代,许多破坏性的力量正发挥着作用。工业化及都市化的发展已经达到了一个临界点:气候变化、天然可再生资源的供应减少、空气和水源质量下降,这些状况必须得到有效遏制,否则会对地球上的生命带来不可挽回的恶果。假如我们不采取切实有效的行动,海水水位会在21世纪内上升1米或更多,会对低洼岛屿及人口稠密的沿海低洼地区构成威胁,这些将更甚于受堤坝保护的荷兰上千年来所面对的困境。超过70%的世界人口居住在沿海平原,世界上15个最大的城市中有11个位于沿海海湾。这正是我们需要立刻全面关注环境保护及可持续发展的原因。在基本的生理意义上,人类生存依赖于环境的可持续性。这就是景观建筑师对生存的解释。

城市公共开放空间可缓和热岛效应,即区域性夏季高温,缓冲疾风从而改善微气候,补充氧气,减少空气污染,改善水质,控制暴雨径流,补充地下水位,控制洪水,提升植物的多样性,为野生生物提供栖息地,包括陆生及水生生物、哺乳类动物、鸟类及微生物。在我们的新世界里,这类城市开放空间不仅是都市之肺,同时也是城市的肾脏——从水分中排除毒素,及城市的心脏——以稳定速度注入含氧及净化的水分并输送到自然环境内。虽然这些绿色基础设施不能单独运作,但它们在城市公共开放空间内发挥着显著的作用。在设计及管理时是不能忽略它们的基本生存作用的。本质上我们需要解决的问题是如何设计真正对环境无害的景观。

源于20世纪90年代的绿色基础设施运动大胆地断言:最终将会是自然环境主宰我们的都市化,而不是由利润驱动力决定环境的存亡。的确,道路的过度铺砌被植树所取代一度受到媒体的密切关注,体现巨大的新闻价值,而且这种潮流还在继续,但如今更具新闻价值的却是一些已然消失的环境的起死回生。所谓的"棕地"，比如废弃的工厂或垃圾堆填区，必须加以净化、封盖，并补充表土层。自然生态系统网络是所有景观概念的中心所在，为我们提供了不可或缺的生命维持功能。基于此事实，"景观都市化"这一术语现被用于描述"重新整合设计领域，让景观取代建筑设计成为都市化的基本组件"。[4]

从可持续发展的角度满足地球和人类的需求

全球气候调节：

保持大气气体的平衡使其维持在历史水平，创造可供呼吸的新鲜空气，隔离温室气体。

当地气候调节：

通过绿荫、蒸发扩散、防风林等方式调节当地的气温、降水、湿度。

空气与水净化：

消除和减少空气和水中的污染物。

水供应及管理：

利用河流和地下蓄水层贮存和提供水资源。

侵蚀产沙控制：

确保土壤拥有一个良好的生态系统，防止因侵蚀和淤积而造成的损害。

减灾：

避免或减少因洪水、风暴潮、自然火灾和干旱所导致的损失。

授粉：

为农作物或其他植物的繁殖提供授粉品种。

栖息地功能：

为动物、植物提供相应的庇护和繁殖栖息地，从而有助于保护生物多样性和遗传多样性，并推动进化过程。

废物的分解和处理：

分解废物与循环养分。

人类的健康和福祉：

增强自身体质与心理健康，完善社会关系，与自然界形成良好的互动。

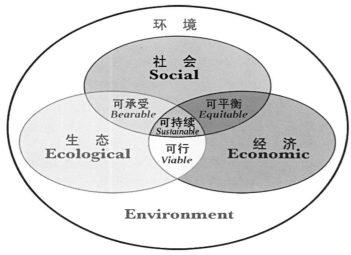

图 2.3
维恩图解三种可持续性
Venn Diagram of the Three Sustainabilities

如果我们短期忽略其中的任何一个，我们就不能长期维持环境质量、社区和谐和经济可行性。
We cannot sustain environmental quality, community harmony and economic viability over the long term if we neglect any of them in the short term. (Credit: Andrew Sunray)

the world's 15 largest cities are on the coastal estuaries. That is why environmental protection and sustainability demand the undivided and immediate attention of all of us. Human survival in the basic physiological sense depends on environmental sustainability. That is what landscape architects mean when they talk about survival.

Urban public open space benefits or services include reducing local summer temperatures in what are now called heat islands, improving the microclimate by buffering high winds, replenishing oxygen, reducing air pollution, improving water quality, controlling storm water runoff, replenishing the water table, controlling floods, contributing to floral biodiversity, and providing wildlife habitat, terrestrial and aquatic, from mammals and birds to microbes. In our brave new world, such urban public open spaces can be thought of as not only the lungs of the city, but also the kidneys of the city, removing toxins from the water, and the heart of the city, "pumping" aerated, purified waters into the environment at a steady rate. Although it cannot do the job alone, the green infrastructure of urban public open spaces can play a significant role, and it must. To ignore that fundamental survival role in their design and management is no longer acceptable. The issue is very much how to design a genuinely environmentally-friendly landscape.

The green infrastructure movement originating in the 1990s boldly asserts that, at long last, it is the natural environment that should dictate how we urbanize, rather than the profit motive that dictates what environment is to die, and what to live. Indeed, it used to be newsworthy when excessive road pavement was replaced with planting, and that welcome trend continues, but more newsworthy nowadays are the environments that have died and are now being brought back to life, so-called "brownfields" such as abandoned factory sites or landfills which must first be decontaminated or sealed and then have their topsoil replenished. The network of natural ecosystems—the heart of any notion of landscape—provides essential life-support functions. Recognizing this fact of life, the term landscape urbanism is now used to describe "the disciplinary realignment [whereby] landscape replaces architecture as the basic building block of urbanism." [4]

Meeting Earth's Needs and Human Needs from a Sustainability Perspective

Global climate regulation

Maintaining balance of atmospheric gases at historic levels, creating breathable air, and sequestering greenhouse gases.

Local climate regulation

Regulating local temperature, precipitation, and humidity through shading, evapotranspiration, and windbreaks.

Air and water cleansing

Removing and reducing pollutants in air and water.

Water supply and regulation

Storing and providing water within watersheds and aquifers.

Erosion and sediment control

Retaining soil within an ecosystem, preventing damage from erosion and siltation.

Hazard mitigation
Reducing vulnerability to damage from flooding, storm surge, wildfire, and drought.
Pollination
Providing pollinator species for reproduction of crops or other plants.
Habitat functions
Providing refuge and reproduction habitat to plants and animals, thereby contributing to conservation of biological and genetic diversity and evolutionary processes.
Waste decomposition and treatment
Breaking down waste and cycling nutrients.
Human health and well-being benefits
Enhancing physical, mental, and social well-being as a result of interaction with Nature.
Food and renewable non-food products
Producing food, fuel, energy, medicine, or other products for human use.
Cultural benefits
Enhancing cultural, educational, aesthetic, and spiritual experiences as a result of interaction with Nature.
(Source: ASLA et al., The Sustainable Sites Initiative Guidelines and Performance Benchmarks, 2009. see www.sustainablesites.org)

Attending to the earth's needs is a basic human responsibility to be shared by all of us, above all the responsibility of those in positions of power and influence. As the warnings of the environmental activists like Nobel Prize winner Al Gore in his movie *An Inconvenient Truth* have advanced our thinking about stewardship and sustainability, many have come to embrace the essential truth that environmental sustainability, economic sustainability and social sustainability are interdependent; indeed, for some thinkers they are one and the same thing. (2.3) That is to say, a healthy society cannot flourish and last very long without a healthy environment. Social harmony cannot last very long without environmental harmony. Straying from harmony goes against a core aspiration of traditional Chinese culture. In an increasingly disruptive and urbanized world, urban parks are indeed at the heart of the matter.

Sustainability and Wellness from a LOHAS Perspective

Lifestyles of Health and Sustainability (LOHAS) describes people who are concerned about sustainable living. This demographic includes mostly affluent and well educated consumers in the West and Asia (Japan, Singapore and Taiwan) who eat organic and locally grown food, reduce their energy and non-renewable resource consumption, prefer Chinese medicine, and buy fair-trade products. As their numbers grow they will become strong advocates for urban public open space excellence both as a personal matter and as a matter of environmental concern.

For too long, urbanization decisions have been made based on financial analysis that has failed to take into account costs and benefits that are hard to put a price on. The true economic costs of degrading the environment which must be borne by society, if not the lucky developer, are being estimated and forecast with increasing accuracy

食品与可再生的非食品类产品：
生产食物、燃料、能源、医药或其他供人类使用的产品。
文化利益：
加强文化、教育、审美、精神层面的相关经历，以便与自然界形成良好的互动。
（资料出处：ASLA et al., The Sustainable Sites Initiative Guidelines and Performance Benchmarks, 2009. see www.sustainablesites.org）

关心地球的需求是人类的基本责任，应由我们所有人共同承担，这远比那些位高权重者的责任更为重要。环保人士，诺贝尔经济学奖得主阿尔·戈尔在他的电影《难以忽视的真相》中启发我们对保护自然和可持续性的思考。许多人开始慢慢接受这样的基本事实：环境可持续性、经济可持续性和社会可持续性是互相依存的；事实上有一些思想家认为它们就是一回事（图2.3）。这就是说，健康社会的繁荣与发展离不开健康的环境。没有环境和谐，社会和谐也不能持久。与和谐背道而驰是违背中国传统文化核心价值观的。在一个日趋纷乱及城市化的世界里，城市公园可以说是问题的核心。

从一个乐活族的视角看待可持续发展和健康生活
（乐活族的）健康观和可持续性的生活方式描述的是那些时刻关注可持续生活的人。这类人群大多是富裕且受过良好教育的消费者，在西方和亚洲（日本、新加坡和中国台湾）都有这样的人群，他们大都食用有机和当地种植食物，注重减少能源和非可再生资源的消耗，喜好使用中药，购买平价产品。随着这类人群数量的增长，他们将成为城市公共开放空间健康发展的有力助推者，使更多人意识到城市公共开放空间的发展并不仅仅是个人行为，更是关乎全人类的环境问题。

一直以来，城市化的决策一直以金融分析为依据，而人们却忽视了，有些成本和收益是无法用金钱衡量的，这在金融分析里自然无法体现。那么这些由环境退化造成的经济成本由社会而不是那些幸运的开发商来承担，成本的估算预测也越来越精确，并将在城市化的决策考量中占到越来越大的比重。相反，生态效益渐趋被理解为可估价的生态服务。随着时间的流逝，这种思想会更准确更广泛地评价公园的经济效益，下面我们将纵观马斯洛经典理论中的所有人类需求层次，以进一步进行探讨。

■ **生理需求**
人类最基本的需求是生理上的。我们都需要呼吸、进食、喝水及睡眠来满足身体机能的正常运转。我们依赖自然资源以满足这些需求。从最基本的层面上看，大部分的城市公共开放空间是一种天然的资源。自然环境可直接或间接为人类提供新鲜空气、食物、水分及阳光，而这些都是我们生存、生育及达到平均期望寿命的基本需要。

甚至在这基本的生理层次上，城市公共开放空间在城市居民的生活中也扮演着重要的角色。它的作用包括补给空气，保护生物多样性、水质，保留更多的地域性未开发用地、水资源及我们赖以生存的生态系统，远离看不见的健康威胁。在这一需求层次，城市公共开放空间作为整体中的一小部分经常被忽略，但当我们提升到更高层次的需求时，城市公共开放空间便扮演着更为关键的角色，来满足我们各式各样的需求，并且更加不能被忽略。

生理需求包括身体的健康。城市公共空间在城市居民生活中扮演着重要的角色。我们都有责任保护好自己的身体，如经常进行慢跑等单人活动。有规律的运动有助于保持身体健康。运动能增强体格、减少持续的疲劳感、增强免疫系统和防止高血压、心脏病、心血管疾病、Ⅱ型糖尿病及肥胖等富贵病。城市公共开放空间为满足人类基本需求创造了理想的环境并扮演着关键的角色（图2.4 - 图2.6）。

■ 安全感

如果吃、喝和睡等基本生理要求已达到，并且身体健康，人们则会关注个人安全问题。不管个人的关注焦点有何不同，其中最基本的都是免除人身、工作、信仰、家庭和财产危险。

如果管理得当，城市公共开放空间可提供没有生命危险的安全环境，如避免汽车噪声所带来的生理不适及有毒气体等带来的无形健康威胁。事实亦证明精心维护城市公共开放空间有助于降低盗窃和侵犯人身安全及违反社会秩序的行为，如骚扰、醉酒、争吵、损坏公物。现在有许多荒废或高犯罪率的大片用地被改造成了管理良好的公园，极大地降低了犯罪率。纽约市的布莱恩特公园经重新设计与适当管理，成为了此类案例的典范。

但是，安全的意义不仅于此。人类的幸福不仅仅只是依赖生存和人身安全，而且还需要良好的心理健康。有规律的运动不仅改善身体健康还可以改善心理健康，有助预防抑郁症及失眠、增强自尊、提高认知能力及减少注意力不集中引起的紊乱。运动也可降低皮质醇水平，而皮质醇是一种令脂肪生长在腹部、令减肥困难的激素。这说明了身体和心理健康是互相影响及互相关联的，而健康需要身心两方面共同决定。

许多人倾向于花钱到健身房做运动。研究指出户外运动具有明显的优势，尤其是在精神上：与室内运动相比，户外运动能有效地提供更振奋的感觉、提升能量及积极的人生态度，同时也能减少压力、困惑、愤怒和抑郁。公园可被视为城市中最大的健身房，而且还是免费使用的。与其他地方相比，在公园内进行散步、

and will increasingly be taken into account in urbanization decisions. Conversely, ecological benefits are increasingly being translated into ecological services which can also be priced. Over time, this kind of thinking will more accurately and inclusively evaluate the economic benefits of parks across the whole spectrum of human needs of Maslow's original framework, which we will now examine.

■ Physiological Needs

The most basic of all human needs are physiological. We all need to take care of bodily functions like breathing, eating, drinking, and sleeping. We depend on natural resources to take care of these needs. At its most basic level, much urban public open space is a natural resource. A natural environment provides human beings, directly or indirectly, with the fresh air, food, water, and daylight that we need in order to survive, to be capable of reproducing, and to enjoy an average life expectancy.

Even at this basic physiological level, urban public open space plays a significant supporting role in the life of urban dwellers. It does its part to replenish the air, and maintain biodiversity and water quality, as part of a larger regional reserve of undeveloped land and water resources as well as part of an essential ecological system on which we all depend. It is free of invisible health risks such as toxins. At this level in the hierarchy of needs, urban public open space may be dismissed as just a small part of this large whole, but as we proceed up the hierarchy of needs, urban public open space takes on an ever more critical role in meeting all the needs considered together, and can no longer be dismissed as expendible.

Physiological needs include physical wellness. This is more basic than engaging in social activities which contribute to psychological wellness. Each of us is responsible for maintaining our own body, often engaging in solo activities like jogging. Regular exercise increases physical fitness, reducing a feeling of persistent exhaustion. It boosts the immune system and helps prevent the "diseases of affluence", such as high blood pressure, heart disease, cardiovascular disease, Type-2 diabetes and obesity. In fulfilling this basic human need, open space provides an ideal setting and plays a key role. (2.4-2.6)

图 2.4 - 图 2.6
生理需求
Physiological Needs

在亚洲的公园里，女性慢跑、做拉伸以及瑜伽来保持体形。
In parks in Asia, women are doing pushops, performing yoga and jogging to stay fit. (Credit: maridav, designpics, Adrin Shamsudin)

■ Safety

If basic physiological requirements such as eating, drinking, and sleeping are met, and people are physically healthy, then they are free to concern themselves with their well-being, that is, their personal safety and security. Whatever else they may wish to focus on, the basics include a life without significant risks to one's person, one's job, one's beliefs, one's family and one's property.

If properly managed, urban public open space provides a safe environment free of physical danger such as from vehicular traffic and physical discomfort such as loud noise. It has been well demonstrated that carefully maintained urban public open space also reduces the likelihood of common crimes such as theft or assault, and anti-social behavior such as harassment, drunkenness, rowdyness, and defacing public property. It is the neglect that attracts antisocial behavior and makes people feel unsafe. There are many examples of how transforming abandoned and crime-infested parcels of land into well maintained parks significantly reduces crime. This happened to Bryant Park in New York after it was redesigned and properly maintained.

But safety means much more than that. Human well-being depends not on mere survival and physical safety, but on good mental health. Regular exercise not only improves physical health, it also improves mental health, helps prevent depression and insomnia, promotes self-esteem, enhances cognitive function and reduces attention deficit disorder. Exercise also reduces levels of cortisol, a stress hormone that builds fat in the abdominal region, making weight loss difficult. The latter demonstrates that physical and mental health are interactive and interrelated. Wellness depends on both.

Many people prefer to work out in a gym, which costs money. Evidence points to the distinct advantages of outdoor exercise, especially psychological: compared to indoor exercise, outdoor exercise affords greater feelings of revitalization, increased energy and positive engagement, together with decreases in stress, confusion, anger and depression. Parks can be thought of as the largest gym in the city, and they are free of charge. Walking, hiking, running, cycling, rowing, *tai chi chuan*, stretching and aerobic exercise courses are all more effective in park settings than elsewhere. Exercise in the form of group sports like baseball, basketball, volleyball and tennis is an important outlet for many people and, apart from expensive private clubs, can only occur in public open space settings.[5] (2.7-2.9)

A major voice for urban parks in the U. S. since 1972, the Trust for Public Land eloquently makes the case that physical activity and exposure to Nature and greenery make people healthier, that access to parks increases the likelihood of exercise and decreases the likelihood of crime, and that physical inactivity goes hand in hand with obesity.[6] (2.10)

Olmsted never heard of the obesity epidemic, which now besets many Western countries and is gaining strength in China. People in earlier times were much more likely to be underfed than overfed. In our world of junk food, soft drinks, mechanized transportation, less physically demanding work, and increased use of labor-saving devices,

图 2.7 - 图 2.9
安全性
Safety

在公园中，孩子们可以从简单到更大的挑战中学会身体上的冒险，这弥补了在很小时候就可能发生的久坐在电脑前所带来的负面结果。
In parks, children can learn to take physical risks, working up from simple ones to greater challenges, offsetting the negative consequences of spending too much time in front of a computer which can start at an early age. (Credit: Dan Provo UT, Cramer, Thomas M. Paine, and Elena Koulik)

爬山、跑步、骑自行车、划船、打太极、伸展及有氧运动等活动将更加有效。团队模式的运动如棒球、篮球、排球和网球是许多人户外活动的重要方式，这些活动除了大型的私人俱乐部，就只能在公共开放空间内进行了[5]（图 2.7 - 图 2.9）。

自 1972 年以来，美国城市公园的主要代言者——公共土地信托会极有说服力地证明了体育运动和沐浴在自然及绿色环境下能使人们更加健康，通过享用公园的机会可以增加运动的可能性，降低罪案发生率和肥胖率[6]（图 2.10）。

在奥姆斯特德的年代从来没听说过肥胖症，但肥胖症如今正不断困扰着许多西方国家，中国也面临这样日趋严重的问题。古代的人一向都食不果腹，而非过量饮食。而现今的社会里充满垃圾食品、汽水、机械化交通，体力工作减少，省劳力设备增加，使肥胖症患者数量飙升。同样地，现代人对手机和网络里的电子游戏和音乐的沉迷使人与人的隔阂、人与真实世界的隔阂日趋严重。人们对虚拟世界的沉迷，导致他们脱离现实，不与跟自己不同的人打交道。人们在年轻时害怕自己与他人不同，长大后他们会开始加入互联网上的兴趣小组，而这些小组往往都会加强人与人之间的意识差异，让人们趋于麻木，没有好奇心和同情心，甚至是缺乏道德洞察力。这种从社会中的抽

图2.10
安全性
Safety

近期世界性的肥胖激增反映出活动的失衡以及不良的饮食习惯。
The recent word-wide surge in obesity reflects an unhealthy balance of activities as well as the effects of poor dietary choices. (Credit: Bruce Rolff)

图2.11
安全性
Safety

在公园里练习瑜伽不仅帮助人们保持健康也帮助人们减轻压力。
Practicing yoga in a park helps people not only to stay fit but also to reduce stress. (Credit: spflaum)

离也同时会影响家庭关系。这群人自然会很少花时间在户外，独自享受大自然或参与具有挑战性的体力游戏。这种损害是双方面的：他们在大自然里会感到不舒服，与不相识的人在一起时会感到不自在。上述的两种情形都是不健康的，要舒缓这两种情况是极其重要的，而公园可有效地缓解这些问题。任何一个阶段开始都不晚，公园甚至对提高环境承载力和社区的活力至关重要（图2.11）。

缺乏公共开放空间和充足绿地资源的城市居民，就像被囚禁在狭窄的世界里，正如陶渊明诗句里的"久在樊笼里"（图2.12）。随着城市密度的上升，更多城市人口居住在超高层的大厦里，越来越多人的健康正受到威胁。近来有人评论："强迫人们留在禁闭及简陋的空间里是一种折磨或处罚。"[7]早在高楼大厦包围中央公园之前，奥姆斯特德就意识到，即使城市居民生活在低层建筑物内，仍会因工作压力及拥挤的居住环境而感到压抑。他们需要体育锻炼、放松休闲及某些城市化不能提供的东西——享受大自然的机会，而这需要城市领导人或有关人士采取行动为大众提供。在现今人口密度极高的社会里，人们对亲近大自然的需求当然更甚。

对于大自然并不陌生的人们沉浸在开放空间里并享受丰富的感官体验，这让他们保持健康并减轻压力。大型公园的宜人环境与城市噪声形成对比，同时也和虚拟世界中的视觉和听觉体验形成对比。公园提供可步行或慢跑的小径、可供人坐下来休憩的长凳或是遮阴避雨的凉亭等舒适的便利设施。这种亲近自然而得到的"修复"功能日益引起人们的关注。[8]自然的景色、声音、气味、质感甚至是味道能刺激并慰藉人们的感官，将其从城市生活的疲倦及高压下释放出来。仅仅是散步、慢跑或是轻松地坐着，已经有很大的舒缓作用。越是多重感官的体验，人就越会被深深触动，精神亦会为之振奋，记忆力也会增强。有谁不愿享受天籁之音？——喷泉、瀑布、风吹过松叶的簌簌声及小鸟的婉转歌声？有谁能不为之心情舒畅？

obesity has skyrocketed. Likewise, in our world of video games and music on our cell phones and the internet, personal isolation has skyrocketed. People are increasingly insulating themselves by tuning out their surroundings in real space, isolating themselves from people not like themselves, and focusing on the virtual world instead of the real world. This can begin in childhood. When still young, people may become fearful of people different from themselves. As they get older they may start to follow internet interest groups which tend to reinforce perceived differences between people and encourage careless thinking devoid of curiosity or empathy, or even worse, deficient in moral insight. This withdrawal also undermines family ties. Those who withdraw spend less time outdoors, let alone enjoying Nature or playing physically challenging games. The damage is two-fold: they feel uncomfortable in Nature, and they feel uncomfortable with people they don't know. Neither condition is healthy; alleviating both conditions is immensely important; and both can be effectively alleviated by parks. And it is never too late to start. The long term viability of the environment and the community depends on it. (2.11)

Without access to urban public open space, urban dwellers with limited resources are, in a sense, imprisoned in a confined world, as Tao Yuanming wrote long ago in his famous poem. (2.12) As urban densities increase, and urban populations increasingly live in high-rise towers, the wellness of more and more people is at risk. Recently it has been observed, "Forcing people to remain in confined, bare settings is a form of torture or punishment". [7] Long before high-rise buildings encircled Central Park, Olmsted himself recognized that even in that low-rise world urban dwellers confined to stressful jobs and crowded living conditions need both physical exercise, relaxation and something urbanization itself does not offer unless leaders and concerned people take action to provide it: access to Nature. The densities in which people now live demand it even more.

For people for whom Nature is no longer an unfamiliar realm, immersion in open space and taking in the full multisensory experience have the power to restore emotional wellness and reduce stress. Large parks offer a welcome contrast from urban noise, auditory and visual, and from the virtual world. They provide amenities—that is, features that contribute to comfort and convenience, like paths where people can walk and jog, benches where people can sit and relax, and pavilions where people can shelter from the sun and rain. The "restorative" value of immersion in Nature has been noted more frequently in recent years. [8] The soothing stimuli of natural sights, sounds, smells, textures and even tastes can work magic on a weary and stressed urban dweller. Walking, going for a jog or sitting and relaxing may be all that it takes. The more multisensory the experience, the more deeply people are touched and their spirits uplifted, and the more powerful the memory. Who does not enjoy the sounds of Nature—a fountain, a waterfall, the wind blowing through pine needles, birds singing—and feel better for it?

Parks accommodate what anthropologists theorize as an innate, almost instinctive human need to be immersed in Nature. A human trait recognized in both Eastern and Western cultures is the deep appeal of contemplation of Nature or quiet reflection on life's challenges and opportunities while being immersed in Nature. Some authorities go so far as to suggest that the typical park landscape of a grove of trees on grass reflects an innate preference for the savannah habitat of our primitive ancestors who felt both sheltered and safe there. People are tapping into the archetypal meaning of the forest, grass, stream, hill and valley, as suggested in poetic terms like grove, turf, rill, knoll, and dell. From communing with Nature, people renew their sense of well-being.

Urban public open space is good for employment; regular visits help people do better in their job. According to recent research, the office worker of today who engages in tasks calling for sustained focus can benefit significantly from a walk in the park. The sights and sounds of Nature appear to be especially beneficial for re-energizing our minds. Dr. Marc Berman and associates at the University of Michigan reported that performance on memory and attention tests improved by 20% after people took a break for a walk through an arboretum. When these same people were sent on a break to stroll down a busy street in town, no cognitive boost was detected. [9] Doing well at a job is what leads to the safety of accumulated resources.

Parks also help people overcome isolation and social anxiety and feel comfortable with other people. U.S. architecture critic Sarah Williams Goldhagen puts the issue like this: society in many countries is at risk of becoming "more an archipelago than a nation, increasingly balkanized into ethnic, class, faith, and interest groups whose members rarely interact meaningfully with people whose affiliations they do not in large measure share". All these social ills, she argues, can be alleviated by urban public open space that brings people together rather than keeps them apart in their own illusory dream worlds. Urban public open space should be designed for interactions

人类学家提出亲近自然是人类与生俱来、近乎本能的一种需求，而公园正可以满足这种需求。东方和西方文化中共有的一种人类特质是沉浸于大自然，或在自然中思考并探索自然本身以及生命中的机遇和挑战。一些权威人士会进一步提出有代表性的草地小树林的典型公园景观设计方案，因为从我们的原始祖先开始，就发现这样的草地树林能够给人安全感并提供庇护场所。人们推敲森林、草地、溪流、山丘及山谷的内涵，就如同诗词语言所描绘的：佳木葱茏、丰草绿缛、溪流潺潺、层峦叠翠、幽谷山涧。人类与大自然水乳交融，可以将他们的幸福感提升到新的境界。

城市公共开放空间有利于工作，经常去开放空间可提升工作效率。最近的研究指出，对于现今长时间高强度工作的上班族，去公园散步可以使他们的工作效率提高。大自然的风景及声音可令我们精力充沛思维活跃。密歇根大学的马克．博曼博士和他的同事在一项研究中表明，在植物园中散步小憩后，人们的记忆力及注意力集中水平可提高20%。而当同一群人被送到市内一条喧嚷繁忙的街道休憩时，数据则没有显著的提高。[9]而同时，把工作做好有利于财富的累积，从而使我们获得安全感。

公园同时也可以帮助人们克服孤独感、社交焦虑症及与他人相处时的不适感。美国建筑评论家莎拉·威廉姆斯·古德哈根把上述情况归纳如下：许多国家的社会"更像一个群岛而不像一个国家，它们逐渐分化成不同的种族、阶级、信仰及利益团体。某一类别的成员很少会和不同于自己的群体成员交流"。她认为城市公共开放空间可将不同的人从他们的虚幻梦想世界带到现实空间中来，有助于改善这种社会病态。城市公共开放空间的设计应考虑到互动，而这种互动"并不是有组织性或特定目标的，因

图 2.12
陈洪绶（1598-1652 年）陶渊明画像
Cheng Hongshu (1598-1652), Tao Yuanming

早在我们面对都市压力的几个世纪之前，陶渊明就指出了城市居住者会在压力中沦陷。
Centuries before the urban stresses we face now, the sage Tao pointed out that urban dwellers can feel trapped in a stressful world. (Credit: Honolulu Academy of Arts Collection)

图 2.13
情感
Love

在纽约的布莱恩特公园,人们晨练太极拳。
People in New York City practice morning *tai chi chuan* in Bryant Park. (Credit: Carl from New York; http://www.flickr.com/people/7811769)

为人类在有既定目标的情况下往往都会将注意力集中于他们的目标,只有目标被移除的情况下,他们才会看见先前看不见的人(或者是他们选择忽略的人)"。[10] 在公园中的偶然邂逅与发现不仅能消除人与人之间的隔阂感,而且通过经常对陌生人的礼貌与友善的态度,可以提高个人的道德修养。在游乐场,儿童通过共享游戏设施可以获取终身受益的社会交往经验。

公共健康对社会意义重大。很明显,城市公共开放空间在帮助城市人口保持健康中扮演着重要的角色。它为我们带来更有生产力与活力的工作环境,可减少社会医疗保健费用的开支。总有一天,城市公共开放空间的公共健康效益将转化成能用经济价值计算的公众健康服务。

物业安全似乎和城市公共开放空间没有太大的联系,但考虑到经济效益时两者却有着密切关系。公共土地信托会的卓越城市公园中心对城市公共开放空间的经济价值成效做出了一系列粗略的估量:除了可提供洁净的空气和水源(可直接使用,改善公共卫生)之外,还可提升邻近地产的物业价值,提高社会的物业税收和对开发商的吸引力,刺激市民对市中心的投资欲望,提高旅游消费及社区凝聚力。[11] 以芝加哥的千禧公园为例,该公园使旅游消费一年增长 1.9 亿美元,同时在邻近社区促成了 10000 多套全新单元的建造。[12] 甚至是树木也能促进零售消费的增加:相对没有种植树木的市中心地区,美国的消费者会在有明显的树冠覆盖下的市中心多购买 9%-12% 的有形和无形的商品。[13] 艺术,

that are "unstructured and non-goal-oriented, because humans, wired to concentrate on goals when goals are set before them, will focus on people whom they might not otherwise see (or whom they might otherwise choose to ignore) only if the pursuit of concrete goals is withdrawn." [10] Not only does the feeling of isolation dissipate, but the random encounters and observations which one makes of other people in parks is beneficial to personal moral growth, not least in learning to be friendly and polite to strangers. Children sharing play equipment in a playground are learning lifelong lessons about social interaction.

Public health matters very much to the community. Clearly, urban public open space plays a critical role in helping an urban population stay healthy, lead more productive, energetic lives in the workplace, at less health care cost to society. The public health benefits of urban public open space will someday be translated into public health services for which economic value will be regularly calculated.

The safety of property may seem to have little to do with urban public open space until one considers the economic benefits. The Trust for Public Land's Center for City Park Excellence has come up with rough measurements of the economic value of benefits of urban public open space in addition to the benefits of providing clean air and water, direct use, and improved public health: increased property value to nearby real estate, increased property tax revenues to the community, increased

图 2.14
情感
Love

成都的人们在春节的假期在一个公园的草坪上休息。
People in Chengdu relax on the grass in a park during Chinese New Years Holiday. (Credit: Qi Feng)

attractiveness to developers, increased incentive for citizens to invest downtown, increased tourism expenditure, and community cohesion. [11] As one example, Millennium Park in Chicago has increased tourism expenditures by $190 million a year and led to the construction of many of the 10,000 new housing units in nearby neighborhoods. [12] Trees even make a difference in retail expenditures: in the U. S. shoppers spend 9% to 12% more for goods and services in downtown districts with a significant tree canopy than one without trees. [13] The arts, including public art, make an even greater difference in attracting local investment and enhancing the local economy; in the U. S., a public investment of less than $3 billion in public art generates $134 billion annually. [14]

In summary, a community's economic health and security and its citizens' personal health and security are closely interrelated, and together constitute human safety. Urban public open space makes significant contributions to both.

■ **Love and Belongingness**

People whose basic physical needs are met and who feel safe and secure are then free to focus on forming healthy personal relationships with others, to be connected intimately with others, to grow with them and share meaningful life experiences, and to feel a strong connection to the broader community. There is no more vivid example of this than the large

包括公共艺术，可以更有效地吸引本地投资，从而带动本土经济；在美国，公众付出不到 30 亿美元的公共艺术投资可带来 1340 亿美元的收益。[14]

总之，一个社会的经济健康和安全与市民的个人健康及安全是休戚相关的，并且它们共同构成了人类的安全性。城市公共开放空间为上述两点做出了重大的贡献。

■ **情感和归属感**

已经满足了生理需求并获得了安全感的人们可以自由与其他人建立健康、密切的关系，一起成长，分享有意义的人生经验，并与大范围的社区中的人和物有密切的联系。中国的公园里，周日清晨一大群人聚在一起唱歌或打太极拳，就是再鲜明不过的例子了（图 2.13 - 图 2.14）。当我们提到"社区"一词时，就自然而然地想到中心焦点或核心利益，这个定义可追溯到城市公共开放空间历史早期。社区的概念，不仅仅是形容一个场所，更是对人类本质意义的诠释。城市公共开放空间比以往任何时候都扮演着更关键的角色，它重新诠释社区意识，就像第一章说的，加入了归属感，或是一种共同命运，一种跨越种族和阶级的交往。在拉丁语中，这被称作 communitas（指无特定结构的社区）是一种可以共享公共空间的生活模式。社区是一种友善的交际形式，不同种族、信仰、世界观的人聚集在一起，彼此关爱、维持家庭及友谊关系、开拓事业、建立邻里关系及

图 2.15 - 图 2.16
情感
Love

在成都活水公园和杭州西湖，新人们摆姿势拍婚纱照。
Couples pose for wedding photographs at Living Water Garden in Chengdu and at West Lake in Hangzhou. (Credits: Thomas M. Paine)

加入共同目的团体。社区还可以通过关爱和友情的滋养来帮助人们克服过度消费的欲望和与人交流的恐惧感，用平等博爱来取代自我任性。

通过对英国、中国及美国南部不同社会阶层的观察，奥姆斯特德总结出公园必须对所有阶层的市民平等地开放，并提供充满了强烈社交氛围的空间，而非让人远离人群，从而使"各阶层的群体都为共同的目的聚在一起，没有自作聪明，没有相互攀比，没有嫉妒，也没有自以为是，个人的存在融入大家的欢乐中，从而令每个人都更加愉快"。15 换言之，奥姆斯特德心中所想的正是 communitas，可以用"热闹"一词描述——一种欢乐、活跃及人们在喜庆节日里的愉快情绪。

英国建筑师罗亚当自 1996 年起在中国工作，他主张具有亚洲风格主义的社区。他归纳了七种高度可持续性社区的特性，分别是——全民参与、共同愿景、交流讨论（反思）、物质与精神平衡、融合（对抗片段化）、服务以及调整适应（转型）——强调群体的凝聚力及社会的可持续性。"我们所面临的挑战是要围绕爱、诚实、节制、谦逊、好客、正直和团结这些价值观或精神准则重新设计及发展我们的社区。这些有助于提升社会凝聚力，不管一个社会的经济如何繁荣、智力资源如何丰富、技术如何发达，没有上述各种特性，社会也难以长期持续下去"。16

社会学家雷·奥登伯格认为单靠生活场所和工作场所并不能满足人类社交的需求。17 人类需要一个"第三场所"以促进社区的繁荣，一个便利、受欢迎、舒适的地方，最好是免费的。城市公共开放空间提供的"第三场所"可令友谊萌芽，让家庭成员一起享受悠闲时光，让人们相遇并成为朋友甚至堕入爱河。在中国，公园或许是婚纱摄影最热门的场地（图 2.15 - 图 2.16）。关于这些美好体验的记忆甚至可以持续一生一世，亦在感情上加深了与社区的联系。水火节是美国罗得岛州普罗维登斯定期的公众表演节目，广受好评，其创办人埃文斯认为，城市公共开放空间成功的标准之一，是和同样文化背景下的其他空间相比其公共表演的受欢迎程度和影响力。

groups singing or performing *tai chi chuan* together in a Chinese park on a Sunday morning. (2.13-2.14) When we use the term "community" we come to what we may think of as the central focus or core benefit that goes back to the beginnings of the history of urban public open space. Community—the concept as much as the place—is at the root of what it means to be human. More than ever before, urban public open space has a pivotal role to play in renewing a sense of community, in other words, as we saw in Chapter One, a sense of belonging, a common destiny and a bond across all classes that we call *communitas* or shared public life within the community. Community is the convivial, day-to-day gathering of people of different races, faiths, and world-views to engage each other, find their shared humanity, to maintain family and friendships; conduct business; establish neighborhoods; and join in common purpose. It helps replace craving for conspicuous consumption or fear of otherness with the genuine nourishment of connectedness and companionship, more egalitarian than egotistical.

From his own observations of social classes in England, China and the American South, Olmsted came to the belief that public parks must be equally accessible to all citizens from all walks of life and include space not just for getting away from people but the exact opposite, a space for intense social immersion, for "the prospect of coming together, all classes largely represented, with a common purpose, not at all intellectual, competitive with none, disposing to jealousy and spiritual or intellectual pride toward none, each individual adding by his mere presence to the pleasure of all others, all helping to the greater happiness of each". 15 In other words, what Olmsted had in mind was *communitas*. What he had in mind could be described as *re4nao4*, the conviviality or lively and festive mood of people having a good time.

Adam Robarts, an architect based in China since 1996, argues for a more Asian-inspired definition of community. What he calls the seven habits of highly sustainable communities—universal participation, common vision, consultation (reflection), balance of material and spiritual, integration (fighting fragmentation), service, and adaptation (transformation)—emphasize group solidarity, and social sustainability. "Our challenge is to redesign and develop our communities around those values or spiritual principles including love, honesty, moderation, humility, hospitality, justice and unity, which promote social cohesion, and without which no community, no matter how economically prosperous, intellectually endowed or technologically advanced, can long endure". [16]

Sociologists like Ray Oldenburg have argued that home and workplace alone cannot meet this need of people to feel connected to other people. [17] There needs to be a "third place" to facilitate and foster community, a place that is convenient, welcoming, comfortable, and ideally free of charge. Urban public open space provides a "third place", where friendships can blossom, families can share leisure time together, people can run into friends, and even fall in love. In China parks are perhaps the most popular place for taking wedding album photos. (2.15-2.16) The memories of such tender moments can last a lifetime and add a more intense feeling of connection to the community. Creator of Waterfire, a highly successful performance event regularly held in Providence, RI, Barnaby Evans argues that one metric of the success of urban public open space is the incidence of public displays of affection compared to other spaces in the same culture.

The character of the central downtown open space of a destination city also strongly affects the first impression and lasting memory of visitors and tourists. That fact drives a desire on the part of local government to create an iconic park. In crass economic terms, urban public open space capable of fostering personal friendships, spawning couples, celebrating marriage, and impressing tourists earns brand loyalty for a community which is wise or fortunate enough to provide that public amenity. In human terms, the personal stories that owe their storyline to urban public open space can speak as persuasively as statistics as to why urban public open space matters, and matters deeply. It makes perfect sense that people who experience such joy in urban public open space will also love that open space and want to protect it and nurture it like a cherished friend.

Neighborhood parks contribute to family life, to creating stable neighborhoods, and to a strong sense of community. Children in particular benefit from close-to-home park experiences beginning early in life. (2.17) If the next generation is to be healthy and physically adept, outdoor play is an essential part of childhood beginning well before the school-age years and continuing into the teen years. (2.18-2.19) Most people living in cities do not have their own yards. For most children a healthy childhood depends on access to public open space, ideally playgrounds designed for them. The importance of play with one's peers in the formation of values and advancement of social adjustment cannot be overstressed. Particularly important is the role of urban public open space in providing children and youth groups with a set of non-materialistic experiences, [18] and functioning as a "safety valve where children can burn off energy." [19] (2.20-2.24)

图 2.17
百童游玩图，陶瓷花瓶，景德镇，约 1770-1795 年
One Hundred Children Playing, Porcelain Vase, Jingdezhen, c. 1770–1795

与同龄人在一起玩耍奠定了童年时代的友谊，并且建立了信任和团体的意识。
Playing together with peers forges childhood friendships and builds a sense of trust and community. (Credit: Victoria and Albert Museum Collection; photo: Andreas Praefcke)

市中心开放空间的特色可强烈地影响到访客或游客的初次印象及永久记忆。这也推动当地政府去创立标志性的公园。从一般的经济角度来看，城市公共开放空间能增进个人友谊、为情侣牵线搭桥、用于举办结婚庆典。社区有意识或有先见之明地提供公用设施，可为社区赚取品牌忠诚度，也给游客留下深刻的印象。从人类角度来看，一些在城市开放空间发生的个人故事能极具说服力地告诉我们为什么城市公共开放空间重要，而且极为重要。在城市公共开放空间里寻找到快乐的人们必定会钟情于这片开放空间，同时也必定会把它当作一个值得珍惜的朋友去保护、珍惜。

邻里公园对家庭生活的贡献在于它能建立稳定的邻里关系及强烈的社区感。尤其是儿童，他们从小在家旁的公园玩耍成长，获益良多（图2.17）。如果我们希望下一代健康并擅长体育，户外玩耍是一个主要部分，这一部分应该在学龄前开始，一直到他们成长为青少年（图2.18 - 图2.19）。大部分居住在城市的人都没有自家的院子。对大部分的小孩来说，一个健康的童年取决于能否自由地在公共开放空间玩耍，尤其是为他们而设计的理想游乐场。与年纪相当的儿童玩耍的重要性在于价值观的建立及社会适应力的提高。特别重要的是，城市公共开放空间扮演着为儿童及青少年提供一系列非物质经验的重要角色，[18] 它们是"儿童释放过剩精力的安全空间"[19]（图2.20 - 图2.24）。

图 2.18
情感
Love

上海的家庭在延安路的中心绿地上沉浸于自然。
Shanghai families enjoy immersion in Nature on Yan'an Road Downtown Greenspace. (Credit: Thomas M. Paine)

图 2.19
情感
Love

一个三世同堂的中国家庭一同在公园散步。
Three generations of a Chinese family are out for a jog in the park. (Credit: Cathy Yeulet)

图 2.20
情感
Love

在芝加哥千禧公园的云门下,人们彼此分享着他们变形的镜像。
Couples lie down to share the experience of their distorted reflection in Cloud Gate at Millennium Park, Chicago. (Credit: Thomas M. Paine)

图 2.21
情感
Love

周日,成千的人们坐在阿姆斯特丹的冯德尔公园的草地上。
On Sunday, people visit Vondelpark in Amsterdam and relax on the grass by the thousands. (Credit: © Jorge Royan / http://www.royan.com.ar)

Chapter Two   Why Urban Public Open Space Matters More Than Ever   **083**

图 2.22 - 图 2.23
情感
Love

在静谧之美国罗得岛州的普罗维登斯市,陌生的人们一起分享沉静音乐和水中篝火倒影结合在一起的体验。水中篝火的创造者把这个组合体验命名为"市民共同体"。
The creator of Waterfire in Providence, RI, USA, gave the name *communitas* to this group experience of calm music and fires reflected in water, shared by strangers in a state of serenity. (Credits: Thomas M. Paine and WF Providence)

图 2.24
情感
Love

波士顿的人们积聚在罗斯·肯尼迪绿道举行纪念阵亡将士的活动。
People in Boston share a Memorial Day holiday in the fountain at Rose Kennedy Greenway. (Credit: Thomas M. Paine)

■ Esteem

People who have reached the stage of experiencing friendship, love, and starting a family may also experience feelings of admiration or respect for others even as they earn the same for themselves. As people gain confidence in their abilities in the workplace and in the community, they may deservingly be blessed with achievements that have won praise and enhance their self-esteem. Increasingly emulating others, perhaps they win greater respect by others. The drive to feel useful to others and win their esteem or even their loyalty is strong in many cultures. Competitiveness plays a strong motivating role, but so does the self-directed desire to be a better person regardless of the opinion of others.

The reader may well ask what esteem has to do with urban public open space. In some spaces, memorials honor heroes or thinkers worthy of our admiration or respect, inspiring visitors to aspire to do more themselves. In some spaces, public art alludes to ideas that inspire or provoke or challenge us to be better people, more sensitive to issues that truly matter in the world. Some spaces may be large enough to provide comfortable and convenient settings where public

■ 尊重

到了经历友谊、爱情以及成家立业的人生阶段，人们会对他人产生钦佩及尊重之情，也同样会从别人那里获得尊重与认可。除了从工作岗位及社会中汲取自信心，人们还可通过获得成就而得到别人的赞许并增强自信心，也可以通过不断地同他人竞争，赢取尊重。在很多文化中，"努力成为对他人有价值的人和赢得别人的尊重和信任"，这样的动力非常强。虽然竞争力是一种强大的驱动力，但不管他人意见、自我导向地成为一个更优秀的人的愿望也具有很强的激励作用。

读者或许会问，尊重与城市公共开放空间的关系是什么。在一些空间，纪念碑会给英雄以荣耀或对思想家致以敬意，激励来访者做得更好。而有些空间，公共艺术会间接影射一些思想，从而激发或挑战我们的思维，令我们成为更优秀的人，同时对世界上至关重要的问题更加敏感。有的空间大到可以为公众提供舒适及便利的环境，使偶像和粉丝可以聚会互动，或是举办演出，使人们的尊重意识在多元社区中到达新的高度（图2.25）。

如果我们将目光转向社区和其集体自尊，城市空间的价值就显得更为突出。中国的很多城市开始意识到，在城市入口交会处建设夺人眼球的入口开放空间及摆放公共艺术品，能够增强城市人民的自豪感，让游客感受到城市的光辉形象，这些改变体现出极大的价值。显然，优秀的设计会赢得人民的敬意，同时也能增强社区的自豪感并加强社区的竞争力。

有人认为城市公共开放空间可为社会创造情感价值，其地位不亚于一个社会的经济价值。在美国，2008-2010年度的盖洛普社会调查得出这样的结论：人们对其社会的情感依附越多，本地的国民生产总值就越高。这项调查同时也发现了对一个地方的热情和忠诚与当地经济的健康度之间的联系。在很多国家，市民自发组织起来，作为"公园之友"或志愿者，帮助清理公园（图2.26）。假如市民对他们的城市公共开放空间附上情感上的投资，他们会更健康、更快乐、更有工作效率。[20] 越来越多的公职人员意识到，对于面临居住和工作选址抉择的居民或企业来说，如果市中心和社区的公共场所直观上不能赢得他们的青睐，就很难把他们留住。一个社区的"中心"之所以称之为"心"是由多种有力的原因共同决定的，包括人们在社区所能感受到的情感联系，特别是对社区起决定作用的开放空间。

■ 自我实现

有自信心的人，可以顺利地实现自我，即达成自己的愿望。通常，寻找生命更深层的意义往往是在努力满足基本需求之后。对于某些人，自我实现来自对社会的接触和回馈。多数能够获得上述深层体会的人都身处大城市。无论是在城市里出生或是为了改善生活移居城市的人，在人口稠密的环境里比在小社区里更容易结识改变自己人生的人。当别人使自己的人生改变，自己也想对别人的生活产生影响。这些人会积极参与社区活动，甚至让世界变得更美好。

人们的密切关系开始于一对一的偶遇。早在古希腊集市，人们就通过在公共空间偶遇的机会产生联系，发掘共同的兴趣，建立关系并激发创意。两个互不相识的人可能只是前一刻挨坐在公园长椅上观看盛会而下一刻便一起畅所欲言。他们或许会与对方

图 2.25
尊重
Esteem

在加利福尼亚州圣迭戈的巴尔博亚公园里庆祝海军荣耀日。
An event at Balboa Park in San Diego, CA, marks U.S. Navy Chief Petty Office Pride Day. (Credit: U.S. Navy 110907-N-ZZ999-005)

图 2.26
尊重
Esteem

志愿者自豪地身着纪念衫集体来到开罗的解放广场打扫场地。
Volunteers who came together to help clean up Tahrir Square in Cairo proudly wear their souvenir T Shirts. (Credit: Sherif9282)

图 2.27
自我实现
Self-actualization

不同文化的人们聚集在麦迪逊广场花园进行冥想。
Meditating at Madison Square Garden brings together people from many cultures. (Credit: Beyond My Ken)

gatherings may bring the admired together with their admirers, or where performances can inspire a diverse community to a level of esteem that makes a difference. (2.25)

If we turn to the community and its collective esteem, the justification for urban space becomes even stronger. Many cities in China have seen the value of investing in ambitious attention-getting gateway open space and public art at major gateway intersections announcing to all visitors that the city was taken its place proudly among great cities. Clearly, they represent an opportunity for design excellence, something that can enhance community pride, and strengthen a community's competitive position.

It has been argued that urban public open space creates emotional value that is as important as economic value to the community. In the United States, the Gallup Soul of the Community Survey from 2008 to 2010 found strong correlations between peoples' emotional attachment to the communities they live in, and higher levels of local GDP. The survey also found a link between passion for and loyalty to places, and the health of the local economy. In many countries, private citizens organize themselves as Friends of the Park or volunteer to help clean up the park. (2.26) If citizens are emotionally invested in their urban public open space, they are healthier, happier, and more productive.[20] Increasingly public officials realize that when people have choices about where to live and work, cities cannot compete for residents or businesses if they fail to make their downtowns and neighborhoods more appealing by creating public places that people intuitively like. The "heart" of the community is called heart for a powerful combination of reasons, including the emotional connection people feel for it, and in particular for the open space that defines it.

■ **Self-actualization**

People who have self-esteem are free to achieve self-actualization, which is synonymous with self-fulfillment. Finding a deep meaning in life usually comes only when the struggle to master more basic needs is no longer a distraction. For some, self-fulfillment comes from reaching out and giving back. Most people who have been fortunate enough to achieve that do so in large cities. Whether they were born in the city or they moved there to improve their lives, in the dense agglomeration of people they were far more likely to meet people who could make a difference in their life than they would have met in a small community. People made a difference to them, and now they want to make a difference to others. They may be open to active engagement to make the community or even the wider world a better place.

People's engagement may begin with a one-on-one encounter. Since the days of the Greek agora, chance encounters of people in public space have led people to make connections, discover common interests, build relationships, and spawn creativity. One moment two strangers can be sitting near one another on a park bench enjoying the human pageant, and the next moment they can be talking. Perhaps one of them shares a dream with the other, who then responds by sharing something that the first person had never thought of before. Suddenly from that chance encounter a wonderful new idea is born. With all our technological innovations, there is still no real substitute for the power of face-to-face encounters. People meeting face to face can judge trustfulness, can sense how best to communicate, and communication can be far more efficient, nuanced, and accurate than by phone calls or texting. It is the power of authenticity found in face-to-face encounters that works the magic of spawning creativity. (2.27)

Self-actualization can also be a solitary matter. In parks people are also free to seek personal enlightenment, renunciation and spiritual transcendence, or shed falsity and be truly authentic. For still others, what matters is the freedom to create, to innovate, "to go boldly where no one has gone before", to be open to fresh inspiration. Here is where parks shine: for at their best, they offer visitors a rich immersion experience, freeing the senses and the soul to become lost in the otherness of the experience, free of distraction, when one is lost in thought, experiencing "flow", at a moment of epiphany when one suddenly experiences a burst of insight, or rearranges all the pieces of the puzzle that have been festering in the subconscious and suddenly imagines something wonderful.[21] (2.28-2.29)

The versatile and bilingual Chinese writer Qian Zhongshu puts that moment of epiphany like this:

> *When we are totally absorbed in an object or situation so that our mind is for the moment emptied or purified of all other cares and assertive self-consciousness, when the concentration of attention brings about the unity of the subject (because all its energies and impulses are focused on a single point) as well as the unity of the subject and its object (because the former loses itself in the contemplation of the latter), we are in a state akin to the ecstasy or exaltation of the mystics:*
> *In body, and become a living soul;*
> *While with an eye made quiet by the power Of harmony; and the deep power of joy, We see into the life of things.* [22]

Parks are places for such Eureka moments, or simply helpmates who are always there for us, come what may, freeing us to become what Abraham Lincoln called the "better angels of our nature." In the long evolution of our nature, those better angels surely include, in no particular order, tolerance, honor, creativity, the pursuit of beauty, sportsmanship, leadership, courage, honesty, humility, tenderness, passion, empathy, curiosity, generosity, forgiveness, humor, respect, reverence for Nature, and of course spirituality. (2.30-2.31)

This contemplative, compassionate, even spiritual benefit of urban public open space complements and crowns all the others. Large parks, and the occasional gem of a small park, celebrate the enduring legacy of Eastern spiritual and ethical thought like Confucianism, Taoism, and Buddhism, and Western spiritual and aesthetic thought linking the sublime (like Mount Everest or Yosemite), the beautiful (like

图 2.28
自我实现
Self-actualization

人们可以在公园里而非办公室或家里自由地使用电脑。
One can find inspiration in a laptop opened not in an office or even at home, but in a park. (Credit: 123rf limited, HK)

图 2.29
自我实现
Self-actualization

一位艺术家沉浸于中央公园的美景中并把它画在了他的画布上。
An artist immersed in a creative act in Central Park makes the park his canvas. (Credit: SpyON)

分享自己的梦想或谈论从来没有思考过的事情。这瞬间的偶遇或许就会促成一个精彩想法的诞生。即使技术不断创新，也仍旧无法取代面对面的相遇所能产生的影响力。面对面让人能够更好地建立信赖感，并判断出什么才是最好的沟通方法，从而使得沟通更有效率、更微妙，比用电话通话或发短信更准确。面对面的邂逅同时也具有产生创意的魔力（图2.27）。

自我实现也可以是一种孤独的状态。人们在公园里可自由地寻找自我启发、自我超脱和精神上的超越，去掉虚伪，坚守正义。对于人们来说，最重要的是有创作与创新的自由，"大胆地去走前人没有走过的路"，去寻求新的灵感。这正是公园的魅力所在：它让访客沉浸在丰富的体验中，将感官及灵魂释放出来并陶醉于独特的体验中，让人迷失于自己的思想中，经历"意识流"，顿悟出突破性的见解，或下意识重新思考所有让人痛苦的难题并憧憬美好的事物[21]（图2.28 - 图2.29）。

才华横溢、通晓多国语言的国学大师钱钟书先生这样描述顿悟的瞬间：

> *当我们全身心地投入、关注某个对象或陷入某种情境时，思维会在此刻将所有杂念屏蔽或净化，注意力的集中除了产生主客体统一，也产生主体自身的统一（因为所有的精力及冲动都集中在一点上），也是主体客体的统一（因为主体在对客体的沉思中失去了自身）时，我们便处于一种类似神秘主义的迷狂或振奋状态：*
> *入睡，灵魂鲜活地醒来；*
> *和谐的力量，欣悦而深沉的力量，*
> *使我们睁开宁静的慧眼，*
> *把万物的纷纭看穿。*[22]

图 2.30 - 图 2.31
自我实现
Self-actualization

大自然的奇妙激励着人们去探索愉悦以及意志。
The wonder of Nature inspires people to find joy and purpose. (Credits: Mikhail Dudarev and Marina Pissarova)

公园正是那些醍醐灌顶的瞬间产生的场所，或者至少也是我们身边陪伴帮助我们的伙伴，从而使我们变成林肯所说的"我们心中的天使"。在天性的漫长演变过程中，我们心中的天使包括：宽容、荣誉、创意、对美的追求、体育精神、领导能力、勇气、诚实、谦逊、柔情、热情、同情心、好奇心、慷慨、宽恕、幽默、尊敬、对自然的敬畏，当然还有灵性（图 2.30 - 图 2.31）。

公共开放空间补充及涵盖了其他空间，它促人沉思、使人有同情心，甚至能带来精神心灵上的益处。大型公园和偶尔点缀的小公园是颂扬东方精神和道德伦理的不朽遗产，如儒家思想、道教及佛教，而西方精神及美学思想则与庄严（如珠穆朗玛峰或优胜美地国家公园）、美（如西湖）和风景如画（如丽江）的景象相联系。曾经有大量文字记载上述各地名胜。上述的每种哲学思想对应一种全新设计表现的兴起，给人们带来了不同程度的复杂或简约、对称或不对称，但所有这些都追求内在的统一与和谐，从而最终产生美。在我们这个时代，表达方式和振奋人心的形式可能会变，但渴望容纳人类精神的潜在欲望却不变。美是永恒的主题。

同样的逻辑也可应用在社会的许多方面；领导要求建造一个宜居社区，也同样会期望把社区建设带到另一个更高的水平上。而创意和创新反过来会形成一股更强的力量，让社区的竞争力胜过其他社区，使得自己的社区变得更好。有哪个城市会拒绝能容纳上千市民休闲聚集的场所吗？拒绝这一机会就如同搬起石头砸自己的脚。不提供这些场所的"机会成本"实在太大。与此同时，延迟当地可持续性环境发展的机会成本也非常大。马斯洛人类需求层次理论的金字塔最上层的自我实现表明，我们应该给予地球需求更多的关注，同时社会领导者必须共同地注意这一现状而我们的讨论才会圆满地达到初衷。

如何把城市开放空间改善人们生活状态的贡献附上一个经济价值？我们不能因为分析上的复杂性而跳过它。无论分析在难度上如何具有挑战性，如果我们可以在可持续性的实践上加以经济价值，智力及精神价值等无形资产也迟早会被列入计算之内。尽管如此，智力及精神价值可以说是超越了"纯粹"的经济利益。

West Lake), and the picturesque (like Lijiang). Much has been written about all of these. Each of these schools of thought gave rise to a style of design expression of varying complexity or simplicity, symmetry or asymmetry, but all have sought an internal unity and harmony, and out of that, beauty. In our era, the means of expression, and the forms that excite us may change, but the underlying desire to accommodate the human spirit remains the same. Beauty will always matter.

The same logic applies many-fold to the community as a whole; leaders entrusted with the livability of their community will likewise aspire to take that community to the next level of excellence. That creativity and innovation, in turn, has the power to make the community a better place and contribute to its competitive advantages over other communities. What city would willingly deny itself places where people at leisure and not in a rush can have that chance to meet, and by the thousands? Denying itself that opportunity would be like shooting itself in the foot. The "opportunity cost" of not providing such places is too great. At the same, the opportunity cost of postponing real progress on making the local environment sustainable is likewise too great. From the top of the pyramid, then, the self-actualized person realizes, above all, that he or she must also attend to the earth's needs, and the community leaders collectively realize the same, and so our discussion has come full circle.

How can we begin to put an economic value on this kind of contribution by urban public open space to the betterment of the human condition? We cannot ignore the matter simply because of analytic complexity. If we can place economic value on sustainable practices, however challenging for economic analysis, so too, intangibles like intellectual and spiritual value must sooner or later be included in that calculation. Nevertheless, intellectual and spiritual value arguably transcends "mere" economics. Indeed, for most of us who are blessed with the luxury of thinking about it and attempting to

图 2.32 - 图 2.34
没有生机的广场
The Lifeless Plaza

每个国家都会有无聊的空间设计。我们随机选取了托格、瑞典、巴西利亚和深圳的广场作为例子。
No country is immune to spaces designed for boredom. We select at random examples from Torg, Sweden, Brasilia, and Shenzhen. (Credits: Jerzy Kociatkiewicz of Colchester, UK, LeComteB, and Wengchuksyiua)

live it, intellectual and spiritual value is what really matters in this world and is the ideal focus of a good life.

These, then, are the benefits of urban public open space to meet the needs of people and to the environment. All of course are deeply interconnected, with soft boundaries between them. Indeed, the levels of need are experienced as a continuum. Of course, not all urban public open spaces answer the same needs. Different open space types answer different needs to different degrees. Community size makes a difference in how ambitious its aspirations for urban public open space should be. Cities with a population of over a million and with adequate financial resources typically will consider creating park systems linking a variety of parks together. Large parks may typically be more focused on the more basic levels of need. Centrally located parks may be more focused on the higher levels. We will discuss this in much more detail in the next two chapters.

In summary, if we add up all the ways in which urban public open space matters, we can see how not only the local economy and environment but also the quality of life in cities depend very much on a community that is rich in urban public open space. The need for urban public open space excellence is too great to be dismissed as an unaffordable luxury.

■ **Some Challenges**

With so many ways for communities to benefit from open space, we may ask how well they have been doing lately in providing excellent urban public open space design. Before we discuss what makes excellent design, it is well to acknowledge that cities do face some broad challenges in the way urban public open space has been designed in recent decades. No country is immune from the challenges highlighted here.

实际上，对于我们大多数人来讲，能思考和实践是一件幸福的事，智力和精神价值举足轻重，同时亦是美好生活的关键。

这些就是城市公共开放空间的好处，它满足了人类与环境需求。所有过程相互之间都有着深刻的关联，彼此没有清晰的边界线。实际上，不同层次的需求是连续的统一体。当然，不是所有城市公共开放空间都能满足相同的需求。不同的开放空间以不同程度去响应不同的需求。社区的大小有助决定城市公共开放空间对人们及环境带来的益处的多少。人口超过100万、拥有富足资金的城市一般都会考虑建立公园系统以连接不同类型的公园。大型公园一般会关注比较基本的需求。位于中央位置的公园可能会注重较高层次的需求。我们将在后续两章有详细的论述。

总之，通过总结城市公共开放空间各方面的重要性，我们可以清楚地知道，一个充满公共开放空间的社区，不但能改善当地经济和环境的状况，同时也能提高人们的生活质量。我们对于优质的城市公共开放空间的需求是如此强烈，不能把它当作是无法负担、可有可无的奢侈品。

■ **一些挑战**

随着社区通过开放空间获得多种益处，我们也想进一步知道社区如何更好地为大众提供优良城市公共开放空间。在讨论杰出设计的成因之前，我们首先要承认在过去数十年间城市公共开放空间普遍面临着一些挑战，任何国家都会经历下列各种的挑战。

1. 没有生机的广场

有人说，通往地狱之路是由善意铺成的。从巴西利亚到波士顿的大会堂广场，很多昔日获奖无数的广场已变成干涸了的沙漠。它们在设计之初旨在成为城市绿洲，最终却沦为缺乏生命力、死气沉沉的广场，并且欠缺妥善维护，冷清孤立，没有情感和灵魂（图2.32 - 图2.34）。这些广场本来是表达良好意愿的，可惜最终浪费了宝贵的机会。公共空间设计师斯蒂芬·卡尔及马克·弗朗西斯对上述问题感到痛惜："具纪念意义的空间设计本意是想令人印象深刻并具有'深远意义',但事实上它并没有将这个意图传递给使用者,而是让他们在步行穿过广场时感觉到不自在而已"。²³ 在这之后的 20 年内,公共空间设计没有什么改变。不幸的是,中国许多城市都为了提供公共基础设施而匆忙兴建了这类空间。或许这些都是在错误的地方建造的错误的公园,它们与场地不符。没有绿荫的大面积混凝土铺装,大量的长椅设置在广场上,冬天冷风飕飕,夏日暴晒炎炎,这些是对设计者设计初衷最具嘲讽意味的背弃。在人口稠密的中国,很多门户公园的使用率很低,公园外马路上热闹无比,公园内却无人问津。这不是社区内公共生活的景观,毫无愉悦感可言,更不用说社区自豪感了,这是一种乏味的景观。乏味的设计不能带来惊喜或振奋,所以也不能引起共鸣或与人建立密切联系。大面积的铺砌广场会导致热岛效应,同时也是不可持续的。穿越空旷广场的人们情愿站在边缘地带或者是更人性化的子空间里。使用者喜欢停留在固定物上,比如边缘、长凳、观赏对象等,但却不愿意处于一个面积过大而又缺乏生气的广场中。通常能够吸引人们视线的是复杂一点的、能给人带来视觉愉悦的事物,当然还有大自然。

打着极简主义旗号的这些开放空间昔日获奖无数,如今虽然屹立不倒,但实际上是在浪费宝贵的公共资源;如果当初的设计具有丰富的想象力,不偷工省事,人们便有享用多重体验的机会了。

另一方面,过度的复杂性也未必是好事。公园项目充满过多的概念式的设计和设施可能会导致失败,因为公园中的"大自然"已被"都市化"取代,混乱代替了安宁,公园不再是一个安静避世的场所。"过度刺激"的设计跟平淡索然的设计一样都不代表着优秀设计。优秀的设计增一分略繁,减一分则不足,是为刚好。对优秀设计的评判着实是一项难得的技能。

2. 干涸的湿地

可持续性发展逐渐成为一种趋势并吸引了大量的仿效者。湿地的作用是净化污水及控制水流,而取代天然湿地的人工湿地,虽然外观上和真正的湿地无异,刚建成的时候似乎也有湿地功能,却不能真正净化水质。它们需要定期的维护或采用水泵等高能耗的人工水循环系统才能继续运作。真湿地尚且可能丧失功效,何

图 2.35
干涸的湿地
Thirsty Wetland

在我们匆忙地想达到可持续性的过程中,许多无效和失败却又十分吸引眼球的湿地要比那些难看的但是正在自然地干涸的河床更有迷惑性。
In our rush to be sustainable, the inefficiency or failure of many artificial wetlands which appeal to the eye is more deceptive than an unsightly but purely natural dry lakebed. (Credit: Paulus Rusyanto)

1. The Lifeless Plaza

It has been said that hell is paved with good intentions. The many-award winning plazas of yesterday—from Brasilia to Boston's City Hall Plaza—are deserts where an oasis was intended. They ended up as lifeless run-down plazas, poorly maintained, aloof, unfeeling and without soul. (2.32-2.34) Good intentions squandered the opportunity. Public space designers Stephen Carr and Mark Francis deplore the problem: "a monumental space consciously designed to be impressive and thus 'meaningful', may convey little to its users—apart from the discomfort of walking across it". ²³ In the twenty years since that was written, nothing has changed. In China, many cities have unfortunately created such spaces in their rush to provide public infrastructure. Perhaps they are the wrong park in the wrong place, at war with the site. Treeless expanses of concrete with regimented benches windswept in winter and blisteringly hot in summer are perhaps the most flagrant violation of the sacred trust imposed on the designer. In crowded China many gateway parks are sadly empty of people and mostly experienced from the road rather than from inside the open space itself. It is not the landscape of *communitas* or joy or community pride. It is the landscape of boredom. Boring design cannot surprise you or excite you, so it cannot engage you or inspire you. It is also unsustainable. Huge paved plazas are heat islands. People pass through emptiness to attach themselves at sides or in human-scaled subspaces. Users like to attach themselves to "anchors"—edges, benches, objects to look at—but not raw space in a huge lifeless plaza. People are drawn to intricacy and visual delight, and to Nature.

The underlying spirit of minimalism that inspired these formerly award-winning projects is still alive and unrepentant, still squandering precious public resources, and still denying people the opportunity to experience many layers of experience that could have been theirs if only the design had been imaginative and inclusive and not so lazy.

On the other hand, excessive complexity may not be much better. The temptation to crowd a project with design ideas and facilities may fail because the "Nature" in the park has been replaced with "urbanization," serenity replaced with chaos, and that cannot serve the fundamental purpose of providing a soothing retreat. "Overstimulated" design is no closer to design excellence than anemic design is. Excellence lies in the median between the two extremes of too little and too much. Excellence of design judgment is truly a rare skill.

2. The Thirsty Wetland

Sustainability has become a movement that has attracted imitators. In place of genuine wetlands purifying polluted waters and regulating water flow is the potential for fake ones that look genuine and perhaps perform like real ones when first constructed but probably do not truly purify the water as they are intended and cannot survive without maintenance or energy-consuming artificial water circulation systems like pumps and aerators. Even those that are not fake can fail. All they do is squander resources or lull us into thinking that they are net contributors to environmental enhancement rather than net detractors. They are especially deceptive if they are simply trophy projects, exceptions to the rule, alone without followers whose combined presence can have a more significant positive impact. They are especially tragic when they replace far more complex and efficient natural wetlands, which are disappearing at an alarming rate. (2.35)

3. The Neglected Landscape

In the West it is called "bait and switch": a retailer lures us into the store with the promise of one product, but when we get there they say they are sold out and we end up with something inferior. Developers do this to buyers, when they shut off the fountains that were flowing until the units were sold. Local parks authorities can end up doing much the same thing to the community, when park maintenance fades away. (2.36)

The neglected neighborhood, lacking easy access to a neighborhood park if not larger open space, is in need of environmental equity. Providing poorly designed or poorly maintained open space does not solve the problem of providing environmental equity, according to the Commission for Architecture and the Built Environment (CABE), the British government's advisor on architecture, urban design and public space.[24] There is no escaping the need for design excellence.

4. Climate Change Denial

Although there is widespread acknowledgement that the global climate is changing, that the polar ice caps and Tibet Plateau glaciers are melting, and that the sea level is rising, there has been scarcely a change of attitude in how to design projects near the sea and close to sea level. Green infrastructure should be on the increase in this coastal zone, but the rate of allocation has not increased. The prospect of short-term profit in coastal real estate development blinds speculators and urban planners to the long-term devaluation that rising sea levels will bring. The rise of the sea level is still occurring too gradually to trigger a massive wake-up call. Sadly, only the most imaginative and forward-looking urban planners and developers will have the foresight to work with designers to take rising sea levels into full consideration in at-risk areas. (2.37-2.38)

5. "There is No 'There' There."

Whether an urban public open space is paved or planted or in between, it is possible to miss every opportunity to create a place where people want to be, and want to return to time after time. (2.39) A space like that is not a place. It is mediocre. Mediocrity arises when imagination and competence are in short supply.

图 2.36
被忽视的景观
The Neglected Landscape

建造一个景观简单，但是维护起来很难。这个高雄的例子可以发生在任何国家。
Building a landscape is easy, but maintaining it too often proves to be too difficult as in this example from Kaohsiung, which could be in any country. (Credit: Thomas M. Paine)

况是装模作样的假湿地。湿地若仅仅是以徒有其表的门面工程作为项目卖点，则更是误导大众；倘若名副其实，定能带来积极影响。最令人痛心的是用这些人工湿地替代天然湿地。结构复杂的天然湿地有生态功效，却正在迅速消亡（图2.35）。

3. 被忽视的景观

西方有句与"挂羊头，卖狗肉"意思相近的谚语——"先误导上钩，再偷梁换柱"，指商家以不正当的手法引来消费者，当消费者决定购买某产品时，商家便会声称此商品已经售完并以品质低劣的货品代替。开发商会以同样手法瞒骗买家，当所有的房售出后，开发商便会关掉喷泉装置。当公园日久失修，地方公园管理者最终可能会以同类手法对待当地社区（图2.36）。

缺乏可达社区公园的居民区同样需要环境上的平等。英国政府建筑、城市规划及公共空间顾问——英国建筑及建成环境委员会（CABE）表示：为这些居民区提供的设计不合理、维护管理不善的开放空间并不能解决环境均享的问题。[24] 大家都有对优秀设计的需求，不能逃避这一点。

4. 拒绝气候变化问题

众所周知，全球气候变化，两极冰川和青藏高原冰川融化，海水水位上升等气候变化问题日趋严重，但同时大家对沿海和靠近海平面的项目如何设计并没有任何态度上的转变。我们该增加沿海地区的绿色基本建设，但可惜基建分配率并没有得到相应提高。沿海地区房地产发展着眼于眼前短期利益，会令投资者及城市规划者忽视海水水位上升所带来的长远负面影响。水位上升这个问题仍然持续，但现在的严重性并不足以唤醒大众对问题的关注。很可惜，只有最远见卓识、最富有想象力的城市规划者和开发商才会携同设计师，努力周详地去考虑高危地区的水位上升问题（图2.37 - 图2.38）。

5. "那里哪里也不是"

不管一个城市开放空间是只做了铺装或是种满植栽，或是两

图 2.37 - 图 2.38
拒绝气候变化
Climate Change Denial

越来越严重的海上风暴和海平面上升都在要求我们要更多地关注低洼沿海区域的技术革新。
Both increasingly severe sea storms and rising sea levels are going to demand a lot more attention to innovative design for low-lying coastal areas globally. (Credits: tedkerwin and Nigel Chadwick)

图 2.39
那里哪里也不是
There is no "There" there

尽管布拉格的霍多夫北部公园有些大胆的设计想法，但是缺少了场地的感觉以及使用者。每个国家都有像这样被浪费的土地，但并不应该这样。
Despite some bold design ideas Chodov North Park in Prague lacks a sense of place, and users. Every country has wastelands like this. No country should. (Credit: Packa, a user of Czech Wikipedia)

者兼而有之，都有可能会错失将这个地方打造成令人流连忘返的场所的机会（图 2.39）。这样的空间不能称为场所，而只是普通平庸的地方。平庸之作是缺乏想象力和能力的结果。

> 在功能以外，设计师需要做很多选择，这通常被看作他们的审美特权。特定形状或材料的选择及它们之间的关系，对使用者的体验有着深远的影响。他们决定一个场地如何跟文化期许和使用者的渴望产生共鸣，这决定了随着时间的流逝，这个地方是否受人喜爱和丰富多彩。在更深层次上，创造一个具有意义的环境是公共空间设计的最大挑战。[25]

一些设计师用"富有意义"一词形容空间时比较迟疑，倒更喜欢用"难忘的"来形容，但两者都是指与使用者有紧密的积极联系。不幸地，对他们有纪念意义的难忘事物可能不会对大部分人留下深刻印象。比简单创造空间更高的境界是"场所营造"。单是提出"场所营造"这个术语就预示着会带来美妙的结果，正如打上可持续设计标签的项目就应该不负众望地带来可持续的效果。也许这样解释场所的定义更好：即一个能"自我实现"的空间。

因此，长期被忽略会导致场所逐渐失去其最精华的特质，慢慢堕落至平庸，远不如从前。这正是为什么美国诗人格特鲁德·斯泰回到家乡加州的奥克兰时说，"那里没有了那里的面貌"。这句另类的评语言简意赅，令人印象深刻。设计师们常常用它来形容不再有场所感或是从来没有场所感的空间。

今天已经不再是可以在平庸景观上浪费资源的时代了，哪怕是可再生资源，更不用说稀缺资源或我们的血汗和泪水了。这个时代已经开始将优秀的城市公共开放空间设计作为基本的社区需求，而将没用的、疏忽的、平庸的城市公共开放空间设计作为一种犯罪。

6. 空降设计师

常常有一些设计师从很远的地方空降，对项目选址的接触时间尚浅，更谈不上融入社区生活，以至于对当地文化和自然环境相当陌生，这导致他们设计的空间水土不服。更糟的是，他们可能从心底轻视当地传统，想当然地用他们充满文化优越感的想法凌驾于场地状况之上，甚至以此为使命。之所以有这样的错位，是因为当地的官员可能觉得从海外引入的设计是真正有权威的。然而，这并不代表人民的意愿。这些设计师的骄矜傲慢与当地官员的盲从偏见

Beyond function, there are many choices to be made that are often thought of as the artistic prerogative of the designer. These choices of specific forms and materials, and the relations among them, greatly influence the user's experience of associative meaning. They determine how a place resonates with the cultural expectations and longings of users, and have a great deal to do with whether it is loved and enriched over time. To create settings that are meaningful in this deeper sense is the greatest challenge of public space design. [25]

Designers who hesitate to use the term "meaningful" may prefer to say "memorable", but clearly both terms speak to making a deep positive connection with the user. Unfortunately, what is memorable to the designer may not make so deep an impression on most people. Designers who understand the higher aspiration than mere space-making call it "placemaking". The term supposes that wonderful results will follow from merely invoking that term, just as sustainable design is supposed to result inevitably from a project that is so labeled. It might better be said that a "place" is a space that is so apt that it has "self-actualized".

So, too, over time, by neglect, a place can gradually lose its essential qualities and descend into a mediocrity which it did not have originally. That is what the American poet Gertrude Stein meant when she returned to her childhood hometown of Oakland, California, and said, "There is no 'there' there." The oddly memorable phrase stuck, and designers frequently invoke it to describe spaces that are no longer places, or never were.

Today is no longer the time for investing even renewable energy, let alone scarce resources or blood, sweat, and tears, on yet another banal landscape. The time has come to treat urban public open space design excellence as a basic community need, and unused, neglected and mediocre urban public open space as a crime.

6. The Airlifted Designer

Time and again, designers airlifted in from afar barely touch the ground, let alone immerse themselves in the community long enough to understand the cultural and natural context which the spaces that they design should evoke. Worse, they may feel secret contempt for the traditions and feel it their duty to liberate the site with their culturally superior idea. They find themselves in this position because local officials may feel that true prestige will be theirs if they can bring home a design from overseas. However, they may not be representing the feelings of the people in doing so. This arrogance is unworthy of the position of responsibility and trust held by both the designer and the local officials. (2.40)

Nevertheless, it would be wrong to dismiss out-of-hand the notion that local sourcing produces the best results, and foreign sourcing produces the worst results. All depends on the intentions, values, and sensitivity of the particular individuals entrusted with the responsibility of providing the people with excellent urban public open space.

■ Conclusion

The design of urban public open space is too important to be done halfway or haphazardly, which is sadly too often the case. People deserve excellent design in the public spaces which they share. Instead of spaces they hardly want to linger in, people deserve spaces which they never want to leave, but return to again and again. People deserve outdoor spaces where their dreams can take wing, friendships blossom, a childhood experience becomes a lifelong memory, wildlife and people live in harmony, people linger to relax, come to be inspired, to find peace, watch a performance, fall in love, hatch a brilliant new idea, make a tough decision, feel connected and reconnected. People who are able to enjoy places as good as that will feel proud to be part of such a community. And they will help assure that places as good as that will long endure. In such common connection is the sustainability of the urban public open space. Places as good as that stand a good chance of inspiring the community to form an emotional bond enduring over generations. They are only sustainable with that connection. It takes a community to sustain its urban open space.

Tourism Value of Urban Public Open Space

Social networking is providing useful information about the value of urban public open space. Maps of tourist photographs document the role of parks.

Mapping-extraordinaire Eric Fischer has mapped the pictures taken by local residents versus tourists, and the results reinforce the belief that parks can be major contributors to a city's tourism.

A look through the maps of 60 different cities shows concentrations of web-posted tourist pictures in parks and key public spaces. The map of New York City appears to show Central Park as the dominant visitor attraction. In Chicago, Millennium Park is a huge draw, in Paris the boulevards and Jardin du Luxembourg, in Berlin the Tiergarten, in Austin, Tex. Zilker Park and Barton Springs and in Oslo the Frognerparken.

Parks can be a travel destination and be real drivers of economic value from visitor spending. For cities looking to draw in outsiders, a quality public realm may do more than any convention center or expensive museum.

使其不值得委以重任进行公共开放空间的设计（图2.40）。

尽管如此，我们不能矫枉过正，一竹竿打翻一船人，草率武断地认为本土的力量会产生最好的效果，而外来的和尚就不会念经。一切都取决于那些负责为人们创造优秀城市开放空间的人物个人意图、价值观和悟性。

■ 结论

城市公共空间的设计非常重要，所以不能草率行事或半途而废，只可惜，这样的案例经常发生。优秀的公共空间设计是大众应该享用的。这些空间应该令他们流连忘返，愿意再次光顾，而不是一刻也不愿停留。人们需要这样的户外空间，在那里，他们可以插上梦想的翅膀，让友谊愈加深厚，使童年的经历成为一生的回忆，实现野生动物与人类的和谐共处，人们在此可以徜徉放松，获取灵感，寻求内心的平和，观看演出，堕入爱河，孵化新思维，做出艰难的决定且不再会感到孤独。可以在社区空间里感受到这样的人，会为自己是这个社区的一份子而感到自豪，同时他们会帮助维护这些场所，使其得以长久维持下去。这些基本的联系是城市公共开放空间的可持续性发展的保障。拥有上述优良条件的场所可鼓舞社区形成一个可持续几代人的情感纽带。通过这种联系，这些场所才能维持下去。一个城市的开放空间的持续需要整个社区的支持。

城市公共开放空间的旅游价值

社会网络给我们提供关于城市公共开放空间的有用信息。游客照片地图体现出公园的重要性。

卓越的地图绘制者艾瑞克·费切尔用当地居民与游客照片编绘成地图，其结果显示：公园确实是城市旅游业的主要支撑。

通过60多个不同城市的地图显示网络发布的游客照片场景主要集中在公园和公共空间（游客照片用红色显示，居民照片用蓝色显示）。纽约市的地图显示，中央公园是主要的游览观光地。在芝加哥，千禧公园是一个游览胜地。此外还有巴黎的林荫大道和卢森堡公园，柏林的蒂尔加藤公园，奥斯汀的济尔克公园和巴顿泉，奥斯陆的维格朗雕塑公园。

公园可以成为旅游目的地，而且通过促进游客消费为当地带来经济价值，对那些寻求吸引外地游客的城市而言，一个高质量的开放空间可能会比会议中心或博物馆带来更大的经济效益。

图2.40
空降设计师
The Airlifted Designer

也许第一位知名设计师或"建筑师明星"，毫不谦逊的弗兰克·劳埃德·赖特把大胆的西方理念带给了那些崇拜他的亚洲客户和他的建筑师同仁们。如这张图片所示，1911-1912年间他与日本建筑师新远藤和林爱在一起。空降设计师的问题在于民众所需要的可能并不是明星设计师所提供的，尤其当设计师没有充分的时间去掌握当地文化的第一手资料的时候。
Perhaps the first celebrity designer or "starchitect", the never humble Frank Lloyd Wright brought bold Western ideas to his adoring Asian clients and fellow architects, as he is shown here with Japanese architects Arata Endo and Aisaku Hayashi in 1911-12. The problem is that what the people need may not be what the star designer provides, especially one who spends little time experiencing local culture first-hand. (Credit: wikimedia)

注释

1. 奥姆斯特德.《公共公园与城镇的扩大》，1870
2. 克莱尔·库珀·马库斯，卡罗琳·弗朗西斯.《人性场所》（纽约：约翰威立合作出版社，1998）
3. 伊丽莎白·梅耶尔.《外观表现的恒久之美》, http://www.arch.virginia.edu/lunch/print/territory/sustaining.html, July 12, 2011
4. 查尔斯·瓦尔德海. 哈佛大学设计学院景观设计学系主任, 在其撰写的《景观都市主义读本》中提到的"景观都市主义"（纽约：普林斯顿建筑出版社，2006），35
5. 汤普森·库恩，博迪·斯泰恩，维尔·巴顿，迪普拉吉.《环境科学与技术》."在户外锻炼身体比室内锻炼更能带来身心愉悦吗？一个综合的评论"，2011 45（5），1761-1772
6. 保罗·雪拉. "公园的优势，为什么美国需要更多的城市公园和开放空间"；以城市公园的优点为中心，公共土地信托，2005 http://cloud.tpl.org/pubs/benefits-park-benfits-white-paperl2005.pdf；"开放空间,娱乐休闲设施以及适于步行的社区设计的经济效益，" 2010年5月，www.activelivingresearch.org
7. 斯蒂芬·卡尔，马克·弗朗西斯，李安尼·里弗林，安德鲁·斯通.《公共空间》（纽约：剑桥大学出版社，1992），134
8. 瑞秋·卡普兰，史蒂芬·卡普兰. "修复的经验：周围环境治愈的力量"；M·弗兰西斯和R·赫斯特编辑，《花园的意义》（剑桥：麻省理工出版社，1990），238-243. 一些认真研究的例子证明了这一点，弗兰西斯·果，"公园和绿色环境：一个健康的人的生活习惯的主要组成部分"（美国游憩与公园协会，2011年，摘自：http://www.nrpa.org/uploadedFiles/
9. 雪莱·王. "休息喝杯咖啡？公园里转转？为何休息那么难？"，华尔街日报，2011年8月30日
10. 莎拉·威廉姆斯·古德汉根. "驻足于此"，《新共和》，2010年9月2日，20-25
11. 2003年的报告，关注的7个特性，提供了华盛顿、圣迭戈、波士顿、萨克拉门托和费城这5个城市的经济统计，已发布到以下网站：www.tpl.org
12. 亚历山大·加文.《公园，适宜居住社区环境的关键》（纽约：诺顿公司，2011），131

NOTES

1. Frederick Law Olmsted. *Public Parks and the Enlargement of Towns*, 1870
2. Clare Cooper Marcus and Carolyn Francis. *People Places* (New York: John Wiley, 1998),
3. Elizabeth K. Meyer. "Sustaining Beauty, The Performance of Appearance: A Manifesto", http://www.arch.virginia.edu/lunch/print/territory/sustaining.html, July 12, 2011
4. Charles Waldheim. Chair, Department of Landscape Architecture, Harvard Graduate School of Design, "Landscape as Urbanism", in Charles Waldheim, ed., *The Landscape Urbanism Reader* (New York: Princeton Architectural Press, 2006), 35
5. J. Thompson Coon, K. Boddy, K. Stein, R. Whear, J. Barton, M. H. Depledge. "Does Participating in Physical Activity in Outdoor Natural Environments Have a Greater Effect on Physical and Mental Wellbeing than Physical Activity Indoors? A Systematic Review", *Environmental Science & Technology 2011* 45 (5), 1761-1772
6. Paul M. Sherer. *The Benefits of Parks, Why American Needs More City Parks and Open Space*, Center for City Park Excellence, Trust for Public Land, 2005, http://cloud.tpl.org/pubs/benefits-park-benfits-white-paperl2005.pdf, "The Economic Benefit of Open Space, Recreational Facilities and Walkable Community Design", May 2010, www.activelivingresearch.org
7. Stephen Carr, Mark Francis, Leanne G. Rivlin, Andrew Stone. *Public Space* (New York: Cambridge University Press, 1992), 134
8. Explore_Parks_and_Recreation/Research/Ming%20%28Kuo%29%20Reserach%20Paper-Final-150dpi.pdf, June 15, 2011) R. Kaplan and S. Kaplan. "Restorative experience: the healing power of nearby nature", in M. Francis and R. Hester (eds.), *The Meaning of Gardens* (Cambridge: MIT Press, 1990), 238-43. For examples of rigorous studies supporting the claim see Frances E. (Ming) Kuo, "Parks and Other Green Environments: Essential Components of a Healthy Human Habitat" (National Recreation and Park Association, 2011, in http://www.nrpa.org/uploadedFiles/Explore_Parks_and_Recreation/Research/Ming%20%28Kuo%29%20Reserach%20Paper-Final-150dpi.pdf, June 15, 2011)
9. Shirley S Wang. "Coffee Break? Walk in the Park? Why Unwinding Is Hard", *Wall Street Journal*, August 30, 2011
10. Sarah Williams Goldhagen. "Park Here", *The New Republic*, September 2, 2010, 20-25
11. Report of 2003 focusing on seven attributes, providing economic calculations for Washington, D.C., San Diego, Boston, Sacramento, and Philadelphia, as reported in www.tpl.org
12. Alexander Garvin. *Public Parks, the key to Livable Communities* (New York: W. W. Norton & Co., 2011), 13

13. "绿色城市，健康". http://depts.washington.edu/hhwb/Thm_Economics.htm，2011年6月1日

14. 奥姆斯特德. 《公共公园与城镇扩建》. 1870年在波士顿的演讲, 萨顿编辑,《创建文明的美国城市，论城市景观／奥姆斯特德》（马萨诸塞州，剑桥：麻省理工出版社，1971），75

15. 奥姆斯特德. 《公共公园与城镇扩建》. 1870年在波士顿的演讲, 萨顿编辑,《创建文明的美国城市，论城市景观／奥姆斯特德》（马萨诸塞州，剑桥：麻省理工出版社，1971），75

16. 发表于《走向可持续的发展：创造环境的角色》，由房地产委员美国商会主办，上海，2007年9月11日

17. 雷·奥登伯格.《为第三个地方庆祝：关于社区中心"伟大的好地方"的启发故事》纽约：马洛公司出版社，2000

18. 贝琳达·余恩."新加坡社区公园的使用和经验". 休闲研究期刊，28（4），1996年，293-311

19. 安娜斯塔萨·卢开拓·赛得利斯，埃立特·斯铁格利茨."洛杉矶公园中的儿童，对社区公园中公平、品质和儿童的满意度的研究". 城镇规划评论，2002，467

20. 彼得景山，"为什么我们不有情感的集中建造城市？"发表在：http://www.infrastructurist.com/2011/05/17/why-arent-we-building-emotionally-connected-cities-a-guest-post/, May 18, 2011. 景山是《为了城市的爱：人们和他们居住的空间的爱的故事》的作者（创意城市产品出版社，2011）

21. 米哈里·契克森米哈赖.《最佳心理体验》（纽约：哈珀·柯林斯出版集团，1990）

22. 钱钟书,《还乡》,《书林季刊》. 1947年，第17-26，摘自诗人威廉华兹华斯《丁登寺》，转载于《钱钟书英文文集》（北京：外语教学与研究出版社，2010），359-360

23. 斯蒂芬·卡尔，马克·弗朗西斯，李安尼·里弗林，安德鲁·斯通.《公共空间》（纽约：剑桥大学出版社，1992），265

24. 建筑与建筑环境委员会（CABE）."社区绿色环境". 2010年5月1日，http://webarchive.nationalarchives.gov.uk/20110118095356/http://www.cabe.org.uk/files/community-green.pdf

25. 斯蒂芬·卡尔，马克·弗朗西斯，李安尼·里弗林，安德鲁·斯通.《公共空间》（纽约：剑桥大学出版社，1992），265

13. "Green Cities, Good health", http://depts.washington.edu/hhwb/Thm_Economics.html, June 1, 2011

14. Frederick Law Olmsted, "Public Parks and the Enlargement of Towns", speech given in Boston in 1870, in S. B. Sutton, ed., *Civilizing American Cities, Writings on City Landscapes/Frederick Law Olmsted* (Cambridge, MA: MIT Press, 1971), 75

15. Frederick Law Olmsted, "Public Parks and the Enlargement of Towns", speech given in Boston in 1870, in S. B. Sutton, ed., *Civilizing American Cities, Writings on City Landscapes/Frederick Law Olmsted* (Cambridge, MA: MIT Press, 1971), 75

16. Presented at *Towards Sustainability: the Role of the Built Environment*, organized by the American Chamber of Commerce Real Estate Committee, Shanghai, September 11, 2007

17. Ray Oldenburg, *Celebrating the Third Place: Inspiring Stories about the "Great Good Places"* at the *Heart of Our Communities* (New York: Marlowe & Company, 2000)

18. Belinda Yuen, "Use and experience of neighborhood parks in Singapore," *Journal of Leisure Research*, 28(4), 1996, 293-311

19. Anastasia Loukaitou-Sideris and Orit Stieglitz, "Children in Los Angeles Parks, a study of the equity, quality, and children's satisfaction with neighbourhood parks", *Town Planning Review*, 2002, 467.

20. Peter Kageyama, "Why Aren't We Building 'Emotionally Connected' Cities?" in http://www.infrastructurist.com/2011/05/17/why-arent-we-building-emotionally-connected-cities-a-guest-post/, May 18, 2011. Kageyama is author of *For the Love of Cities: The Love Affair between People and Their Places* (Creative Cities Productions, 2011)

21. Mihaly Csikszentmihalyi, Flow: *The Psychology of Optimal Experience* (New York: Harper and Row, 1990)

22. Qian Zhongshu, "The Return of the Native", *Philobiblon*, 1 (1947), 17-26, quoting from poet William Wordsworh, *Tintern Abbey*, as reprinted in *A Collection of Qian Zhongshu's English Essays* (Beijing: Foreign Language Teaching and Research Press, 2010), 359-60

23. Stephen Carr, Mark Francis, Leanne G. Rivlin, Andrew Stone. *Public Space* (New York: Cambridge University Press, 1992), 265

24. Commission for Architecture and the Built Environment (CABE), "Community Green", 2010, http://webarchive.nationalarchives.gov.uk/20110118095356/http://www.cabe.org.uk/files/community-green.pdf

25. Stephen Carr, Mark Francis, Leanne G. Rivlin, Andrew Stone. *Public Space* (New York: Cambridge University Press, 1992), 265

Chapter Three
第三章

规划原则
Planning Principles

向香港人提供优质的公共空间和公共设施，是政府不能推诿的责任。当局应该拿出诚意，多虚心聆听社会各方的意见，还我们一个公道，让大家都能真真正正享用属于我们的公共空间，享受一个有质素的生活。
It is the government's clear responsibility to provide quality public open space and public facilities to the people of HK. The government should demonstrate sincerity and humbly listen to the views of the community. It should be fair to us and let us really enjoy public open space that belongs to us so that we can all have a better quality of life.
—— 梁家杰，香港立法会议员，2008 年 [1]
Alan Leong Kah-Kit, Hong Kong citizen, 2008 [1]

我们的目标是让每个纽约人从家走出来 10 分钟就能到达一个公园。
Our goal is for every New Yorker to live within a ten-minute walk of a public park.
—— 迈克尔·布隆伯格，纽约市市长，2010 年 [2]
Michael Bloomberg, Mayor of New York, 2010 [2]

我们对于城市公园的愿景是到 2020 年，美国城市的每个人从居所都能轻松步行到整洁、安全和充满活力的公园。
Our vision for urban parks is by 2020 everyone in urban America will live within walking distance of a park that is clean, safe and vibrant.
—— 城市公园联盟使命宣言，2011 年
City Parks Alliance Mission Statement, 2011

鉴于城市公共开放空间能提供和满足人类及环境的需求，人们具有使用开放空间的权利好像并不是具有争议性的主张，而是一个基本常识。人们对城市公共开放空间的权利是多方面的。1948年，联合国颁布了世界人权宣言，其中包括"和平集会和结社自由的权利"。显然，户外公共空间比室内空间更公平地提供了上述权利。

西方最近发生的多起事件，正好反映出了城市公共开放空间在实践言论自由中的恰当的角色。在西方，集会的人聚集在公共开放空间发表自由言论，这一方面受到很多人嘉许，但另一方面也剥夺了其他人对开放空间的使用权，并最终把场地弄得脏乱不堪。中国是不会容许这种情况发生的，政府对社会动荡的关注是可以理解。所有社会都有社会骚乱、恐怖主义及暴力等风险，最好的解决办法是通过对合理规章的非暴力执法来平衡言论自由的权利和公众安全，而非禁止民众进入城市公共开放空间。社会动荡的风险会因为得以进入维护良好的公共开放空间而得到缓和，而非激化。人们会更倾向于参加高兴的集会，而不是愤怒的抗议。为人民提供可达的开放空间是政府兑现提高人民生活水平的承诺。

在人权问题上的全球对话更加深入。新兴的"第三代"人权中包括享有洁净的环境、清新的空气、安全的饮用水及远离泥土中有害的化学物质的权利。城市公共开放空间在提供这种权利方面发挥着关键作用。

公共空间设计师斯蒂芬·卡尔在20年前阐述了对城市公共开放空间的进入、使用、活动及修改的权利。城市公共开放空间需要对所有人开放，回应人们所表达的需求，保障用户群体的权利，方便所有群体使用，并提供活动的自由。这种权利不包括反社会行为或暴力，但它包括不会煽动暴力、激发仇恨或推翻政府的言论自由。卡尔论证了另一个原则：城市公共开放空间必须是具有意义的，这又把我们带回马斯洛的指导性框架中。卡尔的主要告诫大概如是："当设计不是基于对社会的理解时，它们可能会退回到相对稳妥的几何构型中，偏爱使用用途和意义显而易见的怪异设计……公共空间设计有义务理解并服务于公众利益，而审美仅仅是其一部分"。[3]

无论人权问题是否符合当前形势，优良的城市公共开放空间需要坚守某些高尚的原则，并努力实现某些关键基准或指标。透过这些指标，政府官员可以获得一种衡量成功的真正标准。政府官员和市民领导在这些原则的指导下为人们提供他们应享有的开放空间和绿色空间，而对于究竟哪些空间指导原则应该被采用的最新思考就是本章的主题。鉴于不同地区的城市面临不同的地域环境，倘若每个城市都亦步亦趋遵从同一规划过程未免太过牵强。本章提供的总体原则常常被规避，然而，它们却是任何特定的规划过程取得成功的关键导则。许多已在使用的规划系统不妨明确采用以下部分或全部原则，定会从中获益。

■ **首要原则**
1. 可达性：在整个城市或大都市圈内提供均衡分布的城市公共开放空间

一个社区如果只提供一个"纪念品"式的大公园给大众，而没有其他开放空间的话，是远远不够的，因为大部分人肯定在远

Given the depth to which human and environmental needs are accommodated in urban public open space, the assertion that the people have a right to open space access may not seem too controversial, but really a matter of common sense. The peoples' right to urban public open space is many-fold. Over sixty years ago, in 1948, the United Nations promulgated the Universal Declaration of Human Rights, which includes "the right to freedom of peaceful assembly and association". Clearly outdoor public space accommodates this right more equitably than private or indoor space.

In light of recent events in the West it is well to reflect on the proper role of urban public open space in the exercise of the right of free speech. In the West, the occupation of public space as a matter of free speech, while admired by many, deprives others of their right to use the space, and eventually degrades the site into an unsanitary condition. China would not allow this; its concern about social unrest is understandable. The risk of social unrest, terrorism and violence that all societies face is best dealt with through non-violent enforcement of reasonable regulations balancing the rights of free speech and public safety, not by denying access to urban public open space. The risk of social unrest is alleviated, not exacerbated, by access to well maintained and well looked-after urban public open space, where people are much more likely to engage in conviviality than angry protest. Providing accessible open space projects the government's commitment to a better life for its people.

The global dialogue on human rights goes even further. Among the emerging "third generation" of human rights is the right to a clean environment, clean air, safe drinking water and freedom from toxic chemicals dumped into the soil. Urban public open space plays a key role in providing that right.

Two decades ago public space designer Stephen Carr made the case for the rights of access, use, action, and modification of urban public open space. Urban public open spaces need to be open to all and responsive to the people's expressed needs in that they protect the rights of user groups, are accessible to all groups and provide for freedom of activity. That right hardly extends to antisocial behavior or violence, but it does include free speech that does not condone or incite violence, hatred or overthrow of the government. Carr argued for another principle: urban public open spaces must be meaningful, which brings us back to Maslow's instructive framework. Carr's most cautionary advice may be this: "When designs are not grounded in social understanding, they may fall back on the relative certainties of geometry, in preference to the apparent vagaries of use and meaning…Public space design has a responsibility to understand and serve the public good, which is only partly a matter of aesthetics." [3]

Whether or not the language of human rights is appropriate to current circumstances, urban public open space excellence requires adhering to certain high-minded principles and striving to achieve certain key benchmarks or metrics. It is by these that public officials gain a true measure of success. The latest thinking on what principles should guide public officials and civic leaders in providing the people with the open space and green space which they deserve is the subject of this chapter. Given the variety of local circumstances facing cities across disparate regions, describing a single step-by-step planning process for all cities to follow would be presumptuous. The overarching principles which this chapter provides are often glossed over but nevertheless are essential guides to success in any particular planning process. Many planning systems already in use could benefit from explicitly adopting some or all of the following general principles.

■ First Principles

1. Accessibility: Provide equitable distribution of urban public open space across the entire city or metropolitan area.

It is not enough that a community provides one large "trophy" park and no other open space for all the people, most of whom will have to live or work far away from the park. Park policy must take into account the principle of environmental equity, that is, providing open space as much as possible based on a policy of non-discrimination of neighborhoods, that is, a policy of reducing the disparity between one neighborhood and another. Though some critics will point out that precise parity is absolutely impossible, that is no excuse for not working to reduce disparity significantly. Clearly, no city is free of particular circumstances that preclude an idealized balanced plan with perfect equity of access and accommodation of the maximum variety of experiences. In the real world, public officials must use their limited resources wisely for the improvement of the entire community (3.1).

Access is ideally on foot, as part of a walk for pleasure or on local errands. Access to open space on foot is the most equitable, natural measure, and as such, the question is how far people of average health, including children, are willing to walk to a park, not how long they are willing to sit in a bus to get there. Access by public transportation should be considered only as a last resort.

Cities should establish accessibility goals for the average citizen. How far is too far from the nearest park, for each park type? What is the maximum such allowable distance, and for what percentage of the population should this rule apply? Such goals are a matter of choice; the maximum allowable distance may vary depending on cultural preferences, safety, climate, and other factors. Willingness to walk varies greatly with age, health, time availability, and the environmental quality of the surroundings. But common sense helps determine a reasonable range of choices available to the locale.

Accommodating the citizen of average health and mobility alone is insufficient, in an ideal world. The needs of the elderly or the disabled ought to be accommodated by additional public services as affordable (3.2). Additional considerations of accessibility for those with mobility difficulties will be discussed in Chapter 4.

Minority populations should not to be discriminated against or inequitably treated in the distribution of parks. The principle is that urban planning should rectify inequitable distribution of access to well

图 3.1
可达性
Accessibility

这是一张瑞典斯德哥尔摩的地图，它展示了大多数的居民可以很容易从居住地或者办公室到达相距不远的开放空间。
This map of Stockholm, Sweden, shows that most residents have easy access to open space close to where they live or work. (Credit: Hollger.Ellgaard)

离公园的地方居住或上班。公园的政策必须考虑到环境公平的原则，即基于邻里间的非歧视政策，尽可能提供更多的开放空间，也就是减少各居民区差距的政策。虽然一些评论家指出精确的平等是绝对不可能的，但这并不能成为不去缩小显著差距的借口。显然，没有城市能够置身于不利的特定环境之外，这些不利环境妨碍了理想化的平衡计划，不能确保享用公园的机会平等，不能确保多元体验的最大化。在现实社会中，政府官员必须明智地利用他们手中有限的资源去改善整个社会（图 3.1）。

理想的到达方式是步行，散步消遣或者是出门办点事儿就溜达到了。步行到开放空间是最公平自然的方式，这样一来，要考虑普通健康状态的人（包括儿童）愿意步行多远的距离，而不是他们愿意花多少时间乘坐公共汽车去公园。使用公共交通到达公园应被视为下下策。

城市应该为普通市民树立一系列可达的目标。对每个公园类型而言，距离最近的公园多远才不算太远？什么是最大的许可距离？这些规则该应用于多少人口比例？这些目标在于选择不同：最大许可距离因为文化偏好、安全、气候及其他因素而不同。愿意步行与否，随着年龄、健康、可支配时间及周围环境质量不同而各异。不过，常识可以帮助确定一个场所合理范围内的一系列选择。

在一个理想的世界里，只考虑普通健康状况和活动能力的居民是不够的。年长者及残疾人士的需要应通过额外的可担负的公共服务得以满足（图3.2）。我们将会在第四章讨论对行动不便人士可达性的设计注意事项。

在公园的分布中，少数民族人口不应被歧视或受不公平对待。城市规划的原则是要纠正使用不公平的分配，以制造一个设

图 3.2
可达性
Accessibility

不同年龄和身体状况的人们应该得到同样的权利去亲近开放空间。
People of all ages and physical abilities deserve equal access to open space. (Credit: Stylephotographs, Germany)

designed and maintained open space. Whether or not environmental equity is a right, it is a matter of doing well by doing the right thing, more bluntly a matter of self-interest: a minority population with improving public health due to nearby parks is more productive, happier and less likely to be vulnerable to social unrest.

Using GIS, we can calculate what percentage of housing units in a city are no more than a given distance from a park of some kind. In Manhattan, 94% of housing units are within 400 meters of a park.[4] Judgment will be required to weigh tradeoffs across a varied geography. Quality can partially but not completely compensate for deficient quantity. Calculated distances must take into account barriers such as highways, rail lines, canals or rivers. Traversing a formidable barrier may require pedestrian overpasses or tunnels. Examples of a maximum recommended distances are included in Table 3.1.

Recommendation: the accessibility goal should be that no resident should walk more than 500 meters or 10 minutes to get to the nearest urban public open space. This is consistent with the principle that residents ought to be no more than a 10 minute walk from everything they regularly need, as advanced by many planners going back to the influential Greek town planner Constantinos Doxiadis.[5]

计及维护良好的开放空间。无论环境公平性是否是一种权利，我们都应该把正确的事情做到最好，也是与自身利益直接相关的问题：少数民族人口因为附近的公园而改善健康状态，从而更有生产力，更快乐及较少受社会动荡的影响。

运用地理信息系统（GIS），我们可以计算出一个城市里到公园的距离在一定范围内的房屋的百分比。在曼哈顿，94% 的房屋距离公园 400 米以内。[4] 不同地理位置的权衡需要给出判断。质量可以纳入考虑但不能完全弥补数量的缺陷。已计算的距离必须考虑到公路、铁路、运河或河流等障碍。要横跨一个难以逾越的障碍，可能需要行人使用天桥或隧道（表 3.1）。

建议：可达性的目标是居民步行 500 米之内或 10 分钟内可到达最近的城市公共开放空间。 这符合居民步行 10 分钟内获取日常所需的原则，也是很多规划师提倡的，可追溯到颇具影响力的希腊城市规划师窦加底斯。[5]

Trust for Public Land, *The Excellent City Park System* (2003) Recommendation

Preferably, people are no more than
five minutes *away by foot in densely urbanized areas or*
five minutes *away by bicycle in less densely urbanized sections or*
one-quarter mile (0.4 km) *away or*
six blocks *away*
*from a park at least **one acre (0.4 hm²) in size** and face no major barriers on the way such as a superhighway or rail line.*

公共土地管理，*优秀的城市公园系统*（2003）建议

*倾向于人们距离一个至少 1 英亩（**0.4 公顷**）大小的公园，不超过*
*高密度城市区域步行 **5 分钟**，或*
*中密度的城市区域骑车 **5 分钟**，或*
*1/4 英里（**0.4 千米**），或*
***6 个街区的距离**，*
且在通往公园的路上没有高速路或铁轨等主要障碍物。

最后，公共空间的使用应该是免费的。收取入场费的城市开放空间应成为历史。如果空间不是免费开放，它不应算作城市公共开放空间。

2. 设定一个远大的目标以提供每千人特定公顷数（面积）的城市开放空间

要指导城市的总体规划，必须有城市公共空间总量的目标，以服务现有和预计增长的人口。城市密度是决定提供多少开放空

西方城市公共开放空间步行距离最大期望值　表 3.1
Maximum Desired Distance to Urban Public Open Space in Western Cities　Table 3.1

距离（米）Distance (meters)	距离（英里）Distance (miles)	步行时间（分钟）Walking Time (minutes)	城市	City
200	0.125	4	哥本哈根 芝加哥	Copenhagen Chicago
400	0.25	8	佛罗里达州，迈阿密 明尼苏达州，圣保罗	Miami FL, St Paul MN
530	0.33	10	加利福尼亚州，圣何塞	San Jose CA
800	0.5	16	科罗拉多州，科泉市	Colorado Springs CO
1600	1	30	得克萨斯州，奥斯汀 印第安纳州，印第安纳波利斯 堪萨斯州，维契托佛尔	Austin TX Indianapolis IN Wichita KS
3200	2		佐治亚州，亚特兰大	Atlanta GA

资料来源：彼得·哈尼克. 城市绿地：城市重生的创新园. 华盛顿：岛屿出版社，2010：28.
Source: Peter Harnik. Urban Green, *Innovative Parks for Resurgent Cities.* (Washington, DC: Island Press, 2010), 28.

Lastly, access should be free of charge. Charging admission to permit access to urban public open space should remain a thing of the past. If the space is not open free of charge, it should not be counted toward urban public open space.

2. Set an ambitious goal for providing a specified number of hectares of urban public open space per thousand people.

To guide an urban master plan, there must be a goal of total urban public open space to accommodate the existing and projected population. City density is a key factor in determining how much open space ought to be provided. This is not a decision to make in isolation from the experience of other cities, or without accurate data for existing local conditions. Of course, the relevant municipal department will maintain an accurate base map of existing urban public open space, and even identify possible sites for future acquisition. The best enlightened cities with the resources to do so will also want to collect comparative data nationally and regularly monitor local changes over time. Even in the U. S., national data is spotty on total urban public open space in cities, let alone how that total is divided between parks of various types, but non-profit organizations are working to improve the national database. There is much need for improvement on data, but waiting for perfect data is no excuse for delay in providing urban public open space.

In many cities can be found privately owned outdoor space open to the public free of charge. Commercial office and retail projects may include plazas that function in many respects like a public park, like Rockefeller Center in New York City or Jintai Xizhao in Beijing. These spaces truly enrich the array of open space but are nevertheless separate from what the city itself should take responsibility for. Nor should their presence be used to dilute the statistics. Some cities also have amusement parks. However, privately owned and managed space for which a fee is charged should be considered a separate category.

Cities also benefit from the tree planting in buffer zones surrounding residential blocks and commercial blocks, institutions, factories, power and sewage treatment plants and so on, and from green roofs. Though this green space contributes to sustainable urban development, including it with accessible open space distorts the calculation of total urban public open space. Similarly, common areas in residential projects that are closed to non-residents should not count in calculations of urban public open space.

In U. S. cities, the ratio of existing hectares of urban public open space per thousand local residents ranges from 1 to 6. And these are low by European standards. For example, Stockholm, Sweden, has 11 hectares per thousand local residents.

The current professional standard goal in the U. S. and other Western nations is 4 hectares per 1000 population.[6] Examples are included in Table 3.2.

Local circumstances will dictate what goal is appropriate, realistic 间的一个关键因素。这个决定需参考其他城市的经验，也要参照关于当地现状的准确数据。当然，有关市政部门将维持现有城市公共开放空间的准确底图以及确定未来有可能征用的土地。有资源及开明的城市想这样做的话，它们需要收集全国的相关数据，并随着时间推移而定期监控当地的变化。即使在美国，城市公共开放空间总量的国家数据也是参差不齐，再加上这个总量被划分到不同的公园种类中，令数据更不齐全。一些非营利组织正努力改善国家数据库。我们还需要大量改善后的数据，但并不能以等待完整的数据为借口而延迟提供城市公共开放空间。

在很多城市里，一些户外空间属个体或企业所有，却免费对外开放，供大众使用。商业写字楼和零售商铺的广场在许多功能上与公园相似，例如纽约市的洛克菲勒中心或北京的金台夕照。这些空间可以使一系列的开放空间更加丰富，尽管如此，它们与城市本身应负的责任是分开的。这些空间的存在也不应该降低统计的数据。有的城市设有游乐场，但是私人拥有管理的收费空间应该是作为单独统计的类别。

城市可以在周边的住宅区、商业区、公共机构、工厂、发电厂及污水处理厂等设施的绿化隔离带和屋顶绿化中得益。虽然这些绿色空间可促进城市的可持续发展，但是将其纳入可达的开放空间会导致错误的计算城市公共开放空间的总量。同样地，住宅项目中只供住户使用的公用空间，也不应列入城市公共开放空间的计算之内。

在美国城市中，每千名本地居民对现有城市公共开放空间公顷数的比例为1∶6。按照欧洲的标准，1∶6的比例是很低的。以瑞典的斯德哥尔摩为例，当地的比例是每千名本地居民拥有11公顷城市公共开放空间。

目前美国及其他西方国家的专业标准是4公顷/1000人[6]，表3.2中包括了一些例子。

当地的情况将决定怎样的目标是适当的、切实可行和与文化相关的。高密度的目标相对较低。这似乎违反常理，但我们可留意纽约市这个例子。曼哈顿不可能实现每千人拥有4公顷的城市公共开放空间，除非把绿色空间覆盖整个城市，而这是不合逻辑和荒谬的，特别是考虑到纽约市总面积用于城市公共开放空间的百分比（包括曼哈顿及周围的市区），已经远远超过美国其他主要城市（见表3.3）。

3. 为城市公共开放空间占城市总面积的百分比设定一个远大的目标

随着城市以面积百分比而不是人口设定目标，我们不妨看看几个选定的美国城市的比例以作为基本的衡量标准，这几个城市里，有些拥有"绿化"的美誉，有些则有汽车过量的名声。就算是同一区域，数据也有所不同。现有的最大比例为30%（见表3.4）。一般情况下，未来会决定比例的升幅。中国许多城市凭直觉设定了发展街区及特定用地类型的绿化空间面积的目标为20%-40%。这是斯德哥尔摩等世界上最环保的城市的标准，斯德哥尔摩8500公顷的城市公共开放空间覆盖了城市40%的总面积（图3.3 - 图3.4）。

建议：以城市总面积来计，城市公共开放空间百分比最低不应低于20%。对于有大面积水域的城市，其公共开放空间的百分比应该成比例地升高。

城市公共开放空间的土地公顷量（每1000人）   表 3.2
Hectares of Urban Public Open
Space per 1000 Population   Table 3.2

公顷/1000人 Hectares per 1000 population	英亩/1000人 Acres per 1000 population	总公园面积（英亩）Total Park Acres	城市	City
6.27	15.5	5,864	明尼阿波利斯	Minneapolis
5.22	12.9	7,617	华盛顿特区	Washington DC
4.21	10.04	6,170	西雅图	Seattle
3.36	8.3	5,040	波士顿	Boston
3.11	7.7	4,905	巴尔的摩	Baltimore
3.03	7.5	10,886	费城	Philadelphia
2.83	7.0	5,384	旧金山	San Francisco
2.51	6.2	23,761	洛杉矶	Los Angeles
1.90	4.6	38,229	纽约市	New York City
1.70	4.2	11,860	芝加哥	Chicago
0.61	1.5	34,000	上海	Shanghai

资料来源：皮特．哈尼克．《城市绿化》，2010.
方家，吴承照，杰弗里·沃尔，程励．美国、加拿大和中国居民对城市开放空间的使用和观点的比较．第47届世界规划大会，2011.
Source: Peter Harnik, *Urban Green*, 2010.
Fang Jia, Wu Cheng-zhao, Geoffrey Wall, *Cheng Li, Comparison of Urban Residents' Use and Perceptions of Urban Open Spaces in USA,* Canada and China, 47th ISOCARP Congress 2011.

and culturally relevant. High density requires a lower goal. This may seem counterintuitive, but consider the example of New York City. That city could not possibly achieve the goal of 4 hectares of urban public open space per 1000 population on Manhattan without covering all of Manhattan with green space, a logical impossibility, and absurd when one considers that the percent of New York City's total area (Manhattan and the surrounding boroughs) devoted to urban public open space already exceeds that of any other major U.S. city. (See Table 3.3)

3. Set an ambitious goal for urban public open space as a percentage of total city area.

As cities set goals based on percentage of area rather than population, it may be useful to look at the ratio for selected U. S. cities, some with a reputation for being "green", others with a reputation for being overrun by the automobile, as a general benchmark. The data varies, even for one locale. The maximum existing ratio is 30 percent. (Table 3.4) Generally, the future will dictate that the ratios increase. Already cities in China intuitively realize this when they set targets of 20% to 40% for the green space area within development blocks and specific land use types. They are following the lead of cities like Stockholm, one of the greenest cities in the world, where 8500 hectares of urban public open space covers 40% of its area. (3.3-3.4)

Recommendation: the minimum goal for total urban public open space as a percentage of total city area should be no lower than 20 percent. For cities with a significant percentage of water area, the percentage for urban public open space should be proportionately higher.

图 3.3 - 图 3.4
开放空间的百分比
Percentage of Open Space

这些航拍图恐怕展示了那些看起来最绿的城市，它们有充足的土地资源和水资源。其中斯德哥尔摩为全世界制定了标准，它的开放空间百分比是40%。
Perhaps the greenest of cities, with abundant land and water resources that are most apparent in these aerial views, Stockholm sets a world standard at 40 percent. (Credits: Oleksiy Mark from Ukraine and Windowlicker)

城市公共开放空间土地公顷量每千人目标的范例 表 3.3
Sample Goals of Hectares of Urban Public Open Space per 1000 Population Table 3.3

公顷 /1000 人 Hectares per 1000 population	地点	Location	时间	Date	备注	Notes (1 hectare = 2.47 acres)
14	大部分美国城市的最大限度	Upper limit of most US cities	现在	current	彼得·哈尼克，城市绿地 (2010)	Peter Harnik, *Urban Green* (2010)
7.38	墨尔本	Melbourne	现在	current	澳大利亚	Australia
4.4	美国科罗拉多州，丹佛	Denver CO	现在	current		
4.0	哈尼克的样本的中间值	**Median of Harnik's sample**	现在	current		
3.44	美国华盛顿州，塔科马	Fife WA	现在	current	8.5 英亩公园 + 30 英亩 开放空间	8.5 acre park + 30 acre open space
3.1	美国加利福尼亚州，旧金山	San Francisco CA	现在	current		
2.8	吉尔福德郡	Guilford Borough Council	现在	current	英国	Britain
2.59	美国威斯康星州，沃喀莎	Waukesha WI	现在	current	3.1 英亩社区公园 + 3.3 英亩邻里公园	3.1 acre community park + 3.3 acre neighborhood park
2.5	哥本哈根	Copenhagen	现在	current	丹麦	Denmark
2.5	芬戈郡议会	Fingal County Council	现在	current	爱尔兰	Ireland
2.36	悉尼	Sydney	现在	current	澳大利亚	Australia
2.0	大部分美国城市最小值	Minimum for most US cities	现在	current	皮特·哈尼克，城市绿地 (2010)	Peter Harnik, *Urban Green* (2010)
1.8	纽约市	New York City	现在	current		
1.0	纽约市	New York City	1928	1928		
1.0	巴黎	Paris	1870	1870	法国	France

4. Provide a balance of Civic Plazas, Downtown Parks, Large Parks, Greenways, and Neighborhood Parks.

Three decades ago landscape architect Lawrence Halprin, a great designer and a moving writer, made the case for accommodating a variety of human needs in a variety of urban public open space experiences:

The life of cities is of two kinds – one is public and social, extroverted and interrelated. It is the life of the streets and plazas, the great parks and civic spaces and the dense activity and excitement of the shopping areas…There is, too, a second kind of life in the city – private and introverted, the personal, individual, self-oriented life which seeks quiet and seclusion and privacy. This private life has need for open spaces of a different kind…Our open spaces are the matrix of this two-fold life.

4. 提供市民广场、市中心公园、大型公园、林荫道及邻里公园的平衡

30 年前，著名的景观建筑设计师兼作家劳伦斯·哈普林，阐述了在不同的城市开放空间体验过程中满足人们不同的需求：

"城市生活分为两个方面———一个是公共的、社会的、外向的、有关联的。即街道和广场、良好的公园和市民空间、丰富的活动和令人兴奋的购物环境……另一个方面是私密、内向的、个人的、独特的、自我的，寻求安静和与世隔绝的私密生活。这样的静谧生活对开放空间有别样的需求……开放空间是双重生活的环境基础。我们在其中能寻找到不同的生活体验，使得城市生活具有创造性和鼓舞人心……富有创造力的城市，是指城市具有多样性和自由选择的机会，能够促进人们和城市居住环境的最大交融。" [7]

美国城市公共开放空间占城市总面积的比例

Urban Public Open Space as Percent of Total Urban Area in U. S.

表 3.4
Table 3.4

城市 City	百分比（%） Percentage	资料来源 Source	
新墨西哥州，阿尔布开克	Albuquerque NM	30	彼得·哈尼克，*城市绿地* (2010), 47	Peter Harnik, *Urban Green* (2010), 47
纽约州，纽约市	New York City NY	26	亚历山大·加文，*公众公园* (2011), 59	Alexander Garvin, *Public Parks* (2011), 59
加利福尼亚州，圣迭戈	San Diego CA	21.9	彼得·哈尼克，*城市绿地* (2010), 47	Peter Harnik, *Urban Green* (2010), 47
纽约州，纽约市	New York City NY	19.6	彼得·哈尼克，*城市绿地* (2010), 47	Peter Harnik, *Urban Green* (2010), 47
华盛顿特区	Washington DC	19.4	公共土地管理，2009	Trust for Public Land, 2009
加利福尼亚州，旧金山	San Francisco CA	18	亚历山大·加文，*公众公园* (2011), 59	Alexander Garvin, *Public Parks* (2011), 59
明尼苏达州，明尼阿波利斯	Minneapolis MN	16.7	彼得·哈尼克，*城市绿地* (2010), 47	Peter Harnik, *Urban Green* (2010), 47
马萨诸塞州，波士顿	Boston MA	16.3	亚历山大·加文，*公众公园* (2011), 59	Alexander Garvin, *Public Parks* (2011), 59
俄勒冈州，波特兰	Portland OR	15.7	公共土地管理，2009	Trust for Public Land, 2009
纽约州，纽约市	New York City NY	14	wwwdesigntrust.org/pubs/2011_HPLG.pdf	wwwdesigntrust.org/pubs/2011_HPLG.pdf
亚利桑那州，凤凰城	Phoenix AZ	13.8	公共土地管理，2009	Trust for Public Land, 2009
宾夕法尼亚州，费城	Philadelphia PA	12.6	亚历山大·加文，*公众公园* (2011), 59	Alexander Garvin, *Public Parks* (2011), 59
华盛顿州，西雅图	Seattle WA	11.5	彼得·哈尼克，*城市绿地* (2010), 47	Peter Harnik, *Urban Green* (2010), 47
密苏里州，圣路易斯	St. Louis MO	8.5	公共土地管理，2009	Trust for Public Land, 2009
伊利诺伊州，芝加哥	Chicago IL	8.2	公共土地管理，2009	Trust for Public Land, 2009
加利福尼亚州，洛杉矶	Los Angeles CA	7.9	彼得·哈尼克，*城市绿地* (2010), 47	Peter Harnik, *Urban Green* (2010), 47
堪萨斯州，堪萨斯城	Kansas City MO	6.8	公共土地管理，2009	Trust for Public Land, 2009
密歇根州，底特律	Detroit MI	6.6	公共土地管理，2009	Trust for Public Land, 2009
佛罗里达州，迈阿密	Miami FL	6.0	彼得·哈尼克，*城市绿地* (2010), 47	Peter Harnik, *Urban Green* (2010), 47
内华达州，拉斯韦加斯	Las Vegas NV	5.6	公共土地管理，2009	Trust for Public Land, 2009
佐治亚州，亚特兰大	Atlanta, GA	4.6	公共土地管理，2009	Trust for Public Land, 2009

It is largely within them that we can find for ourselves these variegated experiences which make life in a city creative and stimulating…By creative, I mean a city which has great diversity and thus allows for freedom of choice; one which generates the maximum of interaction between people and their urban surroundings. [7]

Urban public open space needs to be provided in variety. To design all spaces essentially with the same character would fall far short of taking the responsibility seriously to provide excellent urban public open space.

The Center for City Park Excellence at The Trust for Public Land (www.tpl.org) maintains a database of individual parks in the largest U.S. cities — currently at a total of 10,500 parks in the 50 largest cities which yields the following two statistics:

Average Size U.S. City Park: 54 acres (22 hectares)
Median Size U.S. City Park: 5 acres (2 hectares)
Some parks are incredibly large—Phoenix includes 16,000-acre (6,400 hectare) South Mountain Preserve—so the average size skews the number upward to reflect the number and scale of these large spaces. But when looking at the median, or middle value, of the range, the number goes down to two hectares (five acres), which means that half the parks are two hectares (five acres) or less.

As the observations above make clear, a discussion of "parks" in general brings together a range of scales that is too vast for a single generic term to be useful. Rather than using one term, communities must consider a range of types which, taken together, accommodate the variety of needs which we described in Chapter 2, in other words, offer the people the widest possible range of open space benefits. Authorities have classified open space types variously. Clare Cooper Marcus proposed a range of types ranging from the Plaza (a predominantly paved space that designers like to call "hardscape") to the City Square (a rough balance of hardscape and unpaved "softscape") to the Park (softscape or planted area in excess of 50%). *Landscape Architectural Graphic Standards* (2007) divides its discussion of parks and recreation between Large City Parks, Small Urban Parks, Waterfronts and Outdoor Play Areas. There is no one standard typology. For example, large city parks are sometimes called regional parks or comprehensive parks. Neighborhood parks are sometimes called mini-parks or pocket parks. Tot lots are sometimes given a separate category from neighborhood parks. Community parks hover somewhere between large city parks and neighborhood parks. Greenways are sometimes called open space corridors, eco-corridors, green infrastructure, and up to thirty other variations. Undeveloped urban parks are sometimes called urban wilds. City square, civic center, urban square are all synonymous. We have chosen to adapt a five-fold typology that takes into account the full range of benefits for people in the 21st century: *Civic Plaza, Downtown Park, Large Park, Greenway*, and *Neighborhood Park*. These five are the core elements of a park system that should be incorporated in the urban master plan. The following table of examples of urban public open space suggests how the various types proposed by others are subsumed into our typology.

有关当局应该提供各种类型的城市公共开放空间。若所有空间都是相同风格的设计，就谈不上负责任地提供优质的城市公共开放空间。

在公共土地管理中（www.tpl.org）的城市优质公园中心保持着美国各大城市的单体公园数据库——目前，50个最大的城市中有10500个公园，产生了以下两种统计数据：
美国城市公园平均大小：54英亩（22公顷）
美国城市公园中等尺寸：5英亩（2公顷）
一些公园不可思议的大——凤凰城有16000英亩（6400公顷）的南部山脉自然保护区——平均的尺寸大小反映了大空间的尺度和数量。但是在中间值或中间水平，数值下降到2公顷（5英亩），这就意味着公园的一半尺寸是2公顷（5英亩）或更少。

从上述的观察资料可明确地看出，对"公园"的讨论，范围太广，没有哪个通用术语可以概括。社区与其只用一个专门用语，倒不如考虑使用一系列类型的术语，多管齐下，可适应第二章内所描述的各种需求，换句话说，可为人们提供开放空间的最大效益。有关当局已将开放空间划分为不同类型。克莱尔·马克斯建议一系列类型，由公园广场（主要以铺砌空间为主，设计师喜欢称之为"硬质景观"）至城市广场（硬质景观及非铺砌"软质景观"的平衡）至公园（软质景观或种植面积超过50%）等。风景建筑平面规范（2007年）把公园及娱乐空间划分为大型城市公园、小型都市公园、亲水和户外游乐场，但都没有一个标准的类型。以大型城市公园为例，它们有时会被称为区域公园或综合性公园，邻里公园有时候也被称为迷你公园或口袋公园。小型幼儿游乐场从邻里公园里脱离出来，自成门类。社区公园介于大型城市公园和邻里公园之间。绿道有时候也被称为开放空间走廊、生态走廊、绿色基建以及其他30多种名称。未开发的城市公园有时候被称为城市中的荒野。城市广场、市民中心、城市街区都是同义的。考虑到21世纪人们的总体利益，我们选择了5种类型的公园：*城市广场、市区公园、大型公园、绿道及社区公园*。这5个是公园系统中的核心元素，同时被纳入城市的总体规划。以下城市公共开放空间一览表的例子，表明了如何让不同类型的城市开放空间纳入我们的归类中（表3.5）。

Chapter Three　Planning Principles　105

本书中城市公共开放空间的分类
Urban Public Open Space Typology Used in This Book

表 3.5
Table 3.5

5 种开放空间类型 Five Open Space types		范例	Examples	面积（公顷） Area (ha)
城市广场 Civic Plaza				
	城市广场 City Square 城市中心 City Square 露天广场 Civic Center 场所 Place 广场 Piazza 城市街区 Urban Square	独立广场，雅加达	Merdeka (Independence) Square, Jakarta	100
		星海广场，大连	Xinghai Square, Dalian	50
		天安门广场，北京	Tiananmen Square, Beijing	44
		蒙特雷，墨西哥	Macroplaza, Monterey, Mexico	40
		印度门，新德里，印度	India Gate complex, New Delhi, India	30
		阅兵广场，华沙	Parade Square, Warsaw	24
		人民广场，大连	Peoples Square, Dalian	12
		皇家田广场，曼谷	Sanam Luang, Bangkok	11.4
		瓦莱广场，帕多瓦	Prato della Valle, Padova	9
		伊玛目广场，伊斯法罕，伊朗	Naqsh-e Jahan Square, Isfahan, Iran	8.9
		德吉玛广场，马拉喀什，摩洛哥	Djemaa el Fna, Marrakesh, Morocco	8.6
		协和广场，巴黎	Place de la Concorde, Paris	8.6
		帝国广场，里斯本	Praça do Império, Lisbon	7.8
		八一广场，南昌	Bayi Square, Nanchang	7.8
		金日成广场，平壤，朝鲜	Kim Il-Sung Square, Pyongyang, N. Korea	7.5
		塔里尔广场，开罗	Tahrir Square, Cairo	7.4
		宪法广场，墨西哥	Plaza de la Constitución, Mexico	5.8
		自由广场，德黑兰	Arzadi (Freedom) Square, Tehran	5
		宫殿广场，圣彼得斯堡	Palace Square, St. Petersburg	5
		加泰罗尼亚广场，巴塞罗那	Plaça de Catalunya, Barcelona	5
		市民中心公园，丹佛	Civic Center Park, Denver	4.8
		塔克西姆广场，伊斯坦布尔	Taksim Square, Istanbul	3.6
		罗根广场，费城	Logan Circle, Philadelphia	3.2
		三权广场，巴西利亚	Praça dos Três Poderes, Brasilia	2.6
		英雄广场，布达佩斯	Heroes Square, Budapest	2.5
		宪法广场，雅典	Syntagma (Constitution) Square, Athens	2.3
		圣彼得广场，罗马	St. Peter's Square, Rome	2.3
		红场，莫斯科	Red Square, Moscow	2.3
		水坝广场，阿姆斯特丹	Dam Square, Amsterdam	2
		孚日广场，巴黎	Place des Vosges, Paris	1.95
		大教堂广场，米兰	Piazza del Duomo, Milan	1.7
		纳沃纳广场，罗马	Piazza Navona, Rome	1.4
		人民广场，罗马	Piazza del Popolo, Rome	1.35
		圣马可广场，威尼斯	Piazza San Marco, Venice	1.3
		斯坦尼斯罗斯广场，南希	Place Stanislaus, Nancy	1.3
		市长广场，马德里	Plaza Mayor, Madrid	1.2
		特拉法加广场，伦敦	Trafalgar Square, London	1.2
		乔治广场，格拉斯哥	George Square, Glasgow	1.12
		联合广场，旧金山	Union Square, San Francisco	1.1
		老城广场，布拉格	Old Town Square, Prague	0.9
		辟维广场，明尼阿波利斯	Peavey Plaza, Minneapolis	0.8
		大广场，布鲁塞尔	Grand Place, Brussels	0.75
		共和广场，佛罗伦萨	Piazza della Repubblica, Florence	0.75
		时代广场，纽约	Times Square, New York City	0.65
		大教堂广场，康斯坦茨，德国	Münsterplatz, Konstanz, Germany	0.6
		塔林市政广场，塔尔图，爱沙尼亚	Raekoja Plats, Tartu, Estonia	0.55
		十二月党人广场，澳门	Senate Square, Macau	0.4
		卡比托利欧广场，罗马	Piazza del Campidoglio, Rome	0.37
		先驱法院广场，西雅图	Pioneer Courthouse Square, Seattle	0.36
线性广场 Linear Plaza		外滩，上海市中心广场	The Bund, Shanghai	13

续表

5种开放空间类型 Five Open Space types		范例	Examples	面积（公顷） Area (ha)
市区公园 Downtown Park				
公园 Public Garden 公共花园 Public Park 小型城市公园 Small Urban Park		国家广场，华盛顿	National Mall, Washington DC	59
		卢纳塔公园，马尼拉	Rizal (Luneta) Park, Manila	48
		翁代尔公园，阿姆斯特丹	Vondelpark, Amsterdam	48
		海滨公园，路易斯维尔	Waterfront Park, Louisville	29
		杜伊勒里花园，巴黎	Jardin des Tuileries, Paris	25
		卢森堡公园，巴黎	Jardin du Luxembourg, Paris	22
		波士顿公园	Boston Common	20
		王子街花园，爱丁堡	Princes Street Gardens, Edinburgh	15
		古埃尔公园，巴塞罗那	Güell Park, Barcelona	15
		人民广场，上海	Peoples Square, Shanghai	14
		九龙公园，香港	Kowloon Park, Hong Kong	13.5
		纳罗亚公园，檀香山	Moanalua Gardens, Honolulu	10.5
		千禧公园，芝加哥	Millennium Park, Chicago	10
		公共花园，波士顿	Public Garden, Boston	10
		发现绿色公园，休斯敦	Discovery Green, Houston	5
		莱恩公园，纽约	Bryant Park, New York	3.9
		华盛顿广场公园，纽约	Washington Square Park, New York	3.9
		国王花园，斯德哥尔摩	Kungsträdgården, Stockholm	3.5
		华盛顿广场，费城	Washington Square, Philadelphia	3.1
		城市花园，圣路易斯	Citygarden, St. Louis	1.2
		诺曼纳纹莎尔公园，波士顿	Norman B. Leventhal Park, Boston	0.70
小型雕塑公园 Small Sculpture Park		堪萨斯市雕塑公园	Kansas City Sculpture Park	8.9
		明尼阿波利斯雕塑公园	Minneapolis Sculpture Garden	4.5
		奥林匹克雕塑公园，西雅图	Olympic Sculpture Park, Seattle	3.6
大型公园 Large Park				
社区公园、公园、区域公园、城市国家公园 Community Park, Park, Regional Park, Urban National Park		原野之家公园，马德里	Casa de Campo, Madrid	2834
		城堡岛，斯德哥尔摩	Ekoparken, Stockholm	2700
		森林公园，波特兰	Forest Park, Portland	2064
		鹰溪公园，印地安纳波里斯	Eagle Creek Park, Indianapolis	1929
		谢尔比农场，孟菲斯	Shelby Farms, Memphis	1800
		原野之家公园，马德里	Casa de Campo, Madrid	1722
		格利菲斯公园，洛杉矶	Griffith Park, Los Angeles	1707
		费尔蒙特公园，费城	Fairmount Park, Philadelphia	1686
		奥林匹克公园，北京	Olympic Park, Beijing	1215
		佩勒姆湾公园，纽约	Pelham Bay Park, New York City	1119
		凡仙森林公园，巴黎	Bois de Vincennes, Paris	994
		萨顿公园，伯明翰，英国	Sutton Park, Birmingham, UK	970
		里士满公园，伦敦	Richmond Park, London	960
		英国公园，慕尼黑	Englischer Garten, Munich	958
		阿姆斯特丹森林公园，阿姆斯特丹	Amsterdam Bos, Amsterdam	935
		孟山都森林公园，里斯本	Monsanto Forest Park, Lisbon	931
		布洛涅森林，巴黎	Bois de Boulogne, Paris	845
		绿带公园，史德顿岛，纽约州	Greenbelt Park, Staten Island, NY	720
		凤凰公园，都柏林	Phoenix Park, Dublin	710
		石溪公园，华盛顿特区	Rock Creek Park, Washington DC	710
		查普特佩克公园，墨西哥	Chapultepec Park, Mexico	686
		西里西亚中央公园，西里西亚，波兰	Silesian Central Park, Silesia, Poland	620
		普拉特游乐场，维也纳	Prater, Vienna	610
		要塞公园，旧金山	Presidio Park, San Francisco	603
		纪念公园，休斯敦	Memorial Park, Houston	593
		众神的花园公园，科泉市	Garden of the Gods Park, Colorado Springs	534
		城市公园，新奥尔良	City Park, New Orleans	530
		森林公园，圣路易斯	Forest Park, St. Louis	523

续表

5种开放空间类型 Five Open Space types	范例	Examples	面积（公顷） Area (ha)
社区公园、公园、区域公园、城市国家公园 Community Park, Park, Regional Park, Urban National Park	林肯公园，芝加哥	Lincoln Park, Chicago	488
	丛林公园，伦敦	Bushy Park, London	445
	巴尔博亚公园，圣迭戈	Balboa Park, San Diego	442
	金门公园，旧金山	Golden Gate Park, San Francisco	416
	翡翠项链，波士顿	Emerald Necklace, Boston	405
	史丹利公园，温哥华	Stanley Park, Vancouver	405
	百丽岛公园，底特律	Belle Isle Park, Detroit	397
	中央公园，纽约市	Central Park, New York City	340
	德鲁伊山公园，巴尔的摩	Druid Hill Park, Baltimore	301
	郎得海花园，利兹	Roundhay Park, Leeds	280
	天坛公园，北京	Temple of Heaven Park, Beijing	267
	蒂尔加藤公园，柏林	Tiergarten, Berlin	255
	海德公园/肯辛顿花园，伦敦	Hyde Park/Kensington Gardens, London	253
	展望公园，布鲁克林，纽约	Prospect Park, Brooklyn, New York	237
	皇家山公园，蒙特利尔	Mount Royal Park, Montreal	214
	富兰克林公园，波士顿	Franklin Park, Boston	213
	杰克逊公园，芝加哥	Jackson Park, Chicago	200
	法棱公园，明尼阿波利斯	Phalen Park, Minneapolis	200
	斯堪雷公园，匹兹堡	Schenley Park, Pittsburgh	185
	景观公园，杜伊斯堡，德国	Landschaftspark, Duisberg, Germany	180
	南方公园，索菲亚	South Park, Sofia	180
	摄政公园，伦敦	Regents Park, London	166
	高地公园，多伦多	High Park, Toronto	161
	科莫公园，圣保罗，明尼苏达州	Como Park, St. Paul, Minnesota	155
	丽池公园，马德里	Buen Retiro Park, Madrid	142
	国际都市公园，奥斯汀	Zilker Metropolitan Park, Austin	142
	百年纪念公园，上海	Centennial Park, Shanghai	140
	尼迈耶之伊比拉普埃拉公园，圣保罗	Ibirapuera Park, Sao Paolo, Brazil	140
	库本公园，班加罗尔，卡纳塔卡，印度	Cubbon Park, Bangalore, Karnataka, India	135
	肯辛顿花园，伦敦	Kensington Gardens, London	111
	高尔基公园，莫斯科	Gorky Park, Moscow	109
大型雕塑公园 Large Sculpture Park	暴风国王艺术中心，纽约州	Storm King Art Center, New York State	202
	长春世界雕塑公园，吉林	ChangChun World Sculpture Park, Jilin	92
	维格郎公园，奥斯陆	Frogner (Vigeland Sculpture) Park, Oslo	32
植物园 Botanical Garden or Arboretum	邱园，伦敦	Kew Gardens, London	121
	安诺树木园，波士顿	Arnold Arboretum, Boston	107
	莫里斯植物园，费城	Morris Arboretum, Philadelphia	37
	布鲁克林植物园，纽约	Brooklyn Botanical Gardens, New York	21
花园墓地 Garden Cemetery	奥本山公墓，剑桥市，马萨诸塞州	Mount Auburn Cemetery, Cambridge MA	70
水库 Reservoir	栗子山水库，波士顿	Chestnut Hill Reservoir, Boston	60

绿道 Greenway			长度（km） Length(km)
草场、步行道、散步道、滨水小径 Esplanade, Path, Promenade, Waterfront Trail	查理士河自行车道，波士顿	Charles River Bike Path, Boston	37
	滨水小径，芝加哥	Lakefront Trail, Chicago	29
	西湖步行长廊，杭州	West Lake Promenade, Hangzhou	15
	水景步行街，圣安东尼奥	Riverwalk, San Antonio	4
公园绿化带，林荫大道 Boulevard, Parkway	布朗士河公园大道，纽约市	Bronx River Parkway, New York City	25
绿道，开放空间廊道 Greenway, Open Space Corridor	翡翠项链，波士顿	Emerald Necklace, Boston	10
	肯尼迪玫瑰绿道，波士顿	Rose Kennedy Greenway, Boston	2.4
	大通公园，札幌市，北海道，日本	Odori Park, Sapporo, Hokkaido, Japan	1.5

续表

5种开放空间类型 Five Open Space types	范例	Examples	面积（公顷） Area (ha)
快速铁路线 Ex-rail line	中城绿道，明尼阿波利斯 Hi-Line，纽约市	Midtown Greenway, Minneapolis Hi-Line, New York City	9.2 2.33
社区公园 Neighborhood Park			
小型公园 Pocket Park	佩利公园，纽约市	Paley Park, New York City	<1
社区花园 Community Garden	人民公园，伯克利	Peoples Park, Berkeley	1.1
游乐场 Playground			<1
狗公园 Dog Park			up to 20
微型公园 Minipark	绿色街道项目，纽约市	Greenstreets Program, New York City	<<1

注：尺寸是大概的；测量的方法和来源也会不同。
Note: dimensions are approximate; measurement methodologies and sources vary.

类型 Typology Snapshot

城市广场 Civic Plaza	市区公园 Downtown Park	大型公园 Large Park	绿道 Greenway	社区公园 Neighborhood Park
> 0.33 公顷 hectares << 100 公顷 hectares	1–50 公顷 hectares	> 100 公顷 hectares >> 20 公顷 hectares	> 1.5 千米长 km long	< 1 公顷 hectares

In China, to a degree found nowhere else, local authorities have the power to implement an urban master plan and as part of that create parks even on occupied sites by virtue of state ownership of all land. Other nations would be hard pressed to assemble large park sites, or create downtown parks where none has existed, other than on post-industrial brownfields sites like abandoned factories and airports, or on fill along the waterfront. In China, assembling large sites is a routine occurrence. Because so many lives are affected by the decisions that local government officials must make, they should take the utmost care to assure a wise outcome to a well informed and deliberative process. Increasingly, in mainland China as in Hong Kong and Taiwan, and many other nations, the public may be invited to attend public presentations and voice its concerns or even participate in the planning process by making suggestions.

That process may well begin in the urban master plan, at the level of deciding on the right mix of Civic Plazas, Downtown Parks,

在中国，地方当局有权实施一个城市的总体规划，包括兴建公园，极具中国特色的是，有些公园甚至可以建在已经被占用的地点上，所有的土地都是国有的，政府有权征用。而其他国家很难集中一块大型公园用地或兴建全新的市区公园，除非是在弃置工厂或飞机场等后工业废地或沿海滨的填海区域。在中国，集中大片用地是件很平常的事情。因为当地政府的决定影响到很多人的生活，所以他们应该小心翼翼，以确保一个深思熟虑的决策过程和一个明智的结果。在中国大陆、香港和台湾地区和其他许多的国家和地区，更多的公众可被邀请参加公共辩论来发出他们的诉求，甚至可以用提出建议的方式来参与规划。

上述的过程可以从城市总规划开始，决定城市广场、市区公园、大型公园、绿色通道及社区公园的适当配置。每个城市或大都市至少应该有一个城市广场、市区公园及一个大型公园。一些城市有幸能拥有两种以上的公园类型。上海就是一个很好的例子。外滩是一个城市广场，人民广场是一个市区公园，而鲁迅公园、中山公园及浦东的世纪公园都是大型公园。除了这些主要"财富"外,上海还有更多的小型公园,例如静安公园、淮海公园和襄阳公园。黄浦江两岸都是绿道,以及沿延安高架路的带形延中绿地都是市中心的绿色空间。上海市政府更是设立了以下标准(表3.6):

社区公园的分布并不取决于景观特色,而是取决于公平分配以确保每个人都能有合理的使用。在表3.7给出了小型美国社区的公园系统总体规划的例子,并提议一个可比较的方法,就是一系列的公园类型及分布的目标(定义为服务区域)。这个例子更多是一个有用的方案,而非放诸中国皆通的标准。请注意它不包括连接元素。

Large Parks, Greenways, and Neighborhood Parks. Every city or metropolitan area ought to have, as a minimum, a Civic Plaza, and Downtown Park, and a Large Park. Some cities will be fortunate enough to have even more of these two types. Shanghai is a good example. The Bund is a Civic Plaza, Peoples Square (*Renmin Guangchang*) is a Downtown Park, and Luxun Park, Zhongshan Park, and Centennial Park in Pudong are Large Parks. In addition to these major assets, Shanghai is fortunate to have many smaller local parks like Jing'An Park, Huaihai Park or Xiangyang Park. Both banks of Huangpu River are Greenways, as is the linear park along Yanan Elevated Road Downtown Greenspace. The Shanghai municipal government has set up standards as shown in Table 3.6.

上海城市开放空间种类
Urban Open Space Typology in Shanghai — 表 3.6 / Table 3.6

上海分类 Shanghai Classification		本书的分类 Classification in this book		推荐尺度(公顷) Preferred Size (hectares)	服务区域(不超过的距离,千米) Service Area (distance not to exceed, km)	服务人口(千人) Service Population (1000s)
城市综合公园	Urban Comprehensive Park	大型公园	Large Park	>10	3	270 – 450
行政区综合公园	Borough Comprehensive Park	大型公园	Large Park	>10	2	120 – 200
社区公园	Community Park	市区公园	Downtown Park	2 – 4	1	30 – 50
邻里公园	Neighborhood Park	社区公园	Neighborhood Park	0.3 – 0.8	0.5	7 – 15
幼儿游乐场	Tot Lot	社区公园	Neighborhood Park	0.04 – 0.08	0.25	1 – 3

资料来源:方家、吴承照、杰弗里·沃尔、程励. 美国、加拿大和中国居民对城市开放空间的使用和观点的比较. 第47届世界规划大会,2011。
Source: Fang Jia, Wu Cheng-zhao, Geoffrey Wall, Cheng Li. *Comparison of Urban Residents' Use and Perceptions of Urban Open Spaces in USA, Canada and China*. 47th ISOCARP Congress, 2011.

小城市的城市公共开放空间规划奥福尔伦,密苏里州(人口:80,000)
Urban Public Open Space Plan for a Very Small City O'Fallon MO (population: 80,000) — 表 3.7 / Table 3.7

城市公共开放空间类型 Urban public open space type		标准 英亩/1000 Standard acres/1000	标准 公顷/1000 Standard ha/1000	服务半径(英里) Service Area Radius (mi)	服务半径(千米) Service Area Radius (km)
都市公园	Metropolitan Park	4	1.6	5	8
行政区公园	District Park	2.5	1	3	4.8
邻里公园	Neighborhood Park	1.5	0.6	1	1.6
迷你公园	Minipark	0.25	0.1	0.1	0.16

资料来源:福尔伦公园和娱乐休闲总体规划,2010,http://www.ofallon.mo.us/ParksandRec/pubs/3-PLAN_ANALYSIS.pdf。

The distribution of Neighborhood Parks depends less on landscape character and more on equitable distribution to assure that everyone has reasonable access. In Table 3.7, an example from a park system master plan for a small U. S. community suggests a comparable approach, that is, a range of park types, and goals for distribution (defined as service area). This example is useful as an approach rather than suggestive of a standard that might make sense in China. Note also that it does not include linkage elements.

5. Protect (and if necessary restore) significant natural and cultural resources.

While beyond the focus of this book, significant natural and cultural resources in the community that are not already protected at the national or provincial level warrant documentation and protection by local authorities to safeguard the natural environment and for the benefit of the people. Communities may contain remarkable natural resources like wetlands, lakes, waterfalls, or mountain forests, or significant wildlife habitat. These resources should be protected and only made accessible to the extent that access does not degrade their quality and viability, or endanger wildlife habitat. Communities may also have significant historic sites that have remarkably survived the turmoil in the previous two centuries and hold deep cultural meaning for the people. For example, a cultural site could be a memorial, the birthplace or home of a famous person, a rare example of traditional architecture or garden design, or an archaeological site. Such cultural sites should be restored authentically and meticulously as research indicates and resources allow, and be accessible to the people. The presence of such cultural artifacts, especially if their significance is explained in an interpretive marker, instills a feeling of connection to place. (3.5-3.7)

It may seem illogical to include manmade structures in a landscape-centered framework such as an urban public open space system, but the sites that they occupy together with green infrastructure sites form a common urban no-development zone whose longevity will define the community for generations and which are at the core of its environmental and social sustainability. Sites in this category can also contribute significantly to the local tourism economy. Important natural and cultural sites clearly belong in the overall urban public open space master plan and deserve to be linked to the more typical elements that are the focus of this book.

Such precious natural and cultural resources are places where design excellence is more appropriately a matter of restraint, even invisibility, than is the case in the design of parks.

6. Link open spaces in a logical system to enhance equitable access, increase park use, intensify park experiences, shape urban growth, and enhance biodiversity.

The more connected the major park elements are, the better, and the more there is a chance for synergy, and optimization of accessibility and usage levels, and for survival of plant and animal species in a

图 3.5
保护文化场地
Preserving Cultural Sites

法纳尔大厅,自由的发源地,是波士顿很受欢迎的历史古迹,坐落在城市广场的中心,广场被称作法尼尔厅集市。
Faneuil Hall, the Cradle of Liberty, a much visited historic site in Boston, is in the heart of the civic plaza known as Faneuil Hall Marketplace. (Credit: Infrogmation)

5. 保护（并在必要的情况下复原）重要的自然及文化资源

虽然超出了本书的重点,但值得一提的是:社会中未被国家或省级保护的重要的自然及文化资源应授权地方当局来归档和保护,以确保自然环境和人民利益。社区可能会包含湿地、湖泊、瀑布或山林等值得关注的自然资源或是重要的野生动物栖息地。这些资源应该受到保护,他们的开放程度应受限制,不能降低其质量和活力,或危害野生动物栖息地。社区也可能会拥有两个世纪前的社会动荡遗留下的重要历史遗址,而这些遗迹对人们也有深厚的文化意义。比如一个文化遗址的纪念馆,名人的出生地或故居,罕见的传统建筑或园林设计,甚至是考古遗址。我们应该根据研究的成果及资源的允许,真实地细致地修复这种文化遗址,并方便人们参观。这些文化文物的存在,特别是加上标注来解释其重要性,让人们感受到与当地联系的情感纽带（图3.5 - 图3.7）。

把人造结构融入以景观为中心的城市公共开放系统似乎有点不合逻辑,但它们占有的用地可以结合绿色基础设施的用地形成一个共有的城市非发展区域。这个地带的持久性会成为该社区的几个时代的特色,并成为当地环境及社会可持续性发展的核心。这个类别的选址也可为当地的旅游经济作出显著的贡献。重要的自然和文化遗址可归到城市公共开放空间总体规划整体中,并应与本书关注的更多典型要素相联系。

面对如此珍贵的自然及文化资源,设计要学会合理地隐去痕迹,而不是彰显雕琢,这不同于一般公园的设计。

6. 把开放空间连接到一个逻辑系统里,以提高公平的使用权利、增加公园使用率、加强园区的体验、塑造城市的发展并加强生物多样性

公园主要元素的连接越多、越好,越有协同作用的机会,如果可达性和使用水平的越容易达到最佳化,也更便于更多的动物、植物在压力环境下的生存。假如一个空间只有水景而缺乏大草坪,而另一空间只有草坪而缺乏水景,把它们联系起来可以互补不足,

Chapter Three　Planning Principles　**111**

图 3.6
保护文化场地
Preserving Cultural Sites.

孙中山先生的住所,位于上海的文化遗址,与其相邻的博物馆有很多参观者。中山先生的雕像成了景观中画龙点睛之笔。
The house of Dr. Sun Zhongshan, a cultural site in Shanghai, and its adjacent museum welcome many visitors. A statue of the Founder of Modern China graces the surrounding landscape. (Credit: Thomas M. Paine)

图 3.7
保护文化场地
Preserving Cultural Sites.

就像"自由发源地"与"法纳尔大厅集市"相连,中国共产党的发源地也与"新天地"相连。"新天地"由"法纳尔大厅集市"启发而建,现在是一个很受欢迎的目的地。
Just as the Cradle of Liberty lies next to Faneuil Hall Marketplace, the Birthplace of the Chinese Communist Party lies next to Xintiandi, which was inspired by Faneuil Hall Marketplace and is a popular destination. (Credit: Thomas M. Paine)

并加强公园的体验。一种强烈的视觉和功能连接在都市化区域是最难实现的,那里也许只有窄通道。相反,宽阔的绿色连接通道在未城市化地区的城市总规划里是可行的。

联系应该利用自然资源的优势。作为自然资源数据收集的一部分,每个城市应该清查及评估它的天然河流及溪涧、湿地、有成熟植物的池塘地区以及人工湿地、洼地、运河等有人造特征的资源。湿地、池塘及溪涧或河流有时候也被称为"蓝带",来区别陆地上的"绿带"。每种类型都具有绿道的潜在作用,以提供破坏性最少的休闲小径或可能是一条景观大道(或绿化道路),给予大型公园间多形态的联系,这在中国还不常见。多形态是指容纳散步、慢跑、休闲自行车、甚至是在机动车道旁溜旱冰等活动。在严重都市化的环境中,比如中国华东很多地方,绿色基础建设缺乏或落后,城市规划过程应该认真考虑利用天然材料并引入绿色基础建设,而非土木工程中一贯的钢筋混凝土结构。比起缺乏植被的混凝土结构,这种柔性的方案可使公园使用更高效且便利(图 3.8 - 图 3.9)。

绿色基础建设已经越发被当成是塑造城市发展的原则。它包括城市公共开放空间系统及以废水循环和地下水补充为中心的非公共绿色空间。以绿色基础建设为城市的核心的理念始于围绕城市的绿带。自然水景或排水渠道的不规则排列,不被视为是给四通八达的完美网格化街道带来不便的障碍,反而成为界定不同地区界限的愉快插曲。正是在街道的不规则布局和在由楔形绿地分隔的小区域的定义中,一个城市得到它的吸引力和独特性。不管路边的建筑如何各具特色,看似无穷无尽网状的笔直街道很快就会变得似曾相识,而千篇一律并不能促进城市的发展。通过绿道扩展城市公园系统是塑造城市发展的最好方法。

stressed environment. If one space has a water feature but lacks a large lawn, and another has a lawn but no water feature, linking them helps make up for the deficiencies in each, and intensify the park experience. Opportunities for a strong visual and functional connection are hardest to achieve in urbanized districts where a narrow corridor may be all that is feasible. Conversely, wide green corridor links are feasible elements in the urban master plan for not-yet urbanized districts.

Linkages should take advantage of natural resources. As part of its data collection for its natural resources, each city should inventory and assess its natural rivers and streams, wetlands, ponds, areas of mature vegetation, as well as manmade features like constructed wetlands, swales and canals. Wetlands, ponds and streams or rivers are sometimes called the "bluebelt" as distinct from the dry land of the "greenbelt". Each type has a potential role as a Greenway providing a minimally damaging recreational trail or perhaps a scenic roadway (parkway), still unusual in China, offering multi-modal linkage between larger parks. By multi-modal is meant accommodation of walking, jogging, recreational biking and perhaps even rollerblading alongside motor vehicles. In heavily urbanized environments such as are found in much of eastern China, where green infrastructure is lacking or degraded, an urban planning process should seriously consider the reintroduction of green infrastructure using natural materials rather than rigid structures typical of civil engineering. The soft approach can simultaneously accommodate park use much more effectively than can concrete structures devoid of dense planting. (3.8-3.9)

图 3.8
连接作用
Linkage

波士顿的公园道路系统和绿道系统互相连接着市区公园和大型公园，就像这幅查尔斯河盆地的航拍图中所示，公众公园在右侧远端，左侧远端是芬威球场，滨河绿道与河平行并且一直延伸到高层建筑后面的低层住宅区之中。
Boston's system of parkways and greenways interconnect with Downtown Parks and Large Parks, as suggested in this aerial view of the Charles River Basin with part of the Public Garden at the far right, the Fenway at the far left, and greenway along the riverfront and parallel to it down the middle of the low-rise neighborhood beyond the high rise buildings. (Credit: anonymous)

Increasingly, *green infrastructure* is invoked as the principle for shaping urban growth. It includes urban public open space systems as well as the system of non-public green space focusing on grey water recycling and groundwater replenishment. Green infrastructure brings to the core of the city the same kind of thinking that began with the idea of a greenbelt ringing the city. The irregular alignment of natural water features or drainage canals is not perceived as an inconvenient interruption to a perfect grid of streets extending infinitely in all directions, but instead as a welcome interlude that helps define the boundary between one district or land use and another. It is in the irregularity of the street layout and in the definition of distinct small districts separated by green wedges that a city derives much of its attractiveness and distinctiveness. No matter how differentiated the roadside architecture, straight streets on a seemingly endless grid soon start to look alike, and monotony is no way to shape urban growth. City growth is best shaped by extending the city park system using Greenways.

A key ecological benefit of open space linkage is often overlooked. Connecting large natural open spaces by Greenways also promotes local biodiversity by allowing wildlife and plants to migrate from one habitat area to another through the alien urban environment. The increasing fragmentation and isolation of habitats brought on by pervasive urbanization diminishes the chance of survival of threatened species. Landscape connectivity allows species migration, promoting species survival in a time of climate change.

图 3.9
连接作用
Linkage

像得克萨斯州圣安东尼奥滨河步道那样的绿色基础设施，提供了让人们划船的地方。这成了一种愉悦而又轻松的体验城市迷人风光的方式。
Green infrastructure such as the Riverwalk in San Antonio, TX, can include boat rides which provide a pleasant and relaxing way to experience the charms of the city. (Credit: Zereshk)

Chapter Three　Planning Principles　113

开放空间连接的关键——生态效益经常被忽视。用绿道来连接大型的自然开放空间也可同时促进本地的生物多样性，使野生动物、植物可通过外在的城市环境从一个栖息地迁移到另一栖息地。普遍的城市化所带来的日益增加的栖息地分裂和隔离，使受威胁物种的生存机会大为减少。景观连通性允许物种迁徙及促进气候变化中的物种生存。

7. 颁布公共政策以确保连续性、一致性及用户满意度

在中国，不少司法管辖区的官员每两年的轮换或升迁使政策的连续性难以确保。决策者缺乏时间去作出全面的决定，或以政绩作为评价标准，都是与提供优良的城市公共开放空间背道而驰的，这样就导致开放空间的质量存在风险。城市公共空间系统的规划、设计、建设及管理需要慎重的考虑和超乎寻常的连贯性，而不是隔不了几年就改弦易辙。我们一直强调不断监测及保存记录以衡量进度及了解趋势，不管好与坏，这样在必要时就何以随时介入，年复一年，这是一个长年累月的过程。

显然，这些工作的顺利开展需有充足的资源，保证适当的管理及维护，因为人们的知识更丰富，需求更高，所以更要努力赢

7. Promulgate public policy to assure continuity, consistency and user satisfaction.

The two-year rotations or promotions of officials in many jurisdictions in China put policy continuity at risk. Open space quality is at risk where decision makers have too little time to make sweeping decisions or are evaluated on the basis of performance metrics in conflict with providing urban public open space excellence. Urban public open space system planning, design, construction, and management require careful consideration and extraordinary consistency, not a change of direction every few years. We have stressed the need for ongoing monitoring and record-keeping to measure progress and understand trends, good and bad, so that intervention is forthcoming when required, year after year, over the long term.

Clearly, success will require that the community has sufficient resources to provide adequate management and maintenance staffing and that it works hard to earn user satisfaction as the people become more knowledgeable and demanding. Soon their concerns will look beyond the basics like safety from physical hazards and crime. (3.10)

图 3.10
公众参与
Public Participation

在很多西方国家，当地的居民都会出席公共会议，了解将要发生的改变，共同讨论提出意见。例如康涅狄格州哈特福德的这个例子。
In many Western countries local residents attend public meetings to learn about proposed changes and be given an opportunity to voice their concern as in this example from Hartford, Connecticut. (Credit: Sage Ross)

8. Plan for the long term, incorporate flexibility in management to accommodate change in society.

While abrupt changes of direction in the short term are counterproductive, over time external change is inevitable, and will necessitate a realignment of priorities in the programming, design and management of urban public open space to accommodate changing needs in society and conditions in the environment. Such flexibility should inform a deliberative, open process to maintain excellence over the long term. Process is indeed the key. This realization is more compelling than ever before, as thinking about urbanization aligns more with thinking about ecological process and disruptive climate change, and away from planning for some imagined stasis or permanent final outcome.

9. Prepare for a future of fruitful collaboration between public green agencies and private volunteer support groups.

In many countries, non-governmental organizations (NGOs) work with government officials in the management of public parks. The very first NGO in China, Friends of Nature, was founded in 1993, and was followed by thousands more. Sooner or later in China and other countries where an increasingly well-informed citizenry is motivated to become involved, each locale will have an NGO to assist in the immensely important work of protecting the natural environment and urban public open space. Most Western countries manifest a rich participatory environment. In the U.S., The Trust for Public Land (TPL, www.tpl.org) quoted repeatedly in this book, knows as much about urban public open space trends as any government agency, and makes use of the information in a wholly public-spirited way. The Project for Public Spaces (PPS, www.pps.org) is another source of a wealth of online information about urban space in many countries. PPS likes to say that it takes a community to build its urban public open space, and it takes that urban public open space to build a community. PPS strongly advocates tapping the vision of the actual stakeholders, that is, local residents, to guide the designers rather than the reverse. The Sierra Club (www.sierraclub.org) is one of the oldest environmental advocacy organizations in the U. S., and has been instrumental in raising environmental awareness and protection of endangered natural resources. The Massachusetts-based Trustees of Reservations (www.thetrustees.org) is the world's first land trust, formed to acquire and protect forever natural and cultural landscapes that contribute significantly to local quality of life and landscape character, and open these places to the public for a small fee. They work to supplement what the public sector can provide, but cannot replace urban public open space that is open free of charge. Unlike the government, they have no source of revenue other than donations and fees.

Lastly, many urban public open spaces in the West are supported by private "Friends" groups (Friends of the High Line, www.thehighline.org) or Conservancy groups (Central Park Conservancy, www.centralparknyc.org) which raise funds to support maintenance, restoration, and programming beyond what government funding

得使用者的满意。不久他们的关注将会超出基本的需求，例如杜绝危险物品和犯罪的安全感（图3.10）。

8. 目标要长远，加入管理上的灵活性以适应社会的变化

虽然短期内突然的转向会导致不良的后果，但随着时间推移，外来变化不可避免，而且也会有需要去重新部署城市公共开放空间规划、设计及管理的优先权，以适应社会中不断变化的需求和环境条件。这种灵活性应该将整个协商的过程向公众开放，以保持长期的精益求精。过程确实是关键。这个认识比以往任何时候更具说服力，因为关于城市化要考虑到生态因素、破坏性的气候变化，而不是根据想象中的数据或抱着一劳永逸的心态。

9. 为公共绿色机构及私营志愿者团体的未来富有成效的合作准备

在很多国家，非政府组织（简称NGO）与政府官员共同合作管理公共公园。中国第一个NGO "自然之友"，于1993年成立，随后出现了数以千计的其他非政府组织。在中国及其他国家，在信息化社会，越来越多的公民渴望参与社会事务，每个地区迟早将会出现一个NGO，来协助保护自然环境及城市公共开放空间等极其重要的工作。大部分西方国家出现了一个公众活跃参与的环境。在本书中反复引用的美国公共土地信托会（简称TPL, www.tpl.org），对城市公共开放空间的了解不亚于任何政府机构，并以公开透明的方式使用这些信息。公共空间计划（简称PPS, www.pps.org）是另一个有关各国城市空间的网上信息来源。PPS喜欢说需要一个社区来建立它的城市公共开放空间，同时需要那个公共开放空间来建设社区。PPS极力地主张开发实际利益相关者的想象力，换言之，当地的居民引导设计师，而不是反过来。塞拉俱乐部（www.sierraclub.org）是美国历史最悠久的环保宣传组织之一，一直以提高环保意识及保护濒危的自然资源为己任。总部在马萨诸塞州的保护区信托管理（www.thetrustees.org）是世界第一个土地信托，它的成立是因为要获取一些对当地生活质量及景观特色有重要贡献的永久自然资源及人文景观，并且把这些地方以象征性的收费对公众开放。他们的运作是对公共部门所能提供的公共空间的补充，但绝不能取代免费开放的城市公共开放空间。与政府不同，他们除捐赠及收费外，并没有其他收入来源。

最后，很多西方的城市公共开放空间受到私人"朋友"群体（例如高线之友，www.thehighline.org）或资源保护群体（例如中央公园保护处 www.centralparknyc.org）的支持。他们筹集资金支持维护、修复和规划政府拨款能力范围以外的项目，并以不削弱对公众的开放使用的方式进行。上述都是有可能在中国兴起的同类环保组织的先例，当然中国也可建立其他类型的组织再让西方从中取经。

我们必须强调上述都是非营利机构，这一点是非常重要。与NGO、基金会、朋友群体及其他非营利机构不同，房地产公司等营利机构一般都会看重短期的商业利益而漠视长远的环境保护。但幸运的是，实际情况并非一概如此。最好的营利机构会超越这种思维，在不损害结果的前提下提供他们的资源。

在西方国家,当大块用地被分析以确定哪些地区需要保留,哪些地区最适合发展时,公家和私人的伙伴关系有时候是唯一的解决方案,最后达到两种观点间的妥协。由景观建筑师劳伦斯·哈普林设计的吉尔德利广场,位于旧金山,是由私人资助建成却一直维持非商业化运作——除了摆放具有历史意义的广告牌。

城市公园联盟

"城市公园联盟是城市公园运动的领导者。这个城市公园运动涉及了无数的组织致力于遍布美国的冠军级高质量的城市公园。城市公园联盟对关于公园的公共政策至关重要,这些合理的政策将城市公园作为更强大、更充满生气的城市的不可或缺部分,这些城市的市民可从生活的每一步中都得到滋养。

我们对于城市公园的愿景是到 2020 年,美国城市的每个人将居住在整洁、安全和充满活力的公园的步行到达距离内。"

来　源:http://www.cityparksalliance.org/about-us/mission-and-goals

10. 在政策层面、规划和设计上引入注册景观建筑师的参与

正如在第二章中所讲,我们已经进入了由景观和生态思想指导城市化甚至塑造大楼建筑风格的时代。只有景观建筑师才有资格带领由不同学科专业人士组成的团队,共同为当地政府机构提供仔细考虑过的客观建议。他们可以为当地政府权威人士提供整体的全盘观点,以帮助他们做出富有远见的平衡决策。景观建筑师应该与建筑师和工程师同样有注册资格,并有同等重要性,尤其是在尚未有此先例的国家。他们也应该参与政策讨论,可能作为被邀请的专家,或者是适时加入委员会。

■ 智能城市规划的原则

城市公共空间的规划当然不能在真空中产生,而应在城市规划更大的背景下产生。在引导城市化走向一个更乌托邦世界的原则上,尤其是具有丰富城市开放空间的规划,西方设计专业人士之间似乎有着广泛的共识。

智能城市化原则

1. 与自然平衡(可持续性,接近自然,水域与动物、植物生活环境的保护,环境恢复)

2. 与传统平衡(文化遗产、史迹保护、城镇景观、符合人文环境的设计)

3. 合适的技术("小即是美",本地实践)

4. 娱乐休闲场地(不同级别和尺度社会交往的场地,包括独居、朋友、家庭、邻居、社区和更大人流量的广场、公园、体育场、交通枢纽、人行道或画廊)

5. 高效(高效节能的公共交通和基础设施)

6. 人类尺度(适宜行人和当地居民的通用与无障碍设计;行走方便性与娱乐休闲功能重叠)

7. 机会模式(平等获取工作的机会和基本服务,比如公共健康、安全、福利、教育、给水、资金、多样化选择)

8. 区域整合(乡村和城市核心,与自然的平衡与重合)

9. 平衡的发展(好的基础设施,多种模式的、公平的分配,连通性)

---

permits, and in a way that does not undermine public accessibility. These are examples of the kind of environmental organizations that might spring up in China, which will also undoubtedly create other types that the West can learn from in return.

None of these organizations are for-profit, and that is an essential point to make. Unlike NGOs, foundations, Friends groups and other non-profit organizations, for-profit organizations such as real estate development companies typically put short-term commercial gain ahead of long-term protection. But such is fortunately not always the case. The best for-profit organizations will transcend this thinking and offer their resources without compromising the result. In Western countries, public-private partnerships are sometimes the only solution, wherein a large land parcel is analyzed to identify what areas should be preserved, and what areas are most suitable for development, arriving at a compromise between the two points of view. San Francisco's famed Ghirardelli Square, designed by landscape architect Lawrence Halprin, was funded privately and still manages to feel commercial-free—except for the historic billboard.

City Parks Alliance

"The City Parks Alliance is a leader in the urban parks movement. This city parks movement encompasses the myriad organizations dedicated to championing high quality urban parks throughout the [U.S.] The City Parks Alliance is vital to sound public policy that recognizes city parks as an integral part of forming stronger, more vibrant cities that nurture their citizens from every walk of life. Our vision for urban parks is by 2020 everyone in urban America will live within walking distance of a park that is clean, safe and vibrant."

Source: http://www.cityparksalliance.org/about-us/mission-and-goals

10. Involve licensed landscape architects at the policy level and in planning and design.

As we noted in Chapter 2, we have entered the age when landscape and ecological thinking is guiding urbanism, and even shaping the architecture of buildings. Landscape architects are uniquely qualified to lead the team of disciplines that together will provide local government agencies with carefully considered objective advice. They can provide local government authorities their integrative, holistic perspective essential for far-sighted, balanced decision-making. Landscape architects deserve to be licensed on an equal footing with architects and engineers in countries where this is not yet the case. They also deserve to participate in policy discussions, perhaps as invited experts, or serve on committees as appropriate.

■ Principles of Intelligent Urbanism

Urban public open space planning, of course, does not occur within a vacuum, but within a larger context of urban planning. In the matter of principles that ought to guide urbanization toward a more utopian world, especially one rich in urban public open space, there seems to be wide agreement among design professionals in the West.

Principles of Intelligent Urbanism

1. Balance with Nature (sustainability, access to Nature, watershed and habitat protection, environmental remediation)

2. Balance with Tradition (cultural heritage, historic preservation, townscape, contextual design)

3. Appropriate Technology ("small is beautiful", local practice)

4. Conviviality (places for different degrees and scales of social interaction from solitude, friendship, family, neighborhood, community, and more widely visited plazas, parks, stadia, transport hubs, promenades, or gallerias)

5. Efficiency (energy-efficient public transportation and infrastructure)

6. Human Scale (pedestrian and local resident-friendly, universal and barrier-free design, walkability, overlaps with conviviality)

7. Opportunity Matrix (equitable access to jobs and basic services like public health, safety, welfare, education, water supply, capital, diversity of choices)

8. Regional Integration (countryside and urban core, overlaps with balance with Nature)

9. Balanced Movement (good infrastructure, multi-modal, equitable distribution, connectivity)

10. Institutional Integrity (tolerance, rule of law, honesty, transparency, accountability, competence, participatory, free of fraud, theft, bribery and extortion, quality control of development, density control)

One attractive formulation is the ten Principles of Intelligent Urbanism (above). Many of these principles overlap and meld into others on the list. The principles subsume Smart Growth or Compact City, which focus on high-density growth, public transportation, walkability and sustainability. When an urban public open space master plan takes the Principles of Intelligent Urbanism into account and the results include a substantially equitable distribution of resources, we have achieved a necessary condition, but not a sufficient one, for excellent urban public open space design.

Surveys of the most "livable" cities in the West take many of these principles into account in their metrics. No city deficient in urban public open space is ever considered livable, by which is meant, pleasant to live in. In the past decade, terrorism has elevated the importance of personal safety. In some lists of livable cities, none of the top ten cities are in either China or the U. S., but are in Northern Europe, Canada, Australia and New Zealand. On the other hand, the regions with the most livable cities also have many fewer cities with a population of one million or more, only thirty-six. By comparison, China will soon have one hundred fifty of them. The need for excellent urban public open space globally is nowhere more highly concentrated than in China.

10. 体制完整性（宽容、法治、诚信、透明度、问责制、能力、参与性、免除欺诈、盗窃、贿赂和勒索、发展的质量控制和密度控制）

上述的十项智能城市规划原则十分有意义。这些原则，很多都重叠及与其他融为一体。精明增长及紧凑城市都被归入这些原则中，其中重点包括高密度增长、公共交通、步行适应性及可持续性。当一个城市公共开放空间的总体规划考虑到智能城市规划原则，并且其结果包括了相当多资源的公平分配，那么我们也就为优良城市公共开放空间的设计提供了一个必要条件，但不是一个充分条件。

大多数西方"宜居"城市的调查都以上述各项原则为参考指标。缺乏城市公共开放空间的都市算不上是宜居。宜居顾名思义，是适宜人居，且相与怡然。在过去的十年间，恐怖主义给人警醒，更加让人意识到人身安全的重要性。在一些宜居城市中，前十名都是在北欧、加拿大、澳大利亚及新西兰，没有任何中国或美国的城市。另一方面，这些拥有宜居城市最多的地区，往往人口过百万的城市较少，总共只有 36 个。相比之下，中国很快就要有 150 个人口超过百万的城市。全球对优良城市公共开放空间的需要没有一处像中国这样高度集中。

注释
1. 梁家杰（杰哥）."还我们真正的公共空间《杰哥家书 #35》"（2008），http://lnix.online.com.hk/temp/AL/www.alanleong.com/chi/index.php%3Fcat=4&paged=23.html，五月 2011
2. 彼得・哈尼克.《城市绿地：复兴城市的创新性公园》（华盛顿特区：岛屿出版社，2010）http://www.cityparksalliance.org/about-us/mission-and-goals
3. 斯蒂芬・卡尔，马克・圣弗朗西斯，雷文，安德鲁・斯通.《公共空间》（纽约：剑桥大学出版社，1992），18. 我们不是第一次引用这本书；见克莱尔・马克斯和卡洛琳・圣弗朗西斯，《人物与地点》（纽约：约翰威立，1998），8
4. 纽约大学富曼地产及城市政策中心全年住房报告，2010 春季
5. C.A. 道萨迪亚斯.《人类聚居学》（纽约：牛津大学出版社，1968）
6.《专业的标准》，伦纳德・霍珀.《景观建筑图形标准》（霍博肯：约翰威立，2007），76. 美国国家游憩及公园协会和彼得・哈尼克.《城市绿地：创新的城市公园为复活》，（华盛顿特区：岛屿出版社，2010），14，推荐 10 英亩 (4.0 公顷) /1000 人，哈尼克认为这是一个中间值，跨越了美国的城市
7. 劳伦斯・哈普林.《城市》（剑桥：MIT 出版社，1972），4，11

NOTES
1. Alan Leong Kah-Kit (Kit Gor). "Give Us Back Real Public Open Space" (2008), http://lnix.online.com.hk/temp/AL/www.alanleong.com/chi/index.php%3Fcat=4&paged=23.html, May 2011
2. Peter Harnik, *Urban Green Innovative Parks for Resurgent Cities* (Washington, DC: Island Press, 2010), http://www.cityparksalliance.org/about-us/mission-and-goals
3. Stephen Carr, Mark Francis, Leanne G. Rivlin, Andrew Stone, *Public Space* (New York: Cambridge University Press, 1992), 18. We are not the first to quote this; see Clare Cooper Marcus and Carolyn Francis, *People Places* (New York,: John Wiley, 1998), 8
4. NYU Furman Center for Real Estate and Urban Policy, *Annual Housing Report*, Spring 2010
5. Doxiadis. *Ekistics: An Introduction to the Science of Human Settlements* (New York: Oxford University Press, 1968)
6. "The professional standard" per Leonard J. Hopper, ed., *Landscape Architecture Graphic Standards* (Hoboken: John Wiley, 2007), 76. Both U. S. National Recreation and Park Association and Peter Harnik, *Urban Green, Innovative Parks for Resurgent Cities* (Washington, DC: Island Press, 2010), 14, recommend 10 acres (4.0 hectares) per 1000 population, which Harnik also finds to be the median across U. S. cities
7. Lawrence Halprin. *Cities* (Cambridge: MIT Press, 1972), 4, 11

设计原则
Design Principles

Chapter Four
第四章

首先是生活，然后考虑空间，其次是建筑，其他就不那么见效了。
First life, then spaces, then buildings—the other way around never works.
—— 扬·盖尔，1987 年 [1]
Jan Gehl, 1987 [1]

设计是产品的灵魂，它通过整体的外形来表达出产品所蕴含的生命力。
Design is the fundamental soul of a human-made creation that ends up expressing itself in successive outer layers of the product or service.
—— 史蒂夫·乔布斯，2000 年 [2]
Steve Jobs, 2000 [2]

在城市里的公共空间一定要确保多于仅仅是为了这个一般的活动——娱乐，象征性补偿或者容纳而建成的空间的数量。公共空间作为一种容器，承载集体的记忆和欲望，其次，他们是提供地理和社会的想象力的地方，并且可以扩大新的关系和创造可能性。
Public space in the city must surely be more than mere token compensation or vessels for this generic activity called "recreation". Public spaces are firstly the containers of collective memory and desire; and secondly they are the places for geographic and social imagination to extend new relationships and sets of possibility.
—— 詹姆斯·卡纳，2006 年 [3]
James Corner, 2006 [3]

设计一定要激发、刺激以及愉悦人们的感官。设计需要与人建立起不仅仅是物质而且是思想上的联系。
Design must inspire, delight, excite, and stimulate the senses. Design must connect people to a place in a physical and emotional way.
—— 丹尼斯·卡迈克尔，2007 年 [4]
 Dennis Carmichael, 2007 [4]

前面我们已经讨论过，人们对于公共空间不只是简单的需求，而是更希望能体验到其设计之美。我们还讲述过，场地规划的基本原则应高于设计，同时应忽略设计的不完美性。这一章将分别从整体和部分的角度讲述公共空间的设计原则，包括城市广场、市区公园、大型公园、绿道和社区公园。

在我们这个时代，设计如消费品，应做到品质优良，才能满足需求。我们的场地仅仅拥有基本的功能需求是不够的，而更需要一定的美感，从而吸引人前往。正如上文引用的乔布斯关于时代思潮的看法，他也正是秉承这种思想并将这种思想融入到了iPhone和iPad的设计当中，才得以引领时代的潮流。人们对iPhone和iPad产生了感情。就像乔布斯理解的，与人产生的情感已经不是设计的附属物，而是应该作为完整设计的一部分，是设计的精华。越来越多的企业正在实践一个统一化的理论，即场地设计是公司战略的核心，甚至开始研究设计的美感。公共空间的设计发展能延后很久么？新的优质的城市开放活动空间是让人们在情感上团结，更加整体化，更加人文化。正如丹麦的城市规划师和建筑师扬·盖尔所说，设计的过程必须以人为本："首先是生活，然后考虑空间，其次是建筑，反之则不可行"。这位极具影响力的规划师深深地相信，即使在城市中，建筑设计也应该依从室外空间，而不是室外空间设计依从建筑。也就是说，建筑师在某种程度上应该配合风景园林师的思路，其中包括城市公共户外空间的设计和服务设施方面。

值得一提的是，这一章所提及的设计原则对于成功的设计来说是很必要的。但也只有设计师在同一块场地同一时间综合所有设计元素，创造出一个美妙的空间并吸引群众的关注时，人们才会称赞空间的美，才会惊叹于设计师的才华，这种设计才称得上是"成功"。好的设计将以用户体验为核心，利用想象力综合罗列复杂的设计元素，遵循诸如"均衡性"与"和谐性"的原则，从而赢得用户满意。

显然，这需要设计师极度用心。就像我们在第二章里所说，即使在一个过度拥挤的城市，我们也能找到几个没有被充分利用的公共空间。那些空间往往设计水平差，不足以吸引人们前往。这几乎等同于犯罪，也是对公共资源的一种不可谅解的浪费。

■ 基本设计原则

在我们转到城市公共开放空间设计之前，应该牢记一点，那就是基本设计原则支配着整个设计。无论是城市总体规划，还是随后的细节场地设计，下面的重要设计原则都必不可少：

1. 统一性

统一性是完整性与美学协调性的一种融合。它可以是通过对细节、形式、色彩以及材料的重复利用，营造一个整体的空间体验。有时，这种重复强调了一个好的想法或者主题，有时却又在内容上略显空洞。统一性表现为外表上的天衣无缝，从中难以看出公共艺术及景观设计从哪里开始又从哪里结束。统一性是好的，但是统一并不意味着千篇一律。统一不是简单。

2. 多变性

对称常常不是必要的，对称还可能导致设计的单调，让人产

In previous chapters we laid the groundwork for arguing that people have a need for not only urban public open space access but also urban public open space design excellence. We have described the essential planning principles that must precede design and in disregard of which no design can reach its full potential. In this chapter we describe the design principles that are essential for the urban public open space system as a whole and for all its parts, whether Civic Plaza, Downtown Park, Large Park, Greenway, or Neighborhood Park.

In our era, such design deserves to achieve the same degree of excellence that we expect in consumer products. It is not enough that a product functions on a basic level; it must also delight the user by appealing to the user's innate sense of beauty. Steve Jobs caught the new zeitgeist in the quote cited above and went on to deliver the iPhone and iPad. People connect with their iPhone or iPad on an emotional level. As Jobs understood, connection on an emotional level is not an add-on; it is integral to design excellence. More and more companies in the corporate world are practicing a unitary theory that places design at the core of company strategy, and are even talking about beauty. Can the world of urban public open space design lag behind for long? The new standard of urban open space design excellence integrates people to place on an emotional level, by being holistic and humanistic. In other words, people come first. As the Danish urbanist and architect Jan Gehl puts it, the design process must be human-centered: "First life, then spaces, then buildings—the other way around never works". This influential and much quoted urbanist deeply believes that even in the city, architecture should defer to outdoor space, not the other way around. That is, architects should defer to landscape architects in matters involving the design of urban public outdoor space and the facilities which it accommodates.

A word of caution: adhering to the principles in this chapter is surely necessary for success, but will only be sufficient for success if the design elements come together in one place at one time to form a thing of beauty that "lies gently on the land" and invites us to marvel at the genius of the team that brought such a place into being. Guided by general design principles like "balanced" and "harmonious", excellent urban public open space design synthesizes a complex array of elements in a compelling form that comes out of the imagination and goes straight to the heart of the user.

There is clearly a need to design with great care. As we described in Chapter 2, even in an overcrowded city, we can find examples of underused urban public open space, poorly designed, perhaps in the wrong place, and empty of reasons that motivate people to want to go there, and that is almost a crime, an inexcusable squandering of public resources.

■ Fundamental Principles of Design

Before we turn from planning to the design of urban public open space, it is important to bear in mind the fundamental principles that ought to guide design in general. The following time-honored design principles should never be forgotten during both the urban master planning process and the subsequent site-specific design process:

1. Unity. Unity is a feeling of completeness and aesthetic coherence. Unity may come from repeated details, forms, colors and materials that together impart a holistic spatial experience. Sometimes the repetition underscores a big idea or theme, sometimes it is devoid of intellectual content. Unity also resides in seeming seamless, in not knowing where public art and landscape design begin and end. Unity is good, but uniformity may not be. Unity should not be confused with simplicity.

2. Variety. Symmetry is not always necessary, and may lead to monotony. Boredom is deadly. China values clutter, density is good, and emptiness is not. In some spaces, design complexity encourages people to stay, explore, and return more than does simplicity. As in music, rhythmic variety engages people in ways that a monotone simply cannot. On the other hand, cacophony is disturbing, and disturbing people is hardly the purpose of open space design.

3. Balance. Unity and Variety should peacefully coexist. Properly balanced, together they elicit a dialogue; in other words, they elicit a healthy degree of tension. A little tension is a good thing that engages the human spirit. Too much tension subverts the design and may be disturbing.

4. Scale. Spaces or elements should be neither too large nor too small for their purpose, or their context. What works well at one scale (when we say it is in scale) may look silly when reduced or oppressive when enlarged. In other words, right scale is a matter of proportion. Proportion, in turn, is a matter of right relationships inferred from deep-seated cultural norms and perceptual instincts.

5. Harmony. We come now to the term that may define the essence of Chinese cultural aspiration. In harmony there is both unity and variety in balance and in scale. Harmony resides in artful asymmetry as much as symmetry. Harmony sometimes conveys repose and serenity. Harmony of design in the twenty-first century is not simply a matter of imitation of precedent. It embraces, even celebrates, what is new, while integrating it with what is old. This is fusion design of a new kind. It has the power to move people, and build harmony of community. In that sense, nothing could be more in tune with the essence of Chinese culture than harmony in pursuit of harmony. Current thinking in the West to reject harmony as somehow outdated seems too pessimistic and designed to fail.

■ Caveats for the Design Process

Caveats are "dos and don'ts", more cautionary and task-oriented than principles. Here assembled for the first time are fifteen time-tested caveats to guide the design process in general, regardless of project type. They overlap here and there.

1. Consult the context. Context matters. "Consult the genius of the place in all" is how 18th century English landscape gardener poet

生无聊的感觉。中国的设计善于迂回，在某些空间里，复杂的设计比简单的设计更能引起人们的注意，甚至去探索，使人们有想要再来一次的冲动。就像音乐，往往都是音律节奏变化丰富的乐曲能打动人，而单调的曲子却很难做到这一点。刺耳的乐曲则会引发别人的反感，这种反感是设计师不愿意看到的。

3. 均衡性

统一性与多变性应该共存。两者之间适当的均衡可以激发出设计整体的对话，换句话说，这种设计的对话使游客心中产生愉悦的波动。小波动会激起观众对作品的兴趣，但如果设计的过于复杂，会使这种波动变成一种不安，反而破坏了设计的初衷和想要表达的效果。

4. 尺度适宜

考虑到设计目的以及环境背景，空间以及设计元素应大小适当。当扩大或是缩小某个作品，都会破坏这一作品的美感时，那么这一作品的尺度就恰到好处了。换句话说，尺度适宜是比例协调的问题。再换句话说，比例也是由根深蒂固的文化规范以及直觉本身衍生出的一种合适的大小关系。

5. 和谐性

"和谐"这个术语，可能与中华文化的精髓有着深厚的渊源。统一与多变、均衡和尺度适宜造就了和谐，和谐存在于一种艺术上的同等的对称与不对称。和谐也可以表达休息、静止与宁静的状态。在21世纪，和谐的设计已不再仅仅是对先人的简单模仿，它可以包含，甚至赞美，在整合已有的事物时所创造出来的新事物。这是对新思维的一种融合设计。它拥有可以打动人的力量，同时能创造一种社会的和谐。中国自古以来对和谐的追求过程中所产生的和谐精神便是对中国文化最好的诠释。现在西方国家不认同和谐，并认为其不合时宜，这似乎太悲观了，也可能导致设计的失败。

■ 设计过程中注意事项

注意事项就是"可行与不可行"，它更是一种约束，而非原则。在我们开始优秀开放空间设计细节讲述之前，首先我们总结出15条锦囊妙计，不论什么项目类型，你会发现它们无处不在。

1. 因地制宜

设计环境非常重要，关于这一点，18世纪英国造园家及诗人亚历山大·蒲柏引用了诗句"Consult the genius of the place in all"，强调这一点。在中国，风水一直被强调，这是西方所缺少的一种思想。无论东方还是西方，相邻的外界环境在设计中都是很重要的，其重要性甚至远远超过了部分人的想象。视觉是一种交流方式，对周围的视域表达至关紧要。在中国，这个叫做"借景"。

2. 注重人文

设计环境不是肤浅问题，也不仅仅是视觉的问题。我们应该去试着理解当地的文化以及文化形成的原因。要向当地的环境学习，倾听当地的建议。因地制宜就意味着无论是在学术界还是在别的地方，都要寻找当地居民，聆听他们的所知，和他们交流，并收集他们的建议。尊重当地的文化以及文化多样性，让当地的风土人情、精神与物质去启发设计。中国需要重新找回被遗忘的道教及佛教崇尚的生态共同体的思想，以及许多诸如丽江纳西族等的多文化少数民族聚集地。

3. 建立联系

建立的联系越多，设计就越有吸引力和亲和力。我们要思虑全面。这反映出到中国一种古老的文化精神，即协调性原则，它需要我们延伸更多的思考领域，然后再尽可能深刻地去结合我们的设计。试想一下这些联系形成的一个互动式系统。设计师要多与人沟通，多倾听别人的心声。

4. 建立情感联络

了解人们表达的方式。我们要提供一种令人有安全感的环境，即便是无形的。这种环境应该鼓励不同人之间的交流，包括随意的、随机的、无组织的，甚至是陌生人之间的交流。例如，提供一种不寻常的装置，鼓励从未谋面的两人开始一段对话。

5. 大胆创新

在西方有句话，"没有远见，就没有未来"。多聆听客户和用户的想法，多征求专家的意见，这是对设计负责的表现。同时，也要敢于创新，有大胆的想法。可能，今天的拒绝将造就明天的辉煌。也或者，现今被压迫的好思想，未来可能重新获得认可。同样，今天的大奖得主有一天也可能会遭人唾弃。

6. 避免破坏

这个短语让人联想到西方的希波克拉底誓言（医学用语）。尽一切努力不去制造伤害，保护已经幸存下来的好的东西，恢复已经损坏的，然后重拾早已丢失的。另一方面，担心做的不好可能会导致胆怯，即缺乏了一种想做正确事情的意愿。而这一点可以帮助拯救我们脆弱的地球。

7. 态度认真

就像我们常说的，"好心办坏事"和"三思而后行"。做任何事，不要胆小，但是行事要认真。假如我们出于好意，想要避免任何的错误，那我们将一事无成。避免常规的错误就好，小心谨慎有助于避免浪费资源、防止造成伤害。

8. 择机而行

西方还有句谚语，"完美"是"好"的敌人。设计是反复琢磨出来的，但时间和金钱这样的资源却是有限的。不要一直等下去。让完美体现在别的地方吧。

9. 适时否定自我

要有雅量地承认他人的优点，不要顾面子，慷慨大方一些。感谢身边的每一个人。

Alexander Pope put it. In China, *fengshui* takes into account forces that a Western-trained mind may miss. In both East and West, the adjacent physical world matters very much in design, and matters further afield than some people may think. Visual access is access: the circumferential view outward from the site (viewshed) matters. In China above all other places, it may be called "borrowed landscape", a term invented in China.

2. Tap your inner anthropologist. Context is not superficial. And it is not just visual. Try to understand local culture and why local culture is the way it is. Consult the context of local opinion: lead by listening. Indeed, "consult the genius of the place" may equally well refer to seeking the advice of well-informed, concerned local people, in academia and elsewhere. Engage them. Invite their suggestions. Respect local cultural values and diversity. Let local customs, preferences, spirituality and materiality inspire design. In China, retrieve the lost eco-community celebrated in Taoist and Buddhist thought as well as that of many cultural minorities like the Naxi of Lijiang.

3. Make connections. The more connections we make in a design, the wider its appeal and accessibility. Think holistically. Think laterally. This reflects the principle of correspondence, an ancient Chinese cultural trait, and it needs to be extended more broadly across new realms of thought and then embedded more deeply in our design than ever before. Think of the connections as forming an interactive system. Connect with the people, too: listen to them.

4. Apply emotional intelligence. Know how people behave in the culture. Provide settings that make them feel safe, even when they are secluded. Provide settings that enable and even encourage informal, random, non-structured and non-goal-oriented interactions among strangers, among all kinds of people. For example, provide an unusual installation that encourages two parties who have never met to start a conversation about it.

5. Go boldly where no designer has gone before. There is a saying in the West: "Where there is no vision, the people perish". Having listened to the client and users and solicited expert advice, take charge. Dare to be creative in conceiving a vision, or Big Idea. Sometimes, today's reject will become tomorrow's icon. But not always. Today's "assault on good taste" may someday be redeemed as prescient, and today's award winners may someday quietly fall out of favor.

6. Do no harm. This phrase is commonly associated with the Western Hippocratic Oath taken by medical doctors. By all means, do no harm by way of bad intentions. Preserve the good that has survived, restore what has degenerated, and regenerate what has been lost. On the other hand, the fear of doing harm can lead to timidity, that is, a lack of will to do the right thing. This caveat is gaining new life to help save a fragile planet.

7. Be careful what you wish for. As has been so often said in the West, "Hell is paved with good intentions" as well as "Measure twice, cut once". Do something, don't be too timid to act, but be careful about it. If we wish to avoid any chance of doing harm out of good intentions, we would do nothing. Avoid common pitfalls, like rushing too fast. But being careful helps avoid wasting resources and doing harm.

8. Don't wait forever. There is yet another saying in the West, "The perfect is the enemy of the good". Design is iterative, and resources like time and money are finite. Move ahead with implementation when the very good is within grasp. Let that example inspire the perfect somewhere else.

9. Know when to let ego go. Give credit where credit is due. Avoid issues that would cause others to lose face. Lead by being generous. Thank everyone.

10. Be authentic. Be intellectually honest. Faking it is bad in design as in life. Pass on pastiche and kick kitsch.

11. Don't waste space. Space is precious. Make a small space seem bigger than it is. China knows.

12. Go with the flow. Design for movement along "desire lines", that is, the route people naturally want to take across a space. Grace of movement uplifts the spirit. Choreograph the design.

13. Let things grow. Be patient. Work with time; think for the long-term, not always about instant gratification. The larger the trees at planting, the shorter their duration.

14. Do not build what you cannot maintain. Life-cycle costing includes calculating the total operating costs over a projected useful life and subsequent replacement cost. A design made without considering operating, maintenance, and replacement costs is irresponsible. Responsible design requires a life-cycle plan. In the end, planning and design still need to be one integrated process that looks ahead to efficient, responsible and responsive operation and maintenance, and replacement after a long useful life.

15. Persist, for persistence will be rewarded. One person can make a difference. It may take a strong advocate to win a strong design. It is often impossible for strong design to emerge from a team or a committee where participants abrogate their responsibility to defend their beliefs persistently. Perhaps this book can inspire the reader to be that person who can make a difference.

■ **Urban Public Open Space Design Principles**

We turn now to principles of excellent design for urban public open space in general. Each principle is in the form of a declarative sentence; but the meaning is the same as saying "should".

1. Excellent design begins with considering environmental sustainability best practices.

Environmental sustainability is where "Do No Harm" truly begins. Urban public open space design is evolving into a broad awareness of the need to provide environmental benefits or services as the baseline underlying all design intended to accommodate human activity (see the Maslow framework in chapter 2). Because it is so important, we

10. 真诚相待

要在思想上坦诚。在设计中欺骗和在生活中欺骗一样的可耻，拒绝仿制设计，拒绝拙劣设计。

11. 节省空间

空间很宝贵。让一个小空间看起来更大一点。在中国做设计更应知道这一点。

12. 人性化动线

对人的流动性的设计应该随着人们的需求线而变化，这个需求线就是人们在穿过一片场地时自然而然选择的路线。美丽的动线更具吸引力，并为设计加分。

13. 顺其自然

要有耐心。让时间做主，放远眼光，而不要总考虑眼前的利益。成果显现的越早，它持续的时间也就越短。

14. 低成本维护

产品生命周期成本包括计算总的操作成本、预计使用寿命和随后的更换成本。不考虑操作、维护以及置换成本的设计是不负责任的。负责任的做法是有整个生命周期的维护计划。总结起来就是：规划与设计需要成为整体过程，这个过程可以提前安排有效、负责的管理与维护以及长期使用后的更换。

15. 坚持就有回报

一个人是可以有所作为的，一个好的设计想法往往需要强大而坚定的支持。一个不能坚持自己想法的团队是不可能诞生好的设计的，希望这本书可以启发那些愿意有所作为的读者。

■ **城市公共开放空间设计准则**

现在我们来谈下优秀城市公共空间设计的准则。虽然每个准则都是普通的叙述，但它们无异于祈使或强调，意味着我们应该这样做。

1. 考虑可持续发展是开始一个优秀设计的最佳途径

真正做到"避免破坏"是从环境的可持续性开始。设计师普遍意识到，现代城市公共开放空间需要以环境利益或服务为基础，在设计之初能容纳人们活动的需求（请见第二章的马斯洛框架）。因为它是那么的重要，我们在这里再重复一遍，如果现在没有尽可能保护到已有的自然环境以及野生生物保护区的话，那就要还原或者修复已被破坏的环境（如土壤、湿地、滨水缓冲带）和再造所失去的环境。设计应该结合透水铺装，场地排水，渗透与灰水（家用冲澡、洗碗、洗衣之后的水）净化措施，地下水储存和生态调节系统，不需要外来水源的垂直绿化和屋顶绿化，可提高生物多样性并且不需要灌溉就能成活的乡土或非乡土植物材料。我们再次强调，要优先使用当地的材料资源、可循环材料、快速再生的材料（例如竹子），而不是那些非本地、稀有的、不可再生的或者濒临灭绝的材料，并且减少在施工期间产生的废料。通过增加地面绿植面积或者屋顶绿化来减少铺装面积，而铺装应选择反射率高的材料，这样能帮助减少城市热岛效应（特别是深色的表面会吸收更多的热量）。我们强烈建议使用节能设计和装置，避免过度使用混凝土，而选择太阳能照明或者抽水，避免过度的工程上的照明（光污染），优先选择LED灯，而不是高耗能材料，还有（也许是最古老的可持续的做法了）利用建筑阴影在夏天降低能量消耗。在我们建造过程中使用的能耗越少越好，越是依靠太阳能或者风能就越好。越少破坏自然环境，空间越安全，功能性越强越好。

阳光不仅可用于光伏电板。一些有远见的城市通过控制阳光来控制与其相邻的场所。比如在城市公用空间制造阴影区，或减少对阳光的直接获取来抑制植物的生长。

关于可持续场地设计的最新观点是由可持续场地计划（SITES ™，由美国景观设计师协会、约翰逊总统夫人野生花卉中心及美国植物园赞助）发布，适用于所有公共空间设计。这些设计观点在之后被能源与环境设计及绿色建筑评定系统（LEED®）所效仿，具体评定方法见附录表格。一个项目能够使自然以及建造体系融为一体，在不用透支下一代资源的情况下满足当今的需求，那它就被定义为可持续的。可持续设计可让自然利益和自然调节能力恢复，就像在未被破坏的生态系统中一样，并且还将可持续原则应用到未来的维护和管理中。

为了给公众提供可持续服务信息，城市公共空间设计需要包括引导标志、宣传册和相关信息网站链接。

最好的环境可持续策略需要与景观设计自然融合，例如排水和公用系统（管道和电线）。长期以后，人们将不再把可持续性当作一个设计的中心任务，转而变为最基本的需求，从而允许设计专注于满足人们一系列其他的需求（图4.1 - 图4.2）。

2. 优质的设计呈现美感

美学是一种生命力。精致的形式、材料、模式、质地、颜色、灯光、气味和声音都是能深深地打动我们的因素，并且都能为我们的生活增添色彩。城市公共空间应该很漂亮。如果这种言论听起来虚无，那我们反过来讲，即我们要避免丑陋的无聊的城市公

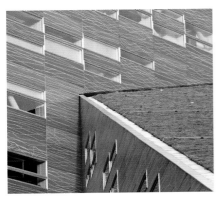

图4.1 - 图4.2

可持续性是在世界的大部分地方的新的最基础标准，需要我们永不抑制设计的创造力，例如屋顶绿化如何启发垂直绿化，或者太阳能电池如何增强公园亭子顶部的坚固性和光的扩散。
Sustainability is the new minimum standard in much of the world, and need not inhibit design creativity, for example in how green roofs inspire green walls, or photovoltaic cells add robustness and light diffusion to a park pavilion roof. (Credits: Rudmer Zwerver and marchcattle)

repeat here that design must restore if not preserve as much of the surviving natural environment and wildlife habitat as possible, restore or rehabilitate what has been damaged (such as soils, wetlands and waterfront buffers), and regenerate what has been lost. Design should incorporate permeable paving, on-site drainage, infiltration and grey-water purification, underground water retention, bio-swales, green wall and green roof irrigation that does not use off-site water, and native or non-invasive plant species that promote biodiversity and do not absolutely require irrigation to survive. We stress again the preference for locally or regionally sourced materials, recycled materials, rapidly renewable materials (bamboo for example), rather than non-local, exotic, non-renewable, rare or endangered materials. Construction practices should focus on reduced generation of waste. Paving areas should be reduced by increasing the area that is planted on the ground or roof, and the paving materials for the reduced paving should be reflective (high albedo) to reduce the urban "heat island effect" (the excess generation of heat by horizontal surfaces, especially those in dark colors). And we strongly recommend the use of energy-efficient design and products, whether the avoidance of the overuse of concrete or the use of solar-powered lighting and pumping, the avoidance of over-engineering of lighting (light pollution), the preference of LED over more energy-inefficient light sources, and (perhaps the oldest sustainability practice of all) the shading of buildings to reduce energy consumption in summer. The less non-renewable energy that is used in the materials and construction of what we build, the better. The more reliance on solar and wind energy, the better. The less structurally invasive practices to provide a safe and serviceable environment to move around in, the better.

Solar access is not just important for photovoltaic cells. Some visionary cities are also controlling adjacent sites to prevent any future development from casting a shadow in urban public open space or decreasing the access to direct sunlight, which would inhibit the healthy growth of plants.

The leading expression of the latest thinking in sustainable site design—which applies to all urban public open space—comes from the Sustainable Sites Initiative ™ (SITES ™) cosponsored by the American Society of Landscape Architects (ASLA), the Lady Bird Johnson Wildflower Center, and the United States Botanic Garden (see www.sustainablesites.org). It is modeled after the Leadership in Energy and Environmental Design Green Building Rating System ™ (LEED®). See the Appendix for the complete list of possible points that can determine a rating. Land practices are defined as sustainable if they enable natural and built systems to work together to meet the needs of the present generation without compromising the ability of future generations to meet their own needs. Sustainable design has the potential to regenerate the natural benefits and services provided by ecosystems in their undamaged state, anticipating sustainable principles in maintenance and management.

To educate the public about the sustainable services being

provided, urban public open spaces should include signage, provide handouts, and be linked to a website providing updated information.

Environmental sustainability best practices need to be carefully engineered and integrated into the landscape as unobtrusively as drainage and utility systems (pipes and wires). The day will come when design does not flaunt sustainability as revolutionary and central to its mission. Sustainable best practices will then simply be expected as the minimum requirement, allowing design to focus on accommodation of the range of human needs. (4.1-4.2)

2. Excellent design achieves aesthetic excellence.

Beauty is a life-force. The exquisite use of forms, materials, patterns, textures, color, light, fragrance and sound have the power to move us profoundly, and make our lives better. Urban public open space should be beautiful. If that statement seems too lofty, consider the converse: urban public open space should avoid being ugly or boring. And yet that is surely the result that comes from designer arrogance, timidity or incompetence, or providing facilities in a random, haphazard, careless or shoddy way. Intense beauty is not easily achieved, to be sure. Nevertheless, beware the temptation to copy high aesthetic achievement found elsewhere. Imitation hardly leads to a feeling of beauty each time a new copy appears; sooner or later the next Campidoglio Square or Lotus Blossom Plaza becomes a boring joke, a cliché. (4.3-4.4)

3. Planting is lush, varied, full of seasonal interest, and well maintained.

Apart from areas preserving native plant communities (wetland parks are an example), or paved areas, urban public open space should typically display ornamental plant materials in profusion. A thinly planted space cannot offset the visual intrusion of the hard, built-up world on all sides, much less the immediacy of hard infrastructure underground, at ground level or overhead which threatens to subvert the Nature that people seek there. However, buffer planting solely of one species is boring and also at high risk of total loss due to pest infestation. Diversity of species is safer, more visually appealing, more natural, and more attractive to wildlife as well. The site with the larger area, greater variety of land and water character, or greater connectivity will usually have greater biodiversity.

Native species are to be preferred to imported species. Indeed, Yunnan Province is blessed with an unsurpassed concentration of native plant species. Although currently only a small fraction of the many native plant species are typically planted in urban open space in China, in the future more diversity will become feasible as more species are offered in the nursery market. Tolerance of urban conditions is important, but plantings in parks have a better chance of survival of urban conditions than do street trees.

Planting design must include consideration of the canopy tree level, the understory tree level, the shrub level, and the groundcover

图 4.3 - 图 4.4

美流淌在从纯自然环境到纯的几何形态的设计范围中。多层次的形式和复杂的感觉可以产生出最震撼有力的美。美感从不是城市公共开放空间设计中一个奢侈的目标。
Beauty resides across the spectrum of design from the purely natural environment (as in Golden Gate Park, San Francisco) to the purely geometric (as in the Tuileries Garden, Paris), and emerges most powerfully when the layering of forms and senses is complex. Beauty is never an expendable goal in urban public open space design. (Credits: Brocken Inaglory and Kimberly Vardeman)

共开放空间。造成不合格空间的原因很多，如设计师自大、胆怯、能力不够，或者随意、无序地安放设施。当然，绝对的美是很难达到的，但抄袭成功之作是毫无意义的。每一个仿制品都很难达到那种原有的美感。下一个坎皮多里奥广场或者莲花广场迟早会被嘲笑，并渐渐被淡忘（图 4.3 - 图 4.4）。

3. 保证植物茂盛，多样化，四季分明，易于维护

且不说保护本土植物群落（例如湿地公园）的场所或者铺装场地，城市公共空间应该具有代表性地展示丰富的观赏植物。一个贫瘠的种植空间不能缓冲建筑边界在视觉上带来的强硬感，更不用说地下、地面或者上空可能威胁人类追求的自然形态的结构。然而种有单一植物的缓冲地带看起来会很无聊，也容易产生病虫害。种类多样化更安全、更吸引人、更自然，同样对野生生物更有吸引力。更大的场地、更丰富的土地和水体因素以及更大的连通性通常更利于生物多样性发展。

本土树种应该优先被考虑。事实上，云南省自然资源丰富，那里本土树种种类之多无可匹敌。尽管现在中国只有少部分典型的本土植物运用到城市开放空间里，但是以后，更多的种类将会

图 4.5 - 图 4.6

出色的种植设计要求对乔木、灌木、多年生花卉、一年生花卉以及地被进行富有想象力的选择与排列，并且有持续的养护。这里表现了英国格罗斯特郡维斯通伯特植物园和杭州西湖的游客们对于色叶植物的喜爱。
Planting design excellence requires imaginative choices for large trees as well as shrubs, perennials, annuals and ground covers, imaginative juxtapositions, and constant maintenance. Here visitors admire the colorful display of foliage at Westonbirt Arboretum, Gloucestershire, Britain, and West Lake, Hangzhou. (Credits: Kimberly Vardeman and Thomas M. Paine)

在苗圃市场上供应。虽然土地的耐受性很重要，但公园里的树种会比街道树种更易成活。

种植设计必须包括树冠层、次冠层、灌木层、地被层，为达到这个目的应该使用落叶树、常绿树、灌木、地被植物以及多年生花卉组成的花带。树冠层的区域供人们聚会，席地而坐，这里需要常绿和落叶树木搭配，夏季可提供荫凉，冬季可引入阳光。从乔木到灌木，从多年生到一年生，设计应该细化到按照生长季节来轮栽每个生长季的花卉，同时花期临近的植物在颜色搭配上要和谐，或者选择单一色调（例如一个全红色的花园）。外部空间的绿篱结合大胆的色彩搭配和图案组合，丰富了设计的视觉效果。在可视性高的空间边缘，修剪整齐的绿篱和植物营造出另一种美丽的画面。中国高速公路两旁被修剪完好的植物同样适用于城市公共空间设计，但更适于行人而非驾车车主。

除了在干旱地带，草坪在种植设计中是很重要的元素。在适

level, using a mixture of deciduous and evergreen trees, shrubs, and ground-covers, and perennial and annual flowers for flowerbeds. At the canopy level in areas where people congregate and sit, a mixture of deciduous and evergreen trees should be selected to provide shade in summer and more sunlight in winter. From trees to shrubs to perennials and annuals, design should include a succession of blooms during each month of the growing season, with the blossom and foliage colors of adjacent plantings selected for color harmony or in some cases monochromatic effect (such as an all-red garden). Green walls that enclose outdoor space can significantly enrich the design with bold color and texture effects. Edges in highly visible locations may benefit from clipped hedges and pruned specimen plants. The rich effects using shaped plantings along urban highways in China are equally applicable in public urban open space, but at a scale and complexity suitable to pedestrians rather than drivers.

Except in arid climates, lawn areas are an important feature in the planting plan. In climates where grass grows vigorously, the public should be allowed to walk and sit on a designated lawn area. While in China many parks currently forbid this, there are exceptions, such as the Shanghai Botanical Garden, where people clearly love the chance to sit on the grass. Norman B. Leventhal Park in downtown Boston opens its tiny lawn to visitors who are permitted to borrow a cushion and sit as long as they want. However, the lawn is closed in the early spring when the grass is emerging from winter dormancy and reestablishing vigorous turf. What is required is adequate maintenance and irrigation. (4.5-4.6)

4. Excellent design makes positive connections on many levels, while avoiding unwelcome intrusions, to meet basic human needs.

Excellent design is intelligent design, connecting to people not only intellectually but also emotionally and spiritually. Excellent design transcends the banal and aloof to deliver a multisensory immersion experience connecting lush nature, meaningful art and smart technology. Such immersion has the power to relax, entertain, inspire, and heal. To deny the people access to this is condescending and timid. At the very least, excellent design should "fill up our senses", to borrow singer John Denver's memorable phrase. If the world is entering what has been called the Conceptual Age, in which the economy is increasingly centered on creativity, culture, and customized services instead of physical and mental rote work, then excellent design should do no less. Indeed, Steve Jobs said, "creativity is just connecting things",[5] synthesizing old things in new ways, in new cultural contexts, for new services, in a new life. Unable to anticipate where inspiration will come from, the creative designer is willing to look everywhere. Creative connection requires an open and courageous mind that thinks "outside the box" and rejects conformity and imitation.

On the other hand, harmful things should not be connected. Intrusions that needlessly undermine the ability of urban public open space to meet basic human needs should be avoided. Just as water pollution is unacceptable, so too are noise and light pollution. On energy conservation grounds alone, noise and light pollution are unjustified, but they also undermine social sustainability; they are disruptive and can even cause health problems. People need quiet and nighttime darkness. Gratuitous noise and light above a minimal level are an imposition. Loud speakers should be used only for emergencies. Commercial audio messages are unacceptable. Light fixtures must be selected to light only the immediate ground area, and by all means minimize brightening the night sky or causing light trespass onto any abutters' property. A lit sign for commercial purposes is an even more unjustified hardship imposed on people. Strobe lighting or lighting that expresses movement with rapidly changing colors and patterns can be particularly annoying to abutters, whose needs and concerns should not be ignored.

If light pollution undermines the quality of the experience of urban public open space, so too do the shadows of high-rise buildings which for too much of the day darken much of the outdoor space. Reduced solar access not only inhibits plant growth but also degrades the experience of urban public open space. The need for shade from the sun is best accommodated by providing modest pavilions, not by allowing high rise buildings to blanket a huge outdoor area in its shadow.

Overarching all these environmental concerns is the people's basic need for access to clean air to breathe when they are outdoors, especially while they are getting exercise and breathing deeply. Progress in reducing air pollution from factories and power plants, particularly the less visible "$PM_{2.5}$" particles, will ideally go hand in hand with increased access to outdoor space in China as it has in many other countries whose air was once equally polluted. (4.7)

5. Enriching urban life with content-rich design is integral to the mission of urban public open space.

Open public open spaces are not only firmly rooted in the soil by planting native plants, but also firmly rooted in their place by planting native ideas, or content. The definition of sustainability is expanding to include social sustainability. At the heart of sustaining a community is the goal of providing the setting for a shared appreciation and understanding of local history and culture. Elitist designers who dismiss content-rich design as decadent or possibly dangerous, as if the space should not be contaminated or cluttered, are denying people access to enriched urban life, including what landscape architect James Corner once called "the associative play of narrative references".

Unfortunately, designers are sometimes tempted to make connections to worldwide trends rather than make local references. For example, the still relatively novel "QR" (quick response) code—black modules randomly arranged in a square—looks like a wonderful idea for a random paving pattern, as does the circuitry in a computer chip. Indeed, architects have already "quoted" these motifs in recent work

合草本植物生长的气候下，草坪应该允许公众在上面踩踏或休憩。但是在中国，现在很多公园禁止踏入草坪，也有例外，比如上海植物园的草坪就允许人们进入，人们会很享受坐在草坪上。波士顿的邮局广场公园有一个小草坪，人们可以借一个坐垫，在那里想待多久就待多久。但是，草坪从冬天休眠期到早春刚发芽的时候是封闭的。足够的养护以及灌溉也是必需的（图 4.5 - 图 4.6）。

4. 优秀的设计能够有效合理地连接各个层面的元素，规避不和谐元素，以迎合人类的基本需求

优秀的设计是充满智慧的设计，在情感与智慧层面都相当吸引人。好的设计超越了陈腐与单调，传送了一种混合着繁茂自然的、有寓意的艺术和智能科技的多重感觉。这种感觉有着放松、娱乐、激发以及治愈人们心灵的力量。否认人们接近这一切则是盲目和畏缩的。一个好的设计，借用约翰·丹佛的名言，应该是"填满我们的各种感觉"。如果说现在我们正在进入一种所谓的概念时代，这个时代的经济增长集中在创新、文化和个性化服务上，而不是体力活和死记硬背的工作上，好的设计也应该如此。事实上，史蒂夫·乔布斯已说过，"创造力就是把各种事物连接起来。"[5] 在新的生活下为了新的服务，在新文化背景下用新手法整合旧事物。有创造力的设计师是无法预知他们的灵感来源的，因为他们时刻准备观察周边一切。有创造力的连接需要一个开放的、大胆的头脑，那就是冲破束缚，拒绝相似和模仿。

另一方面，应摒除不利元素。为了满足人的基本需求，而去对城市公共空间进行不必要的破坏是应该避免的。就像水污染是不能被接受的，噪声污染以及光污染同样不受欢迎。从能量守恒单方面看，声音与光污染是不科学的，但他们更破坏了社会的可持续性，他们的破坏性甚至可能给人类造成健康问题。人们需要安静而避光的夜晚。迫使人们无理由受到噪声污染和光污染是不合理的。扩音器应该只在紧急情况下使用。商业广播也是不受人欢迎的。灯具仅限安装在需要照明的地方，而且应该尽可能地减少夜晚照明，同时尽可能避免打扰邻居休息。以商业为目的的照明标识对人们来说是更加不合理的事。对附近居民来说，闪光灯或者那些颜色和图案快速变化的照明是最让人烦扰的，附近居民的诉求和担心是不应该被忽略的。

如果光污染破坏了城市公共开放空间的质量，同样，拔地而起的高楼带来的阴影也让白天的户外空间显得阴暗。减少阳光的直接照射不仅妨碍植物的生长，同时也破坏了城市公共空间的户外体验。我们需要的是可以适时遮阳的凉亭，而不是高耸建筑的巨大倒影遮挡住大部分的户外空间。

在以上所有的环境诉求中，首要的是人们对户外新鲜空气的诉求，尤其是当人们锻炼和深呼吸的时候。减少工厂和发电站产生的空气污染，尤其是不可见颗粒 $PM_{2.5}$，会一步步地增加中国人与户外空间的接触，就像其他的曾经经历空气污染的国家一样（图 4.7）。

5. 用丰富的设计丰富城市生活是城市公共开放空间的使命

公共开放空间不仅通过种植本土植物根植于土壤，也通过注入本土思想和文化根植于地方。可持续性的定义已经延伸到社会的可持续性。可持续性社区的核心在于，提供给人们一种可以理

图 4.7

把每个层面都积极地连接起来可以在放松和沉思中调动所有的感官以及大脑。这张图片是在波士顿的诺曼·利文萨尔公园里音乐家热情招待中午来访的游客们。
Making positive connections on all levels can involve all the senses, and the mind, in a state of relaxation and contemplation. Here musicians entertain noonday visitors of Norman B. Leventhal Park in Boston. (Credit: Thomas M. Paine)

解并感恩当地历史和文化的环境。部分优秀的设计师不赞同富有文化特色的设计,他们认为这样的设计是甚至是危险的,他们坚持空间不应该被"污染"或者随意分割。这些设计师否认了人们的可进入性可以丰富城市生活,同时也否认了景观建筑师詹姆斯·科纳的理论"设计可唤起相关联的感觉"。

不幸的是,设计师有时更倾向于跟随世界新的潮流,而忽略本地文化。例如,新潮的二维码——黑色的小方块随机地排布在一个正方形中——看起来是铺装的一个绝佳想法,电脑芯片也是如此。事实上,建筑师已经在迪拜和第戎引用过这些图案。新鲜感会很快消退,意义也随之消失。我们应该经常问问自己,一个独特的设计和一个独特的地方之间有怎样的关联?这才是应该被经常考虑的问题。

认识一座城市,就从了解当地的文化开始,而且,这也是用来区分与其他城市不同最有效的方式。庆祝当地的文化与在城市中心进行的国家庆典不同。在过去的世纪里,庆祝当地文化包括给当地英雄人物塑像或安装有历史意义的标识或者奖章,这些标识、奖章通常记录了历史故事以及历史画面。在一些有大墙面的地方,富有想象力的华美的壁画或者浮雕则记录了当地的历史和文化。很多城市的公园里,有些园路以诗人或者哲学家的名字命名,用来纪念这座公园给前辈作家和思想家带来的灵感。这些庆祝方式都是不错的选择,但是他们绝不是唯一的选择。重要的不是数量,而是这种庆祝文化的意义及文化故事的力量。例如,绍兴所需要的就是向世界讲述鲁迅笔下的故事,因为它有如此多文化上的建树值得赞颂。每个地域都有比史册更多的文化以及"英雄故事"值得庆祝。还有好多激动人心的工作要去做!不久,人们对优秀城市开放空间的需求将包括对本地文化的体验。

公共开放空间在现今越来越多地被看做是一个能容纳艺术和历史文物的户外空间。显然,户外的游人数量超过了参观博物馆的人数,也许是因为博物馆是收费的。尽管一些形式的雕塑可

in Dubai and Dijon. The novelty will soon wear off, and the meaning lost. What is the relevance of a particular design for a particular place? That is the question that should always be asked.

Communicating local culture is a powerful way of differentiating a city from all others. The celebration of local culture is distinct from the celebration of national identity which is usually on display in the city center. In past eras celebrating local culture has meant including the obligatory statue of a local hero or heroes, or installing historic markers or plaques that usually inscribe or "tag" the space with text recounting history and sometimes display historic images. In other spaces with large wall surfaces, imaginative and colorful murals or wall sculptures have celebrated local history and culture. In a few cities a park pathway earned the name Poet's Walk or Philosopher's Walk celebrating how parks inspired former generations of writers and thinkers. These remain good choices, but they are hardly the only ones. It is not the quantity that matters, but the means by which the culture is celebrated, and the power of the story. For example, all Shaoxing needs is Luxun to tell a powerful story to the world, and yet it has so much more cultural greatness to celebrate. Each community has more culture and more local heroes awaiting celebration than local officials have yet considered honoring. There is much exciting work to be done! Indeed, before long the case will be made that the people's need for access to excellent urban open space includes the need for access to cultural meaning and the experience of *communitas*.

Public open space is increasingly being seen as the proper place to bring art and historical relics outdoors. Clearly, the audience outdoors is vastly enlarged over those who decide to enter a museum, perhaps having to pay an admission fee. Although many forms of sculpture can survive outdoors, not all can. For them, as well as for other art forms and historical relics that cannot practically be placed outdoors, urban public open space can showcase images or replicas of them. While any type of open space can benefit from such cultural content, the main focus for it is typically the most intensely used spaces such as Civic Plazas and Downtown Parks. (4.8-4.9)

Excellent Design and Big Ideas

Excellent design makes connections of all kinds, not only in the obvious ways of connecting place to place and connecting visitors to each other, but more ambitiously and profoundly, connecting people to ideas. Excellent design celebrates big ideas that may transcend local culture, but in a locally distinctive way. The ideas are not ideas about design or materials per se, but through design and materials the space celebrates big ideas that can inspire visitors to be better people.

Excellent design does not settle for small ideas, it makes connections between big ideas, East and West, North and South, secular and spiritual, ancient and new. Big ideas in one way or another celebrate the better realms of our shared experience—justice, peace, and of course community. They might also exalt cherished gifts like redemption, transformation, metamorphosis, ecstasy, serenity. The appeal of these gifts is as timeless and universal as the mythic traditions that celebrate

图 4.8 - 图 4.9

深入呼应当地环境设计庆祝了当地的文化，甚至当地的历史故事，就像我们在绍兴纪念鲁迅，或者在悉尼的情人港的环境的展示。
Content-rich design celebrates local culture and even local stories, as in the case of celebrity author Luxun in Shaoxing or an environmental exhibition at Darling Harbor in Sydney.
(Credits: Thomas M. Paine)

them. Lastly, they might simply exult in a new understanding of the wonder of the world—its hidden harmonies, its synergies, its sublime order transcending chaos. Perhaps this gift transcends all others. These realms and cherished gifts are the grand themes overarching what urban public open spaces could celebrate, each in their own distinctive and distinguished way.

Excellent design celebrates big ideas suggestively rather than stridently like billboards trying to sell a message. Rather than adding more static statues of dead people, the idea is expressed in design forms and materials, and art installations. And in that richly textured network of deeply "planted" content, users of diverse backgrounds will find something that they can connect with even in a world where the shared lore of symbols and cultural meaning has eroded and become trivialized in slogans, and the usual so-called big ideas have been reduced to the subject of special-effects performances developed by Disney imagineers, like "It's a Small World After All." Truly big ideas are not the themes of Theme Parks.

6. The allocation of space equitably accommodates a wide variety of activities.

The five space types discussed in this book—Civic Plazas, Downtown Parks, Large Parks, Greenways and Neighborhood Parks—form a system and provide settings for a range of social units, from crowds, teams, interest groups, neighbors, to families, couples, and solitude. These terms suggest the basic hierarchy of human needs that we saw in Chapter 2. They are worth keeping in mind even as we focus on specific activities that should be accommodated. Those activities comprise no short list. Not all of them are in demand in all cultures. The demand for some activities may be changing as new ideas are imported. Local authorities may perhaps survey the local population to assess evolving preferences. The process of deciding what activities to accommodate in a particular project, which activities should be closely

以展示在户外空间，但是不是全部的雕塑。对于那些不适合在户外展示的雕塑以及其他艺术形式和历史文物，我们可以展示图片和复制品。尽管任何形式的开放空间都可以在类似的文化产物中获益，最适合做展示活动的开放空间形式是使用率最高的城市广场和市区公园（图4.8 - 图4.9）。

优秀的设计与奇妙的想法

优秀的设计建立了各种各样的关联，不仅是明显的场地与场地之间的关联、游客与游客之间的关联，更强大且更深刻的是，它建立了人与不同想法之间的关联。一项优秀的设计用本土独特的方式表现出可能超越了当地文化的奇妙想法。这些想法不是关于设计或者材料本身，而是通过设计和材料的应用来反映出不同的理念，这些理念可以鼓励人们做得更好。

优秀的设计不满足于平庸的想法，它更联系着各种不同的奇妙想法，从东至西，从南至北，从世俗至精神，从古至今。伟大的想法用这样或那样的方式庆祝我们共同经历的领域——正义、和平，更不用说我们的社会。奇妙的想法也许还颂扬了更珍贵的品质，比如救赎、改变、蜕变、极乐以及宁静。这些品质的吸引力是普遍而且永恒的，就像那些颂扬他们的神秘传统一样。奇妙的想法也许还在不经意间对这个世界做出了新的理解——比如它潜在的和谐，它的协同，它超越混沌的升华，或者某种超越了所有的品质。这些领域和品质，每一个都有自己独特和杰出的方式，亦是包罗万象的城市公共开放空间可以庆祝的宏大的主题。

优秀的设计理念循循善诱，而不是像广告试图出售消息一样刺耳。不同于给那些逝去的人塑立雕像，好的理念存在于设计形式、材料表达以及装置艺术。在"植入"内容丰富的社交活动里，不同背景的使用者总会找到一些可以相关联的信息，哪怕世界符号和文化意义已然被侵蚀和贬低为毫无意义的标语口号，并且那些所谓的奇妙想法已经被降低到迪士尼幻想工程师所开发的特技游艺，比如"小小世界"这个游乐项目。而真正的奇妙想法并不是那些主题公园的"主题"。

6. 为多种活动合理地分配空间

本书中讨论的五种空间类型，城市广场、市区公园、大型公园、绿道以及社区公园形成一个系统，并且为不同的社区团体提供所需的环境，这些团体包括：群体、团队、团体、邻里、家庭、夫妻以及独居者，这些团体类型反映出在第二章中我们提到的基本层级。这几种空间类型很重要，即便是我们专注于适于某些活动的一种空间时，也应该将其全部熟记于心。这些活动各种各样。在所有的文化背景中，并不是所有的活动都被需要，对某些活动的需求可能会随着新理念的植入而变化。决定哪些活动对应哪个特定的项目，哪些活动应该紧密相连，哪些活动应该分开，这一决定过程叫规划。供人类使用的设计始于规划，而规划的开端就是决定哪些活动更适合特定的公共空间。**城市公共开放空间活动表格**（表4.1）包括了从"积极的"或有体力需求的到"消极的"或不费力的许多活动。所有的公共空间类型都应该包含积极和消极两种活动类型。我们这个表格很难彻底地叙述所有的活动，尽管上面按照季节、时间和年龄分类罗列了超过100种的活动。我们将人类活动周期划分成六个年龄组（婴儿、1-4岁的儿童、5-11岁的儿童、12-15岁的少年、16-35岁的成年、36-60岁的中年以及61岁及以上的老年），每个年龄段特定的技能、社交圈及体力都不同。有些设施鼓励使用者相互配合（如给孩子玩的沙坑和喷泉，给成人的花园）；有些甚至可以允许残障人士参与到集体活动中；6 毕竟，人的整个生命周期都需要对健康的挑战和把握。

当然，人们的需求是多样的，在任何时间、任何分区、任何类型的开放空间里，公园最好的提神功效恐怕就是：让不同年龄段、不同职业的人们可以有空间静静地思考和反思，幽静的空间和人群聚集的空间一样可以促进社会的可持续发展。

城市公共空间使用权，男女平等

在运动方面，男女平等一直是西方国家关注的焦点。但这一点还未在关于平等使用城市公共空间的例行讨论中出现。可以肯定的是，在不久的将来，男女平等必会提上城市公共空间设计的议程，并促使城市公共空间的设计者们通过不断的试验来研究如何为男性和女性分配空间才最为合理。这假设了在大多数文化背景下，运动队的组织者不顾对性别歧视的禁止，一味地要求团队性别单一，这种为了进一步解析"人"而将人细分成多组的做法已经在一定程度上发展到了十分荒唐的程度，但是为了达到性别平等仔细思考男人和女人对于公共空间的不同需求还是有一定道理的。考虑到目前对此研究的缺乏，城市公共开放空间活动表决定在任何活动上都不按性别分组。

城市公共开放空间活动表格显示出：哪些活动在之前提到的那五种典型的空间（①城市广场、②市区公园、③大型公园、④绿道以及⑤社区公园）中是合适的。例如，高尔夫、宠物公园、滑板或者植物识别就在①和②中是不合适的。常规演出在⑤中是不合适的等等。设计师要根据场地特点调配项目需求，然后评估每一种活动的空间需求，比较空间可达性，再决定适合每种活动的空间尺度，判定是否有足够的空间给整个项目并且满足所有活动的需求，合理的分配空间。最重要的原则是：不管多少种活动适应这一设计，一定有其他的很多活动因为不适合而被排除。所以当地政府和设计师肩负的首要责任是了解社区居民的最大需

linked, and which ones separated, is called programming. Design for human use begins with programming, and programming begins with determining which activities are appropriate for a particular open space type. The **Urban Public Open Space Activity Table** (Table 4.1) includes activities ranging from "active" or physically demanding, to "passive" or less strenuous. All open space types should accommodate both active and passive recreation. Although our list is hardly exhaustive, over one hundred activities are included, classified by seasonal suitability, night-time suitability, and age-group suitability. We divide the human life cycle into six age-groups (infants and toddlers 1 to 4 years old; children 5 to 11 years old; pre-teen/early teens 12 to 15 years old; young adults 16 to 35 years old; middle-aged adults 36 to 60 years old; and older adults over 60 years old). Each age group has its own level of skill proficiency, socialization, and physical stamina. Some facilities could encourage user manipulation and interactivity (sand play, fountains for the young, gardening for adults); others could allow wheelchair users to engage in team sports; indeed, "opportunities for healthy challenge and mastery are needed across the life cycle" [6] and for all skill proficiency levels.

Of course, people are multi-taskers, and at any time, in any subarea of any open space type, people of any age or any skill proficiency level can lose themselves in contemplative introspection, and that may be the park's greatest refreshment of all. Including space for solitude as well as space for crowds best enhances social sustainability.

Gender Equity

Gender equity in access to sports is an issue in Western countries. While the issue has yet to emerge in the conventional discourse on the equitable use of urban public open space, surely in the not too distant future the issue of gender equity will emerge and cause planners and programmers of urban public open space to examine more closely how best to allocate space for equitable participation by women and men. This assumes that sports teams in most cultures are limited to one gender, regardless of prohibitions on gender discrimination by team organizers. Extending this kind of thinking to further parsing of the "people" into ever more specific subgroups at some point reaches an absurd level, but surely thinking about the varied open space needs of men and women in the interest of achieving the goal of gender equity is not going too far. Given the current dearth of research on the subject, The Urban Public Open Space Activity Table does not assign gender to any activity.

The Urban Public Open Space Activity Table suggests which activities are appropriate in each of the five space types discussed in this book: Civic Plazas, Downtown Parks, Large Parks, Greenways and Neighborhood Parks. For example, golf, dog parks, skateboarding or wildlife appreciation are inappropriate in a Civic Plaza or Downtown Park. Formal performances are inappropriate in Neighborhood Parks. The designer reconciles the program requirements for a specific location and then assesses the space required for each activity, compares it to space available, and decides how much space is appropriate to each, and whether there is even sufficient space for the entire program, given the demand for all activities, and the need to allocate limited space equitably. The fundamental principle to keep in mind is this: whatever user activities are accommodated in the design, it precludes a host of other user activities (the so-called opportunity cost) that could have occurred in that space instead, so it is a matter of great responsibility that falls on the shoulders of local officials and the designer to know what the community really needs the most. Instead, spaces should be flexible (permitting multiple use, changes at night, seasonal variation, and evolution over a generation) to accommodate a diversity of users. For example, a small Neighborhood Park should not be filled by a single tennis court. Given skate-boarding's wear and tear on park furniture, that activity is usually forbidden in paved plazas or restricted to a special location.

Adjacent activities should be mutually compatible. The Urban Public Open Space Activity Table is complicated enough without addressing compatibility between activities. Some activities do not belong next to each other; others can sit side by side with careful design; still others may actually benefit from mutual proximity. The goal should be to accommodate activities with minimal mutual interference. Obviously, noisy activities should not abut areas for passive recreational activities like quiet immersion in Nature. Typically, active sports are best located on the perimeter, on flat ground, away from water features or quiet areas within the center. Conflicts may also arise over the potential for stray balls or other objects. For example, archery should be located nowhere near a playground. The potential conflict of kite flying and field sports may not be so obvious. Mistakes can be avoided simply by drawing a matrix for all activities and evaluating each possible pairing for how compatible the one activity is with the other: either highly desirable to be adjacent, or relatively near, or else far apart. Obviously, this does not take into account possible larger groupings, threesomes or more, but the same logic applies. By going through this discipline, local authorities can both avoid making embarrassing mistakes and take advantage of opportunities for efficiency and synergy. Judgments of visual incompatibility will require the services of a good designer. Lastly, support facilities like maintenance equipment, supply storage or waste disposal should not adjoin public use areas. (4.10-4.31)

7. The main park entrance conveys a sense of friendly welcome.

The main entrance should welcome users while conveying dignity and beauty and setting the tone for what lies beyond. The entrance should clearly announce and identify the space, but avoid making too massive an architectural statement. Where applicable, the main entrance should be near a bus or rapid transit stop or a crosswalk. Some very large parks could benefit from creating an urban arrival experience at the gate with a visitors center, cafes, boutiques, and equipment rentals. (4.32)

求。相反，空间应该灵活（允许多种使用方式，根据夜晚与季节来做相应的变化，空间的发展演变在时间上可以跨越几代人），来满足用户的多样性。例如，小型公园不能单独设网球场；因为滑板运动会磨损公园的设施，铺地广场通常禁止这种活动或仅限于一块特定区域内进行。

相邻的活动应该相互兼容。城市公共开放空间的表格虽然很全面但是没有包括兼容性。有的活动与相邻活动不相关，有的活动可以通过精心的设计联系在一起，从彼此中受益。我们的原则是在满足活动需求的同时不互相干扰，显然，吵闹的活动区不应与自然安静的区域毗邻。通常情况下，运动场地最好选址在整体用地的周边地带而非中心，并且有平整的土地，还要远离水域和安静区。一些潜在的冲突是可能的，比如，箭术场地应该远离任何其他运动场地，而放风筝的场地和运动场地的潜在冲突就不那么明显。通过给所有的活动建立模型以及评估每一种可能的、互相匹配的活动可以简单地避免错误：要么完全接近，要么相对接近，要么远离。显然这还没有考虑到更大的团体，比如三人或者更多人，但是逻辑上同样是适用的。通过使用这些规律，地方政府不仅可以避免出错的尴尬，更可以高效而协同地完成工作。一个好的设计者会在视觉上判断相容性。最后，类似维修设备、储藏室和垃圾箱等配套设施，不应该设在邻近公共使用区域的位置（图 4.10 - 图 4.31）。

7. 公园主入口设计要传递出友好热情的信息

主入口的设计应当庄严优雅，让游客感受到友好热情，同时还应奠定公园内部景色的基调。入口应该明确空间范围，但是要避免大体量的表述。情况允许时，主入口应该邻近汽车站、快速交通站或者人行横道。一些大型公园可以通过在大门口设游客中心、咖啡店、精品店、器材出租店来营造一种城市公园的氛围，同时，公园也可以从中获利（图 4.32）。

图 4.10
公园里的咖啡吧
Park Café

城市公共开放空间可以从提供有户外座椅的小咖啡店中受益，例如波士顿的诺曼·利文撒尔公园。
Urban Public Open space can benefit from the presence of a small café with outdoor seating, as in Norman B. Leventhal Park in Boston. (Credit: Thomas M. Paine)

城市公共开放空间活动表格
Urban Public Open Space Activity Table

表 4.1
Table 4.1

城市公共开放空间类型 Urban public open space type		城市广场 City Plaza	市区公园 Downtown Park	大型公园 Large Park	绿道 Greenway	社区公园 Neighborhood Park	年纪 Age						夜 Night	春 Spring	夏 Summer	秋 Autumn	冬 Winter
							1-4	5-11	12-15	16-35	36-60	61+					
文化活动 Cultural Activities	游行，庆典 Parade, Ceremony	■		■				■	■	■	■	■	■	■	■	■	■
	正式的表演 Formal Performance	■	■	■				■	■	■	■	■	■	■	■	■	
	非正式的表演，卡拉OK Informal Performance, Karaoke	■				■		■	■	■	■	■	■	■	■	■	
	艺术家，演讲家 Artist, Orator	■	■							■	■	■		■	■	■	
	节日，集市 Festival, Fair	■	■	■		■	■	■	■	■	■	■	■	■	■	■	■
	展览会 Exhibition	■	■					■	■	■	■	■		■	■	■	■
	大合唱 Singing Group		■			■					■	■	■	■	■	■	■
	跳舞（秧歌）Dance (Yangge)	■	■	■		■			■	■	■	■	■	■	■	■	■
	展销 Interpretive Markers, Plaques	■	■	■	■				■	■	■	■	■	■	■	■	■
室内活动 Indoor Activities	攀岩 Rock Climbing			■					■	■	■			■	■	■	■
	台球 Billiards									■	■	■	■	■	■	■	■
	乒乓球 Table Tennis								■	■	■	■	■	■	■	■	■
	健身器材，爵士健美操 Fitness Equipment, Jazz Aerobics								■	■	■	■	■	■	■	■	■
	蹦床，体操 Trampoline, Gymnastics							■	■	■	■			■	■	■	■
	拳击 Boxing									■	■	■			■	■	■
	跆拳道 Taekwondo								■	■	■	■			■	■	■
	冰球 Ice hockey								■	■	■	■		■			■
	游泳 Swimming								■	■	■	■	■	■	■	■	
	羽毛球 Badminton								■	■	■	■	■	■	■	■	■
	咖啡厅 Coffee House	■	■	■		■				■	■	■	■	■	■	■	■

续表

城市公共开放空间类型 Urban public open space type		城市广场 City Plaza	市区公园 Downtown Park	大型公园 Large Park	绿道 Greenway	社区公园 Neighborhood Park	年纪 Age						夜 Night	春 Spring	夏 Summer	秋 Autumn	冬 Winter
							1-4	5-11	12-15	16-35	36-60	61+					
户外活动 Outdoor Activities	户外咖啡店 Outdoor Café																
	夜市 Night Market																
	迷你高尔夫 Mini Golf																
	羽毛球 Badminton																
	排球 Volleyball																
	柬埔寨踢排球 Cambodian Kick Volleyball																
	蹦床 Trampoline																
	棒球 Baseball																
	网球 Tennis																
	足球 Soccer																
	英式橄榄球 Rugby																
	橄榄球 Football																
	长曲棍球 Lacrosse																
	儿童足球 Kickball																
	板球 Cricket																
	草地曲棍球 Field Hockey																
	篮球 Basketball																
	草地保龄球, 法式滚球 Lawn Bowl, Bocce, Pétanque																
	推圆盘游戏 Shuffleboard																
	溜冰 Ice Skate																
	散步, 手里拿着登山棍行走 Walk, Nordic Walk																
	自行车, 极限自行车 Bike, Race Bike (BMX)																
	跑步 Run																
	滑板 Skateboard																
	花样轮滑, 轮滑舞 Roller Blade (Inline Skate), Roller Dance																
	街头曲棍球 Street Hockey																
	划船 Row, Paddle																
	游泳 Swim																
	玩模型飞机, 模型船和模型车 Use model airplane, boat, car																
	滑草 Grass Ski																
	模型火车 Miniature Train Rides																
	旋转木马 Carousel																

续表

城市公共开放空间类型 Urban public open space type		城市广场 City Plaza	市区公园 Downtown Park	大型公园 Large Park	绿道 Greenway	社区公园 Neighborhood Park	年纪 Age						夜 Night	春 Spring	夏 Summer	秋 Autumn	冬 Winter
							1-4	5-11	12-15	16-35	36-60	61+					
自然环境中的活动 Usually Natural Setting	瑜伽, 冥想 Yoga, Meditation Setting									■	■	■	■	■	■	■	
	太极 Tai Chi									■	■	■	■	■	■	■	■
	遛鸟 Walk Pet Bird										■	■		■	■	■	■
	攀爬台阶 Step Climb								■	■	■	■	■	■	■	■	■
	飞盘, 飞盘高尔夫 Catch, Frisbee, Disc Golf								■	■	■	■		■	■	■	
	踢球 Kickball								■	■	■	■		■	■	■	■
	风筝 Fly Kite								■	■	■	■	■		■	■	■
	攀岩 Rock Climb									■	■	■			■	■	■
	高尔夫 Golf										■	■	■		■	■	■
	迷你高尔夫 Miniature Golf								■	■	■	■	■	■	■	■	
	户外健身器材 Outdoor Fitness System								■	■	■	■	■	■	■	■	■
	单杠 Chin-up Bars								■	■	■	■		■	■	■	■
	平衡木 Balance Beam, Other Gymnastics								■	■	■	■			■	■	■
	围棋 Go									■	■	■	■	■	■	■	■
	象棋 Chess, Chinese Chess									■	■	■	■	■	■	■	■
	跳绳 Jump Rope								■	■	■	■		■	■	■	■
	跳远 Long Jump								■	■	■	■			■	■	■
	远足 Hike								■	■	■	■	■		■	■	■
	社区花园 Community Garden								■	■	■	■	■		■	■	■
	闲坐, 小睡, 看过往的人们 Sit, Nap, Watch People							■	■	■	■	■	■	■	■	■	■
	浪漫的婚纱摄影 Romance, Wedding Photos										■			■	■	■	■
	儿童探险花园 Childrens Discovery Garden							■	■	■					■	■	■	
	户外课程, 露营, 野外漫步 Outdoor Class, Day Camp, Nature Walk								■	■	■	■	■		■	■	■	

续表

城市公共开放空间类型 Urban public open space type		城市广场 City Plaza	市区公园 Downtown Park	大型公园 Large Park	绿道 Greenway	社区公园 Neighborhood Park	年纪 Age						夜 Night	春 Spring	夏 Summer	秋 Autumn	冬 Winter
							1-4	5-11	12-15	16-35	36-60	61+					
自然环境中的活动 Usually Natural Setting	清理，打扫 Engage in Cleanup, Fixup																
	日光浴 Sunbathe																
	社区网络，电话 Social Network, CellPhone																
	上网 Surf Internet																
	写作和阅读 Read, Write																
	绘画写生 Paint, Sketch, Photograph																
	骑马 Ride Horseback																
	观察，摄影野生生物 Watch, Photo Wildlife																
	钓鱼 Fish																
	滑雪，雪地汽车 Cross Country Ski, Snowmobile																
	雪橇 Sled																
	野餐 Picnic, Barbecue																
	野营 Camp Out																
	狗公园 Dog Park																
游乐场 Playground	玩水 Water Play																
	溜滑梯 Slide																
	秋千 Swing																
	攀爬游乐器材 Jungle Gym																
	冒险活动 Adventure Play																
	沙坑 Sand Play																
需要监护的活动 Need Supervision	铅球 Shotput																
	飞镖需要监护的活动 Darts																
	射箭 Archery																

Passive recreation, heavy use　　高使用率，被动式休闲活动
Passive recreation, light use　　低使用率，被动式休闲活动
Active recreation, heavy use　　高使用率，主动式休闲活动
Active recreation, light use　　低使用率，主动式休闲活动

注 Notes:
不包括环境可持续发展服务。
不包括海洋和湖泊活动（帆船，浮潜，滑水，冲浪，帆板）。
假设照明仅提供于不大于排球场的铺装场地或区域。
Excludes environmental sustainability services.
Excludes ocean and lake activities (sailing, snorkeling, water skiing, surfing, wind surfing).
Assumes lighting only for paved courts or areas up to volleyball size.

图 4.11
公园里的座椅
Park Seating

可移动的座椅增大了在公园里的人们的自由性，例如波士顿的肯尼迪绿道。
Movable chairs add immeasurably to the feeling of freedom in a park, as on Rose Kennedy Greenway in Boston. (Credit: Thomas M. Paine)

图 4.12
公园里的座垫
Park Seating

提供免费座垫会可以让人们坐在草坪上，远离拥挤的人群和座椅，例如波士顿的诺曼·利文撒尔公园。
Cushions made available free of charge allow people to sit on the lawn away from the crowds of pedestrians and chairs, as in Norman B. Leventhal Park in Boston. (Credit: Thomas M. Paine)

图 4.13
在桌子上游戏
Table Games

在中国和许多其他国家，一种受人欢迎的活动就是朋友们聚在一起打牌、玩象棋和其他游戏，例如在衢州的城市公园。
A favorite pastime for friends in parks in China and many other countries is to meet and play cards, chess, and other games, as in this downtown park in Quzhou. (Credit: Thomas M. Paine)

图 4.14
公园"阅览室"
Reading

香港汇丰银行帮助纽约的布莱恩特公园建立"阅览室"，它的座椅模仿了巴黎的卢森堡公园中的座椅。
HSBC bank helps support the "reading room" area in Bryant Park in New York City; the chairs are modeled on those in the Jardin du Luxembourg in Paris. (Credit: Jim Henderson)

图 4.15
供乘骑的游乐设施
Rides

一个在公园里和游乐场里都很受孩子们欢迎的游乐设施就是旋转木马。它同样令旁观者愉悦。图片中的例子是罗马的纳沃纳广场。
A popular ride for children in public parks as well as amusement parks, the carousel is also entertaining to spectators. This example is in the Piazza Navona in Rome. (Credit: Lalupa)

图 4.16
供乘骑的游乐设施
Rides

波士顿的公共公园中标志性的天鹅船有着超过一个世纪的历史。
The Swan Boats in Boston Public Garden have been its signature for over a century. (Credit: Daderot at en.wikipedia)

Chapter Four　Design Principles　**137**

图 4.17
驻足观察
People Watching

坐下来观察其他路人的精彩之处是典型的公园体验，例如本图所示的上海静安公园。
Sitting and watching the spectacle of other people passing by is the quintessential park experience, as in Jing'an Park in Shanghai. (Credit: Thomas M. Paine)

图 4.18
群体活动
Group Activities

没有一个比公园更适合群体舞蹈、太极拳以及大合唱的场所。比如上海的淮海公园。
No place is more appropriate for group dancing, *tai chi chuan*, and singing than a park, as in Huaihai Park in Shanghai. (Credit: Thomas M. Paine)

图 4.19
群体活动
Group Activities

上海淮海公园脚下柔软的草地吸引了三位女性之中的两位跨过了护栏。中国现在对公园草坪的使用可能还是禁止的，但是这个规定可能会被期待很快放松。
The softness of grass underfoot has tempted two of three women to cross the line in Huaihai Park in Shanghai. Park lawns may be typically off-limits to Chinese now, but that rule may be expected to be relaxed soon. (Credit: Thomas M. Paine)

图 4.20
舞蹈表演
Dance performance

法式芭蕾之美为塞尔维亚贝尔格莱德的托普契代尔公园增色。
The beauty of French ballet graces Topčider Park in Belgrade Serbia. (Credit: tdjoric)

图 4.21
游乐场地活动
Playground Activities

游戏设施可以做成许多充满想象力的形式。它们鼓励集体活动,并且帮助建立身体的技能。比如在波兰一个公园中的金字塔形的绳索。
Play equipment that encourages group activity and builds physical skills can take many imaginative forms such as this rope pyramid in a park in Poland.
(Credit: Damian Stawowy)

图 4.22
游泳池
Swimming Pool

有想象力的设计可以使一个游泳池融入公园的自然形式中,就像照片中智利的圣地亚哥,那里的岩石不是一个障碍物,而是一个自然的元素。
Imaginative design can integrate a swimming pool into the natural forms of a park, as here in Santiago, Chile, where the ledge was retained as a feature rather than removed as an obstacle.　(Credit: Alejo Dice)

图 4.23
网球场
Tennis Courts

公共的网球场合适建在有足够空间的公共场地上,同时不会让其他的场地变得太拥挤。
Public tennis courts are justified where there is enough room in the urban open space to accommodate it and not cause the rest of the site to become too crowded. (Credit: M. O. Stevens)

140　第四章　设计原则

图 4.24
射箭
Archery

残疾人应得到平等的机会去参加各种不同的活动，超过了大多数人们可以想象的，就像在马来西亚的残疾人从事箭术，这项活动出于安全原因应设置在远离其他活动的地方。照片中所展示的也许超出了大部分人的想象，在马来西亚，参加人参与到射箭中来。
The disabled deserve equal opportunity to engage in a wide variety of activities, beyond what most people may imagine, as in this example of the disabled in Malaysia engaged in archery, an activity that for safety reasons cannot be located too close to other activities. (Credit: Shariff Che'Lah)

图 4.25
迷你高尔夫
Minigolf

尽管在一个完整的高尔夫球场里练习是不可能的，人们愿意在迷你高尔夫球场上练习挥杆，这样的小球场被放置在公园里，例如在比利时奥斯坦德的利奥波特公园。
Even if a golf course is unavailable, people love to practice their golf putts on a minigolf course, which can fit into a park setting as in Leopold Park in Ostend Belgium. (Credit: Georges Jansoone)

图 4.26
冬季活动
Winter Activity

公园应该在每个季节都能使用。在冬季滑冰者蜂拥至波士顿公园中的青蛙池塘。
Parks deserve to be used in all seasons. Skaters flock to the Frog Pond on Boston Common in winter. (Credit: Daderot)

图 4.27
冬季盛典
Winter Festival

哈尔滨的冬天很冷，但是它的冰雪大世界享誉世界。
Harbin may be frigid in winter but it is famous for its enchanting Ice Snow World. (Credit: 李吉秋)

图 4.28
冬季盛典
Winter Festival

渥太华会举行自己的冰雪节，从字面意义上这个由"冬季"和"插曲"组成的合成词暗示着其中的乐趣。
Ottawa holds its Winterlude, a play on words (winter and interlude) that suggests the fun to be had. (Credit: Jim Lopes)

图 4.29
冬季盛典
Winter Festival

日本北海道的札幌，彩色的冰雪节沿着购物大道一直延伸好几个街区。
The colorful Snow Festival in Sapporo, Hokkaido in Japan extends for many blocks along a mall. (Credit: Eckhard Pecher)

图 4.30
冬季盛典
Winter Festival

日本北海道的札幌，彩色的冰雪节沿着购物大道一直延伸好几个街区。（夜景）
The colorful Snow Festival in Sapporo, Hokkaido in Japan extends for many blocks along a mall, seen here at night. (Credit: 京浜にけ at ja.Wikipedia)

图 4.31
狗在人的公园里
Dogs in People's Parks

在一些没有狗公园的城市里，市民允许牵着他们的宠物狗进入公园，只要保证拴好狗链并且在它们便后进行清理。例如在波士顿的公共花园就是如此。
Where there is no dog park, in some cities owners may be permitted to bring their pet dogs as long as they keep them on a short leash and clean up after them as here in the Boston Public Garden. (Credit: Thomas M. Paine)

图 4.32
公园入口
Park Entrance

新艺术运动建筑家安东尼奥·高迪为巴塞罗那的古埃尔公园设计的入口很好地营造出与公园内容相符的愉悦情感。这个有历史意义的设计是永恒的，并且打破了文化的界限。
Art nouveau architect Antonio Gaudi's design of the entrance to Park Guell in Barcelona establishes the right mood for what lies within the park. The lessons of this historic design are timeless and cross cultural boundaries. (Credit: böhringer friedrich)

Chapter Four Design Principles **145**

8. 公园交通系统应在保证安全及可达性的前提下，设置机动车、自行车、轮椅、个人交通工具、滑板、轮滑以及行人通道

公园交通系统应该指引游客经停主要的景点（主题区或相关设施）。在特定的路径应该引导游客经过并且自行发现潜在的社交或活动场地，而不是强迫他们进入。如果有足够的空间，至少应该有一条路径，最好是一个环路，能够为游客提供一个探索的旅程。这条路径应该充满了不同的景观特色，可以营造出从宁静到惊叹，从惊叹到欢乐的各种氛围。

公园交通系统的冲突很可能会导致碰撞事故。因此我们应该提供被分隔的步行道，以及被分隔自行车道，甚至是单独的自行车游园道。这一原则可以追溯到150年前西方第一个大型公园的建立，这个时间甚至早于自行车的发明。城市公共开放空间应提供短期低价的自行车和个人交通工具租赁，并提供停车场地。

卡车在任何一个城市公共开放空间都不应该允许进入（维护车辆除外），还有，甚至汽车（私家车）也应该被禁止在特殊的风景区干道以及穿越大公园的交叉路口，并且限速。考虑到人们的需求，卓越的城市公共活动空间应该远离噪声、高速行驶、安全风险和空气污染。在有车行城市交通的地方，都应该提供单独的车行道或步行道，用于人们骑自行车、滑板、轮滑、慢跑等，这些道路应该与城市交通线平行，并且用景观带隔离，人行横道应该设有缓坡，方便残障人士使用（图 4.33 - 图 4.36）。

应该限制停车区域的面积，同时需用景观将其隔开。没有什么能比一个大而空的停车场更糟糕的了，这非常的浪费，而且区域内的土壤也因此失去了生态功能。在美国，车辆的数量和可提供的停车面积比例相差非常大。大型公园停车场的供给率从0.25%到20%不等，而在城市广场和市区公园，这个数字趋于零。几个美国大型停车场的地表停车率情况见表4.2。

9. 设计、导示、导览图、游园须知、规章、维护的标准都应建立全套系统

城市公共空间是一个品牌，应该像维护品牌一样管理维护它，如果一个品牌具有高标准，并且无论时间、空间和媒介的变化都一直保持这个高标准，那么人们就会很信任这个品牌，我们的原则就是要保持高标准，而且彰显出的标准一定要友善并且受人欢迎，一些粗劣的设计往往会破坏品牌形象，例如媚俗卡通字符、低俗幽默、华而不实的灯饰以及刺眼颜色的陈设和标识。要传达世界级的印象，应该使用双语标识，为了避免尴尬发生，英语使用要地道、准确。

建立全套系统并不意味着所有公共开放空间形式必须标准化。可以做些合理的调整去反映使用强度的差别、硬质景观与软质景观的比例、种植类型以及设施的范围。社区公园可能不像其他城市公共活动空间类型那么需要标准化，但是标准依旧很高。

入口引导标志的尺寸应该很小（显然不能是广告牌的尺寸）。上面要有即时的信息，包括游园须知和联系人信息，并应加柔光便于夜晚识别。下面是两个关于入口的标识。罗斯肯尼迪绿道公园位于波士顿，是一个由私人资助集团管理的2.4公里长的廊道。邻近纽约公共图书馆的布莱恩特公园，占地3.9公顷，之前疏于管理，曾经轻微犯罪行为猖獗，而现在，它是很热门的去处，并且成为公园管理和项目规划的典范。

8. The circulation system accommodates motorized vehicles, bicycles, wheelchairs, personal transports, rollerbladers (inline skaters) and pedestrians in a fair and balanced manner while addressing safety and accessibility issues.

The circulation system should lead visitors past the major elements (features or amenities). In particular the route should allow visitors to pass by and check out the areas of potential social contact without forcing them to enter. If there is sufficient space, at least one walkway, ideally in a loop, should offer a journey of exploration embellished with diversions and digressions of varied landscape character that appeal to various moods, from serenity to surprise to awe to joy.

Where conflicting modes of circulation are likely to lead to collisions, separate bicycle lanes or even separate bicycle trails, as well as separate foot paths, should be provided. This principle goes back to the first large parks designed in the West 150 years ago, even before the invention of the bicycle. Urban public open spaces should offer the short-term low-cost rental of bicycles and personal transports, and provide parking racks.

Trucks should not be allowed within any urban public open space (except for maintenance of that space), and even cars ("pleasure vehicles") should be restricted to special Greenway parkways and crossroads through Large Parks, and restricted to low speed limits. Honoring the people's need for excellent urban public open space should include accommodating the need to be away from the noise, speed, safety risk, and air pollution generated by vehicular traffic. Wherever roadways enter urban public open space, separate lanes or pathways for bicycles and rollerblading (inline skating), personal transports (Segways), pedestrians and joggers should be provided parallel to the roadway but be separated by a landscape buffer. Crosswalks should have curb cuts for wheelchair users and those who are otherwise physically impaired. (4.33-4.36)

Parking areas should be limited in size and be screened as well as landscaped internally. There is nothing worse than an empty and excessively large parking lot. It represents waste, and soil lost to the ecosystem. In the U. S, which drives and paves far too much, the provision of at-grade parking spaces per hectare of parkland in Large Parks ranges from 0.25% to 20%. The rates approach 0 for Civic Plazas and Downtown Parks. The Table 4.2 shows the ratio of surface parking in select large parks in the U.S.

9. Standards for design, way-finding, graphics, regulations, policing and maintenance are applied system-wide.

Urban public open space is a brand and should be managed as one. People come to trust the brand more if those standards are high and if there is consistency of standards over time, across all spaces in the system, and across all media. The goal should be to maintain a high level of dignity. A friendly, welcoming tone for posted regulations is appropriate. The brand will be undermined by allowing kitsch such as cartoon character emblems, low forms of humor, flashy lighting, bright colors for signage and furnishings. Bilingual signs should be checked by a native English speaker for accuracy to avoid embarrassment and to convey a world-class image.

A system-wide approach does not mean that all open space types must have identical standards. Reasonable adjustments can be made to reflect differences of intensity of use, ratio of hardscape to softscape, planting style, and extent of facilities. Neighborhood Parks may not have identical standards to those of other types of urban public open space, but the standard is still high.

Signage at entrances should be small (certainly not billboard size), informative, current, include regulations and information about whom to contact with concerns and suggestions, and be softly lit at night. Below are two good examples of regulations. Rose Kennedy Greenway is a new 2.4-kilometer park corridor in Boston managed by a private support group. Bryant Park is a 3.9-hectare park adjacent to New York Public Library that formerly was poorly maintained and a haven for petty crime and is now a popular destination and a model of park management and programming excellence.

图 4.33
骑车
Bicycling

骑车对于年轻人和老年人来说都是一个受欢迎的休闲活动,包括在适于专有轮胎的自行车道上骑行,就像图中展示的一样。
Bicycling is a popular recreational activity, for young and old, including trail biking using special tires as shown here. (Credit: Suprijono Suharjoto)

Regulations posted in 2010 on Rose Kennedy Greenway, Boston

Help us keep your public parks clean and safe!

Park hours 7 AM-11PM

No skateboarding, bicycling, or rollerblading while in the parks

Use the provided bike racks; do not secure bikes to trees or park furniture

All pets must be leashed

Pick up after your pets

No loitering after park hours

No organized sports

Vending by permit only

Open fires by permit only

No unauthorized vehicles in the parks including personal transports

Drug and alcohol use is prohibited

In case of emergency call 911

To report a maintenance issue call 617-111-1111

图 4.34
自行车租赁
Bicycle rental

为没有骑车来到公园的游客提供可租赁的自行车以及配套的安全头盔是一项很有价值的服务,倒如纽约中央公园。
Renting bikes and safety helmets is a worthwhile service to offer visitors who do not arrive at the park on their own bike, as here in Central Park in New York. (Credit: Jim Henderson)

Welcome to Bryant Park

Bryant Park is a city park renovated, funded and managed by the Bryant Park Restoration Corporation.

You are Welcome

1. To enjoy the park, including the great lawn

2. To spread blankets on the lawn, but not plastic material or tarpaulins

3. To enjoy the gardens—without entering flowerbeds or picking flowers

4. To use a park chair or one seat on a bench designed for sharing

5. To deposit waste in green receptacles

6. To walk your dog—on a leash—and not on the grass—if you clean up after it

7. To take souvenir photos—commercial photography by BPRC permit only

Park Guidelines Prohibit

• Drug Use

图 4.35
轮滑
Rollerblading

轮滑也被称作直排轮滑,这项运动在西方很受欢迎。一些地方的轮滑者享受着与机动车道相分离的独立滑行道。
Also known as inline skating, this activity has become very popular in the West. In some locations rollerbladers deserve their own lane separated from cars. (Credit: Sonya Etchison)

图 4.36
赛格威骑行
Segway Riding
这种最近由美国人发明的代步工具已经在西方的公园里越来越流行，例如萨凡纳的齐佩瓦公园。
This recent invention of an American inventor is becoming popular in many Western parks as in Chippewa Park, Savannah. (Credit: kmf164)

- Alcohol Use outside the Grill and Café
- Organized ballgames
- Sitting or standing on balustrade
- Entering the fountain
- Feeding pigeons
- Rummaging in trash receptacles
- Amplified music that disturbs others

2010年波士顿罗斯肯尼迪绿道公园游园须知
让我们一起努力，保持公园卫生和安全！
公园开放时间：7：00-23：00
请不要在公园里玩滑板、骑自行车和玩轮滑
请使用自行车停放架，不要把车靠在树上或者公园设施上
请为您的宠物系上链子
请处理好宠物的排泄物

表 4.2 / Table 4.2
选出的几个美国大型停车场的地表停车率
Surface Parking Ratio in Select Large Parks in U. S.

城市公共空间 Urban public open space	公顷数 Hectares	每公顷的停车空间 Parking Spaces/ ha
圣路易斯森林公园 Forest Park, St. Louis	523	15
芝加哥林肯公园 Lincoln Park, Chicago	490	12.3
华盛顿国家广场 National Mall, Washington DC	59	9.6
旧金山金门大桥公园 Golden Gate Park, San Francisco	412	9.1
圣迭戈巴尔博亚公园 Balboa Park, San Diego	486	8.6
纽约中央公园 Central Park, New York	341	0.50
纽约布鲁克林展望公园 Prospect Park, Brooklyn, NY	212	0.24

资料来源：Peter Harnik. *Urban Green, Innovative Parks for Resurgent Cities* (Washington, DC: Island Press, 2010), 148.

图 4.37
公园巡逻队
Park Rangers

大型公园会授权统一着装并经过训练的公园巡逻者来友善地协助与提供有用的信息给游客。例如纽约的中央公园。
Large Parks may warrant uniformed and trained park rangers who offer friendly assistance and provide useful park information to visitors, as in Central Park in New York. (Credit: Maria Azzurra Mugnai)

- Performances, except by permit
- Commercial activity, except by permit
- Obstructing park entrances
- Bicycle riding, skateboarding or rollerblading

New York City's vast park system was recently gone even further in its regulations, banning smoking from all its 11,100 hectares of public plazas, parks, greenways and beaches.

In some heavily used parks in Chinese cities there may be interest in providing park rangers, specially uniformed personnel on duty to assist visitors during open hours. Their presence is friendlier than police, and they are well informed and committed to the broader purposes that the park system serves. The model for such park rangers may be those of the national parks in China or the U.S. National Park Service. (4.37)

10. The spaces in the park system adhere to principles of Universal Design.

Currently, Western countries are increasingly adopting "universal

请不要在开放时间以外进入公园
请不要在这里组织体育活动
未经许可不能进行贩卖活动
未经许可不能生火
未经授权的车辆不能驶入，包括私人交通工具
禁止吸毒和酗酒
如有意外，请拨打 911
上报维修问题，请拨打 617-111-1111

欢迎来到布莱恩特公园

该公园是由布莱恩特公园修复公司负责修复、资助和管理的城市公园

欢迎游园
1. 公园内大草坪对外开放
2. 可以使用毯子，但请不要使用塑料材料以及防水油布
3. 赏花时请不要进入花坛，不要随手摘花
4. 公园设有座椅供休憩使用，请礼让
5. 废弃物请丢入垃圾桶

Chapter Four　Design Principles　149

图 4.38
通用性设计
Universal Design

新加坡建筑与建造局画廊内的感知花园展示了多种的通用设计，比如图片中在道路右侧的艺术装置，道路有特殊纹理，不设台阶，标识清晰可见，这些都方便了轮椅使用者的使用。
The Sensory Garden at the Building and Construction Authority Gallery in Singapore showcases various Universal Design features, such as art installations located right on the pathway, pathways with special textures and without steps, and signage legible from a wheelchair. (Credit: Chenzw)

6. 宠物狗请拴好狗链，并清理好排泄物，请勿带入草地

7. 允许拍照留念，如果是商业摄影，要经过布莱恩特公园修复公司的许

公园禁止条例
- 吸毒
- 在烧烤点以及咖啡店外饮酒
- 组织球赛
- 坐在或者站在栏杆上
- 进入喷泉
- 喂鸽子
- 翻垃圾桶
- 未经许可的表演
- 未经许可的商业活动
- 阻塞公园入口
- 骑自行车、玩滑板或者轮滑

目前纽约市区庞大的公园系统在高标准上又前进了一步，11100公顷的公共广场、公园、绿道以及沙滩已全部禁烟。

中国的一些使用频率高的城市公园都会安排管理员，他们统一着装，在公园开放时间协助游客。他们比警察更友善，消息很灵通，也很心甘情愿地从事公园服务工作，这些公园管理员应该成为中国和美国国家公园管理服务的榜样（图4.37）。

10. 在公园系统中坚持"通用性"原则

目前，西方国家越来越多地采纳"通用性设计"，这就是说，设计的标准不能忽略残障人士。过去的建筑师和景观设计师并没

design", that is, design standards that do not discriminate on the basis of disability or sensory impairment. This is not how architects and landscape architects have thought about design in the past. Before, they designed for people without mobility difficulty or sensory impairment. But in fact, at some point in one's life, everyone has mobility challenges at least temporarily, whether as a baby who cannot walk, an adult with a twisted ankle, or a senior with arthritis, anyone confined to a wheelchair. Recognizing that, universal design rethinks design: it is not a deficiency of people but a deficiency of design that causes a poor fit between a particular user and a particular place. Adding ramps and special paving textures to alert the blind is not enough. Designers need to enrich the experience so that those with sensory impairment are nevertheless stimulated. This thinking is eminently applicable to excellent design in urban parks. This worthy goal assumes the equality of all people, and the need of all people to have ease of access, ease of use, and safety. (4.38)

7 Principles to achieve Universal Design: "Design for Lifetime, not just for Primetime"

(A single design which accommodates all users no matter how impaired their mobility, sight, or hearing):

- *Equitable use*
- *Flexibility in use*
- *Simple and intuitively easy to use*
- *Easily perceived*
- *Tolerant of error*

- Physically undemanding (ergonomically sound)
- Choices for universal design should be made for the way-finding system, traffic signalization devices, curbs and radii at intersections, special parking spaces, paving, sitting, tables, drinking fountains, wide-door building entrances, and elevators, all of which should accommodate baby carriages, wheelchair users and sight-impaired people. Chinese Braille 现行盲文 signage and handouts should be the norm.

See: http://www.universaldesigncasestudies.org, http://en.wikipedia.org/wiki/Universal_design

The Americans with Disabilities Act ("ADA") (1990) declares that accessibility is a civil right, not a human right (a civil right is guaranteed by a particular nation's constitution whereas a human right transcends any constitution). For detailed ADA standards (such as railing dimensions, ramp slopes and lengths between landings) to assure accessibility in the outdoor environment for the people who have a disability, see 2010 ADA Standards for Accessible Design issued by the U. S. Department of Justice. In the U. S. 19% of the people have a disability, 1% use wheelchairs. Universal design goes much farther.

See http://www.ada.gov/regs2010/2010ADAStandards/2010ADAstandards.htm

11. Start with the design vision, not with the budget.

Costs should not drive the decision on something as fundamental as giving cities a heart; quite the reverse: first comes the vision, then comes the reality of finding a way to organize the effort and to pay for it. There are always a range of cost options to achieve the same end, from a minimal level to an optimal one. Money is always too scarce for meeting every perceived need, but that scarcity imposes on decision makers the obligation to spend wisely, not the obligation to refrain from spending, to meet the people's needs for urban public open space excellence. As resources become ever scarcer, squandering of financial and material resources becomes ever more unacceptable, even as the need to share those resources becomes ever more urgent.

注释
1. 扬·盖尔.《建筑之间的生活：使用公共空间》（纽约：Van Nostrand Reinhold 出版社，1987）
2. 财富杂志采访，"苹果公司年薪一元的人"，2000年1月24日
3. 詹姆斯·克纳尔."地球激浪"，查尔斯·沃德海姆编辑，《景观城市主意读者》（纽约：普林斯顿建筑出版社，2006），32
4. 伦纳德·霍珀编辑，《景观建筑图形标准》（霍伯肯：约翰威立出版社，2007），821
5. 史蒂夫·乔布斯与盖瑞·沃尔福的采访，"史蒂夫·乔夫斯：下一个疯狂的大事件"，连线杂志，IV：2，1996年2月
6. 斯蒂芬·卡尔，马克·弗朗西斯，里恩·瑞夫林，安德鲁·斯通.《公共空间》（纽约：剑桥大学出版社 1992），125

NOTES
1. Jan Gehl. *Life Between Buildings: Using Public Space* (New York: Van Nostrand Reinhold, 1987)
2. Interview, "Apple's One-Dollar-a-Year Man," *Fortune Magazine*, January 24, 2000
3. James Corner. "Terra Fluxus," in Charles Waldheim, ed., *The Landscape Urbanism Reader* (New York: Princeton Architectural Press, 2006), 32
4. Leonard J. Hopper, ed., *Landscape Architecture Graphic Standards* (Hoboken: John Wiley, 2007), 821
5. Steve Jobs, interview with Gary Wolf, "Steve Jobs: The Next Insanely Great Thing," *Wired Magazine*, IV:2, February 1996, http://www.wired.com/wired/archive/4.02/jobs_pr.html, March 8, 2012
6. Stephen Carr, Mark Francis, Leanne G. Rivlin, Andrew Stone. *Public Space* (New York: Cambridge University Press, 1992), 125

有想到这一点。之前的设计都基于身体健康的人。但事实上，在人生的某些时刻，一个还不会走路婴儿，一个崴脚的成人，患有关节炎的老人或是坐轮椅的人，每个人都会经历对行动力的不便，哪怕是暂时的。基于这一点的认识，通用性设计对设计作了反思：这不是人的缺陷，而是设计的缺陷，它不能连接一个特定的用户到一个特定的场所。仅仅添加无障碍坡道和铺设盲道是不够的。设计师需要丰富自己的经验，使得设计不再从心理上刺激那些有障碍的人。优秀的城市公园设计需要做到这一点。这个有价值的目标是为了让所有人平等化，以及满足所有人对易于到达、易于使用和安全性的需求（图 4.38）。

达到通用性设计的 7 原则："要做到长期规划，而不能只顾眼前利益"

（一个简单的设计能适应所有的使用者，无论他们的肢体、视觉或者听觉是否有缺陷）：
- 平等使用
- 弹性使用
- 使用简易
- 易于理解
- 容许小的错误
- 符合人体工程学原理
- 足够的尺寸及空间供使用

通用性设计应该应用于导示系统、交通信号设施、十字路口的转弯及路缘、专用停车位、铺装、座椅、桌子、自动饮水器、大门入口以及电梯上，所有这些都应该考虑到婴儿车、轮椅使用者以及盲人的使用。中国盲文的标识和手册也应该成为规范。

资料出处：*http://www.universaldesigncasestudies.org*
http://en.wikipedia.org/wiki/Universal_design

美国残疾人法案（ADA）规定，可达性是一种民权，不是人权（民权受到特定国家的宪法保护，而人权超越任何宪法）。ADA 标准的详细规定（如栏杆尺寸、楼梯平台之间的斜坡率和长度）确保了残疾人对户外空间使用的可达性，更多请参考由美国司法部颁发的《2010 年美国残疾人法通用设计标准》。在美国，残疾人口的比例是 19%，1% 的人使用轮椅，但通用性设计遍及的范围更广泛。

资料出处：*http://www.ada.gov/regs2010/2010ADAStandards/2010ADAstandards.htm*

11. 先愿景，后预算

预算不应该成为一个城市最原本的设计理念的先决条件；恰恰相反，设计应以城市愿景为前提，然后才可计算为此付出的努力和财力。为达到同一个结果往往有不止一个预算方案，从最节省的到最理想的。如果要满足所有需求，金钱永远是不够的。但也正是因为这种财力的缺乏，恰恰监督着决策者要把钱花到实处，把钱花在满足人们对优秀城市公共空间的需求上。如今资源日渐紧缩，浪费财力和物力就变得越来越不可接受，分享这些资源的需求也迫在眉睫。

Chapter Five 第五章

城市广场、市区公园、大型公园、绿道及社区公园的设计指南

Design Guidelines for Civic Plazas, Downtown Parks, Large Parks, Greenways and Neighborhood Parks

场地效应的产生在于当人们迫不及待地进入一个空间后，就再也不愿离开它。
Placemaking occurs when you transform a space you can't wait to get through into one you never want to leave.
—— 凯西·马登，2011 年 [1]
Kathy Madden, 2011 [1]

城市被存在于其中的那些美丽公园，公共空间，大街小巷，还有那些像瑰宝一样装饰着城市街道的平凡楼宇们所定义。
Cities are defined by their great parks, their public spaces, their streets and alleyways, by the ordinary buildings that make up their streets as well as their grand civic jewels.
—— 诺曼·福斯特勋爵，2011 年 [2]
Lord Norman Foster, 2011 [2]

图 5.1
特拉法加广场，伦敦
Trafalgar Square, London

这个广场一直以它的背景建筑、焦点柱式雕塑和双子喷泉闻名，使它变得更加出色的是，周边道路在把这个广场与国家美术馆分开的同时在台阶的顶端转变成一条步行道。
Always memorable for its architectural backdrop, focal column and twin fountains, Trafalgar Square just got even better when a perimeter road separating this Civic Plaza from the National Gallery at the top of the staircase was turned into a walkway. (Credit: "Photo by DAVID ILIFF. License: CC-BY-SA 3.0")

我们现在详细地介绍五种开放空间类型的基本设计原则，这五种类型分别为城市广场、市区公园、大型公园、绿道及社区公园。在这些设计中，每个城市公共开放空间类型都有其独特的设施和活动要求，其中当然避免不了一定程度上的重复。例如，在合适的条件下，绿道和社区公园需要的设施可以相同，因此，绿道也可作为其毗邻社区的社区公园。本章的案例分析涵盖了我们认为近期在中国和全球的优秀设计范例。

■ 城市广场

在所有城市公共开放空间类型中，城市广场对人们的生活最具影响力，并在维护城市的社区可持续性和树立城市的独特个性中起到重要作用。克莱尔·库珀·马卡斯以术语"大型公共场地"来描述城市广场所扮演的角色，这使我们联想到古老欧洲小镇的城市中心广场或露天市场，正如第一章所描述的，它们承担着多功能的用途（城市机构、零售、商业交通等）。在所有城市公共开放空间系统元素中，城市广场以其多样性在最大程度上吸引着当地居民和众多流连忘返的人（图 5.1）。因此城市广场在城市规划决策中占据着重要的位置，同时也极大地影响着公众对整体

We turn now to recommended guidelines specific to the five open space types, Civic Plaza, Downtown Park, Large Park, Greenway, and Neighborhood Park. In their design, each urban public open space type offers distinct facilities and activities, although of course there is some degree of overlap. For example, to the extent that Greenways offer the same facilities as Neighborhood Parks, Greenways can, in appropriate situations, also serve as the Neighborhood Park for the neighborhood adjacent to the Greenway. We include case studies of recent work in China and globally which we regard as examples of design excellence.

■ Civic Plaza

Of all the urban public open space types, the Civic Plaza has the most potential to make a difference in people's lives and in the social sustainability of the urban community as well as project a city's distinctive character to the world. Clare Cooper Marcus captures this role by using the term Grand Public Place, which is suggestive of an old European town square or piazza in the heart of the city, located near diverse uses (civic institutional, retail, commercial, transit), as we

图 5.2
时代广场，纽约
Times Square, New York City

当人行道和座椅取代了车行交通，这个曼哈顿的城市广场无疑变得更好了，这一举措恢复了穿行和停留两种需求之间的平衡。
Truly the Civic Plaza of Manhattan, Times Square got even better when pedestrians and chairs took over space from vehicular traffic, restoring a balance between the competing needs of being somewhere and passing through it. (Credit: Terabass)

saw in Chapter 1. Of all the elements in the urban public open space system, the Civic Plaza attracts users from the greatest distance in the greatest numbers with the greatest diversity, and attracts as many visitors as local residents. (5.1) Therefore its role in setting the "tone at the top" and influencing public opinion about "the brand" of the city as a whole is enormous. The Civic Plaza is where local residents take guests to show off their city. In this sense, the Bund in Shanghai, although elongated, functions more as a Civic Plaza than as a Greenway. As iconic as the Bund, Times Square has been called "New York's agora", a place to gather to await great tidings and to celebrate them, whether a World Series or a presidential election. (5.2) Both the Bund and Times Square have been improved by the recent removal of adjacent roadway lanes. Removal of roadway paving to enlarge Civic Plazas or even create new ones has been a major trend in the West.

Plazas within retail or office or mixed use commercial development projects, often provided in exchange for higher project density, may perform some of the functions of a Civic Plaza and are a welcome supplement, but they are hardly a substitute. No private developer can

城市形象的看法。城市广场是当地居民带领访客游览城市的场所。就此而言，上海的外滩尽管是细长形的，但功能上更像是一处城市广场而非绿道。同样是城市标志的纽约时代广场被称为"纽约的集会"，无论是世界系列赛还是总统选举，人们聚集在此等待并庆祝好消息的到来（图5.2）。最近，与之相连的城市道路车道被移除，这更使外滩和时代广场的可达性得到提升。道路为广场让路在西方已成为一种主流趋势。

一些带有零售、办公或者混合使用的商业发展项目的普通广场，通常可以换来更高的项目密度。这些广场或许具有城市广场的一些功能，也可以成为城市广场的一个有效补充，但很难完全替代。没有哪个私人开发商能够满足一个城市对城市广场的需求，更没有人能够保证永久满足这种需求。普通广场大多缺乏便利设施（有形的或者无形的，可为人们的生活带来舒适和便利的设施）。但是另一方面，这些广场的设计也确有可能包含了一些值得在城市广场借鉴的好的创意。

大城市通常不只有一个城市广场，他们各有千秋。但是都应该具有以下几点特征：

图 5.3
罗斯·肯尼迪绿道的喷泉，波士顿
Fountain, Rose Kennedy Greenway, Boston

整个带状城市开放空间中这个城市广场的部分是以一个大型喷泉为中心，这个焦点吸引了从各个社区来的游客和居民，尤其是在炎热天气中变得清凉是一件很愉悦的事情。
The Civic Plaza portion of this linear urban open space centers around a water jet fountain, a focal point that attracts visitors and people from all over the community, especially on a hot day when getting wet is encouraged. (Credit: Thomas M. Paine)

图 5.4
云门，千禧公园，芝加哥
Cloud Gate, Millennium Park, Chicago

云门是印度裔英国雕塑家安妮施·卡普尔的杰作。很少有雕塑能够达到引人沉思并且互动的层次。云门把芝加哥的天际线变为有着特殊效果的背景，使其成为整个公园的焦点。
Rarely does sculpture invite the level of contemplation and interaction as Indian-born British sculptor Anish Kapoor's masterful form, which turns the Chicago skyline into a special effect backdrop. Its Civic Plaza setting is the focus of the whole park. (Credit: Thomas M. Paine)

1. 城市广场是庄严的，是承载记忆的

在设计城市广场时，要以激发人们的情感和唤起人们的回忆为目的，也就是说，要把空间变成场所。这就是我们说的"一个场所的能量"以及"难忘的场所"。城市广场真正让人难忘和钦佩，是它们所传达的深层含义不仅仅被设计者，更被使用者理解之时；是当这些含义优美地体现在人性化的设计、最持久的材料以及最末梢的细节之时。这个场所应充满了自身存在的自豪感。在城市的中心，我们用心交流。

2. 城市广场是举行纪念活动以及文化交流的场地

城市广场应该举办以文化庆祝而不是商业为宗旨的社区标志性活动、纪念活动、节日庆典、表演、户外的艺术展等等。绝对不能让广场的基调错误地变成有侵略性的商业购物中心。一些庆典和纪念的活动每年举办一次，而其他的活动例如表演或者展览每周、甚至每天都会有，这些活动因天气情况而定，几乎涵盖了全年的大部分日子。一天中举办活动的黄金时间是傍晚，那个时候人们刚下班，如果是夏天，刚好退去了白天的炎热。在意大利，这是"漫步"的时间，这种传统本身几乎就是一种艺术——在傍晚散步，享受欣赏与被欣赏（图 5.3 - 图 5.4）。

活动的策划应该包括有效的管理和宣传，并以活动的重要性来决定规划规模是当地性的还是国家性的。当地政府可能会从与私人团体合作中获益，它可以协调参加重大活动中的文化组织，给供应商和表演艺术家发行许可证，从而来保证更多的日常文化活动。管理团队可以在与其他城市的交流中学习最佳的管理方法，并鼓励有兴趣的当地市民提出建议。

城市广场应当纪念当地的历史，因为这是令人自豪的源泉。当地的成就或重要事件都值得在公共空间庆祝，以此激励后人。当地的英雄以及广受尊敬的本土艺术家、作家和各类人才都应该

afford to provide what a city needs for its defining Civic Plazas, nor promise to provide it forever. Given the developer's focus on profit, many such plazas lack amenities (tangible or intangible benefits that contribute to people's comfort or convenience). On the other hand, their design may include some truly visionary ideas that are worth considering in Civic Plazas.

The largest cities will have not one but several Civic Plazas, and each will have its own unique character. But they should all adhere to the following guidelines:

1. Civic Plazas are evocative and dignified.

In the design of the Civic Plaza, the goal should be to evoke emotion and memory, for that is what transforms space into place. There is much talk about "the power of place" and "memorable places". What makes Civic Plazas truly memorable and admired is when they convey deep meaning that is understandable to users, not just designers, and when that meaning is embodied in beautiful, user-friendly design executed durably in the best materials, down to the last detail. Places as well designed as that have the power to instill pride of place. In the heart of the city, they speak to the heart.

2. Civic Plazas are ceremonial and cultural.

The Civic Plaza should host the community's signature public events, ceremonies, festivals, performances, outdoor art exhibitions in a spirit of cultural celebration rather than commercialism. In no way should the tone be mistaken for the flashy aggressive commercialism of a shopping mall plaza. Some events such as festivals and ceremonies are held annually, other events such as performances or exhibitions take place weekly or even daily, throughout as much of the year as possible, depending on the climate. A key time of day for special programming is from the late afternoon into the early evening, when many people get off work and the air cools down on hot days. In Italy this is the time of *la passegiata*, a tradition which is almost an art in itself, the art of strolling in the evening, and enjoying seeing and being seen. (5.3-5.4)

Events programming should include active management and publicity, both local and national depending on the significance of the event. Local government may find it advantageous to partner with a private support group to coordinate with cultural organizations on major program events as well as to issue licenses to vendors and permits to performing artists for more routine cultural events and activities. The management team will inform itself of best practices in other cities, and welcome suggestions from interested local citizens.

Civic Plazas should commemorate local history that is a source of pride. The achievements and sacrifices of local citizens or important local events deserve to be celebrated in a public space, and inspire generations to follow. Local heroes, and widely respected local artists and writers and talents of all kinds should be considered. Twenty-first century technology allows this highly important public benefit to take on many new forms which explore new ways of touching people on an emotional level.

3. Civic Plazas are known for a signature focus or activity of wide public appeal.

The design should feature a strong focal object. Historically this has meant a commemorative statue, a fountain, sculpture, mural or other public art. Some might argue that in our era representational sculpture and murals are too old-fashioned, static and unchanging to appeal in a dynamic world of digital technology and connectivity with the same intensity they once had. What is displayed in our era could change on a regular basis, for example in a recurrent sequence, or an occasional update or improvement.

Twenty-first century technology provides communities new ways to showcase richly rewarding public art that dares to offer content. In our era of such rich possibilities for projecting or embedding meaning in public space, there is no need anymore for artless public art. Cities do not need to settle for art that is incomprehensible or trivial. Anemic content might do for large outdoor landscapes that are unfrequented, where little harm can be done, but it will not suffice for the Civic Plaza where the opportunity cost of not providing artistic excellence is too great. Providing what designers of urban public open space sometimes call "narrative" and "legibility" is not enough, if these alleged goals elude visitors by being culturally incomprehensible. The strength of the community brand depends on the popular perception of its culture, in which Civic Plazas play a pivotal role.

The Bund offers a spectacular cityscape view across water of high-rise office buildings mounted with LED screens showcasing a

借助这里被铭记。21世纪的高科技使得这一极其重要的公共活动富有了许多新的形式，从而以不同的方式在情感层面上给人以触动。

3. 城市广场应被称为标志性焦点或者有吸引力的公众活动中心

城市广场的设计应塑造出一个强有力的焦点物体。在历史上这通常指的是一个纪念雕像、喷泉、雕塑、大型壁画或其他公共艺术。有些人可能会认为在我们这个时代，标志性雕塑和大型壁画太过时，在当今数字技术的动态世界里太一成不变，并且也不能体现出当时的感觉。那么我们可以让它显示的内容定期更换，例如循环更替，或者偶尔更新或改进。

21世纪的科技提供给社区全新的途径去展示大量有意义的公共艺术。这些公共艺术都敢于表达实际内容。城市没有必要再去新置那些难以理解或是微不足道的艺术品。缺少活动内容的艺术品也许适用于大型户外景观，而且也没什么害处，但是它们不足以满足城市广场，如果城市广场浪费了这个提供优秀艺术品的机会，这个代价是巨大的。另外，为了避免公共空间带来文化上的误解，使游客感到迷惑，城市公共空间就不能仅仅做到设计师所熟知的"叙事性"和"可读性"。社区品牌的感召力取决于对其文化的普遍认可程度，在这方面，城市广场起到了至关重要的作用。

在上海外滩，人们可以领略到壮美的城市景观。一系列的短片或广告语展现在黄浦江对岸高耸写字楼的LED巨屏上。同样的，在美国，横跨五个街区的纽约时代广场是一个被歌剧院、品牌零售店、LED商业显示屏和巨幅广告环抱的公共场所，曾经它是蹩脚的X形交叉路口，现在这是个充满活力又让人视觉兴奋的地方，同时也是举行大型城市庆典的集中地，比如，著名的新年庆典自1907年开始在这里举行，每个跨年夜里，标志性的装饰灯球徐徐落下，吸引着成千上万的观众。当然，这些气氛的营造都要依靠周边私人业主的合作。都要请他们帮忙点亮靠近公共空间一侧建筑的外立面。纽约时代广场的新置巨幅广告由风能和太阳能提供电力。而位于芝加哥千禧公园的皇冠喷泉表面播放着众多芝加哥市民的面孔，这些巨型面孔交替变换的同时，有趣的喷水从面孔中的嘴里喷出，落在方形的浅水广场上，成了孩子们的戏水乐园。

活动的策划应侧重考虑舒适、放松和更深层次的社区需求。使用者想要在此欣赏美景，想在此娱乐，甚至希望在这里被某种文化触动。这就需要一些刻意的引导，让人们去探索，去回忆，去感知。通过这些有激发性的设计，城市变得更有城市气息，而人们变得更加儒雅有礼。

活动策划应该加强当地的独特性。有些活动是充满童趣的，像某些艺术展或者享有盛名的锡耶纳坎波派力奥赛马节，每年由色彩丰富的中世纪游行拉开序幕（图5.5）。其他的活动可以每周一次或者更频繁地开展。提供当地的农副产品、烘烤食品、鲜花、手工制品、艺术品的户外市场是城市广场的主要用途之一，这不应该仅仅局限在一些欧洲城市的传统集市。其他城市广场也可以开展农民市场、手工艺品展销或者二者合一等多种形式的活动，需要注意的是，城市广场并不主要是商业空间，所以市场活动的时间应该被明确的限定，例如周末、特殊的日子或者特定的季节里定期举行。位于德国安娜贝格县的圣诞节市场是个很好的典范，这里所有的小商店都是临时性的建筑，它们被高大的装饰着各种圣诞彩灯的圣诞树包围，不论风雪都吸引很多人前往。街市摊位或者展台不是永久性的固定安装，因为这样可以在集市结束之后将空间提供给其他的活动。当地政府或者相关的团体和机构应该对摊位进行严格的管理：确保空间的合理安排，控制包括遮阳棚和遮阳伞等其他摊位物品的使用，监控每日的卫生环境，禁止工业产品、旅游纪念品、服饰的销售等这些与当地活动没有关系的物品出现。音乐家的表演可以适当地安排到集市中来。

4．确保城市广场具有大面积铺装，减小种植面积

对比其他的城市公共空间形式，城市广场是铺装面积最大、种植面积最少。种植可能仅限于盆栽或池栽。或者自然的元素被仅仅浓缩为一个喷泉。为了体现主导作用，铺装应该使高质量的材料和精湛的工艺。实际上，铺装应该被当做一种艺术来看待。米兰的长方形主教堂广场，1.7公顷的户外空间是非常规整的矩形几何铺装。深灰色的石材与教堂白色大理石的外立面形成了美妙的对比。罗马的坎皮多里奥广场和澳门的参议院广场是图案化铺装的典范，后者在20年前用石材重新铺装成经典的黑白波纹。这种鱼鳞纹理朴素且不易过时。需要注意的是：道路铺装的材料，包括公园中游园道的材料不适合用于城市广场（图5.6 - 图5.9）。

亚洲最大的广场——大连星海广场，稍有不同，它占地面积超过了50公顷（图5.10）。除了中心区域2公顷的铺装，其他地方为种植区，并且禁止人们进入。为了庆祝1999年香港回归而建的华表广场，占地48公顷，同样也是禁止入内，椭圆形的场地被一些反射性的道路穿过，同时也被一条椭圆形的道路一分为二，这种巨大空间感觉不像是室外的围合空间，也不是很有活力。西方的许多空间以前也有过类似的做法。

图 5.5
派力奥赛马节，坎波广场，锡耶纳
The Palio Horse Race, Piazza Del Campo, Siena

这个在城市广场举办特殊活动的著名例子吸引了不计其数的观众。活动时用泥土来盖住砖块铺装，活动后便立刻去除（另见图1.26）。
This famous example of special event programming for a Civic Plaza attracts thousands of spectators. Dirt is brought in to cover the brick paving for the event and promptly removed afterwards.(See also Image 1.26) (Credit: Roberto Vicario)

sequence of images. Five-block-long Times Square in New York City has evolved from an awkward X-shaped street intersection into an outdoor room surrounded by theaters and retail and commercial LED displays and billboards, a dynamic, visually exciting place to visit and a focal gathering place for civic celebrations very much in the spirit of conviviality, notably on New Years Eve with its famous ball drop held annually since 1907 and attracting a million people. In both places, the atmosphere of the space depends on the cooperation of private abutting owners. Both cities require owners to light up their facades facing the public space. A recently installed billboard in Times Square is powered by wind and solar energy. Crown Fountain in Millennium Park in Chicago displays supersized, ever changing faces of ordinary people that animate and playfully spout water from their mouths into the watery plaza area where children delight in splashing themselves.

Programming addresses deeper community aspirations than comfort and relaxation. Users want to watch, be entertained and even be moved. Designed amenities should deliberately provoke active engagement and discovery, tapping memory and emotion. By inspired design, urban becomes urbane, and humanity acquires urbanity.

Programmed activities should reinforce the uniqueness of the locale and its region. Some events are annual, like arts festivals or the famed Palio, the colorful medieval pageant and horserace held in the Piazza del Campo in Siena. (5.5) Other events are held weekly or even more often. Outdoor markets of locally grown produce and baked goods, flowers, local crafts and artworks, are a popular use of some Civic Plazas, and not just traditional market squares in European cities. The Civic Plaza is not primarily a commercial space, so the market activity, whether a farmers' market or craft fair or a combination, should be restricted in hours, perhaps held only on certain days, like the weekend, or in a certain season. For the seasonal Christmas market in Annaberg-Buccholz, Germany, small shops in temporary structures surrounding a tall Christmas tree, all adorned with Christmas lights, attract crowds even in the snow. It is very important that the market stalls or booths not be permanent installations, because that would preclude any other activity when the market is closed. The vendors need to be well regulated, by a local government agency or by an association working closely with a local government agency, to assure fair allocation of space, control the use of awnings, umbrellas and other display design features, assure sanitary conditions and daily clean-up, and prevent the selling of factory-made goods, novelties, clothes and the like that do nothing to celebrate creativity of local people. Some outdoor markets even allow musicians to perform.

4. Civic Plazas are largely paved. Planting may be minimal.

In comparison to all other urban public open space types, the Civic Plaza is the most paved and least planted. Planting may be restricted to planters or pots. Or Nature may be evoked simply by the water of a fountain. Given its dominance, paving should be of the highest quality materials and workmanship. Indeed, the paving pattern itself should be a work of art. In the rectangular Piazza del Duomo in Milan, the 1.7 hectare outdoor room is paved in formal rectangular geometry; its dark grey-stone contrasts nicely with the white marble of the magnificent cathedral. The Piazza del Campidoglio in Rome or Senate Square in Macau are examples of how iconic a paving pattern can be; the latter was repaved two decades ago in its signature black and white waves of stone setts. The fishscale pattern is a more understated and timeless choice. Road paving material, acceptable in other park types for trails, is never acceptable in a Civic Plaza. (5.6-5.9)

There are exceptions. The so-called largest city square in Asia, Xinghai Square in Dalian, covers 50 hectares. (5.10) It is mostly planted and off-limits to people, except for a paved centerpiece of 2 hectares, the Ornamental Column Square, which commemorates the return of Hong Kong to China in 1999. The off-limits 48 hectares form an ellipse crossed by radial roadways and bisected by an elliptical roadway. A space so vast does not feel like an outdoor room, nor is it alive with activity. It projects power and status. Many spaces in the West used to do the same.

5. The perimeter buildings on adjacent parcels are continuous enough and tall enough, to define an outdoor room and not so tall that they overwhelm the scale of the space. The space is human in scale.

The best Civic Plazas are described as outdoor rooms. Scale matters. Larger does not mean better. If the space is too wide or too long, the sense of being a room is lost. People on the far side of the space should not be so far away as to seem like faceless specks. As a general rule, a horizontal dimension in excess of 250 meters is excessive.

Civic Plazas feel like outdoor rooms when the architectural enclosure is nearly uniform in height. Conversely, Civic Plazas lose much of their appeal when the enclosing perimeter is interrupted by large gaps, such as vacant lots, parking lots, or wide roadways. Tall perimeter buildings not only dwarf the scale of the space but also cast long shadows. The visual power of Tiananmen Square in Beijing, the

图 5.7
澳门，参议院广场
Senate Square, Macau

这个有着欧洲元素的迷人的城市广场比罗西奥广场小，但没有机动车，因此吸引了大量的行人。
Smaller than Rossio Square, this charming Civic Plaza with European character is free of vehicles and therefore a magnet for pedestrians. (Credit: Thomas M. Paine)

图 5.8
四方街，大研古镇，丽江
Sifangjie Square, Dayan, Lijiang

这里是联合国教科文组织所定的世界遗产之一。这个城市广场的精致之处在于它谨慎地保留了周边的建筑、石头铺地以及像无广告白色阳伞这样并不突兀的新特色，并且没有车辆。
The exquisite character of this UNESCO World Heritage site Civic Plaza is sensitively preserved in the perimeter architecture, stone paving, unobtrusive new features like the white umbrellas free of advertising, and absence of vehicles. (Credit: Thomas M. Paine)

图 5.6
罗西奥广场，里斯本
Rossio Square, Lisbon

尽管它的铺装样式和澳门的参议院广场相同，它的喷泉也与波士顿公园而来的喷泉类似，但它对周围环境关系的处理是独一无二的。但周边建筑广场之间机动车的介入在某种程度上抑制了人们对广场的使用。
While it shares its paving pattern with Senate Square in Macau, and its fountain design with Boston Common, its contextual intricacy is uniquely its own. The intrusion of vehicles between perimeter buildings and the space somewhat inhibits its use by people. (Credit: "Ceinturion at the English language Wikipedia")

图 5.9
平展鱼鳞式的铺装，奥登广场，巴黎
Slab and Fish scale paving, Place du Panthéon, Paris

即使简单的石铺装也可以体现朴素的优雅和耐久性。平展鱼鳞式的图案的尺度在实际铺设中取决于工人跪在地面上时的臂展。
Even simple stone paving can be a thing of understated elegance and durability. The fish scale pattern in the center of the space is practical, being determined by the reach of the kneeling mason's outstretched arm. (Credit: Freepenguin)

图 5.10
星海广场,大连
Xinghai Square, Dalian

越大并不都意味着更好,将来的某天这个巨大的空间应该被重新设计,使其吸引更多的活动,变得更有可持续性。广场抛光的地面在雨天时会变得很滑。
Bigger does not always mean better. Someday this huge space will be redesigned to attract more kinds of activity and be more sustainable. The polished surface can be slippery in rain. (Credit: Thomas M. Paine)

图 5.11
纳沃纳广场,罗马
Piazza Navona, Rome

这个户外空间的尺度很完美,看起来紧凑但是又足够宽敞,并且让游客们感觉很舒服。他的尺寸是 65 米 ×240 米。周边虽然在细节上不同,但是尺度很统一,因而既避免了混乱又避免了单调。
This outdoor room is perfectly scaled to seem intimate yet roomy, and accommodate many visitors comfortably. It measures 65 by 240 meters. The enclosing architecture is varied in detail but uniform in scale, avoiding both chaos and boredom. (Credit: Ee60640 at en.wikipedia)

5．周边建筑应有足够的连续性和高度来围合中间的室外空间。但是它们又不能太高以至于破坏了空间的尺度感,空间应该是人的尺度感

最佳的城市广场被描述成户外围合空间,周边建筑应该有足够的连续性和高度来围合中间的室外空间,但并不是越大越好;太宽或者太长也会让围合感缺失,不能让位于空间两端的人看彼此只是没有五官的小黑点,通常水平长度超过 250 米,就会偏大了,也不能太高,那样就破坏了舒适的尺度感。

Bund in Shanghai, the Place de la Concorde in Paris or the Piazza Navona in Rome would be much diminished if tall buildings were permitted next to them. (5.11) A city that truly values its Civic Plaza as essential to its identity is not going to give in to commercial pressure to allow high-rise development next door. The urban master plan will set strict height restrictions.

The ideal architectural perimeter provides retail, cafés, and public art at ground level and is permeable, that is, accommodates lots of foot traffic in and out of the perimeter buildings. The best Civic Plazas enjoy a perimeter of pedestrian activity that is at least a block deep, and can be spotted a block away. For example, the Piazza de Duomo in Milan links directly through a huge open archway to the iconic Galleria Vittorio Emanuele II, a 200-meter long retail arcade with vast skylights, open day and night. (5.12) Perimeter buildings that are "dead" and "impermeable" at street level undermine the vitality and visual interest of the Civic Plaza. (5.13)

6. Perimeter transportation should not overwhelm the serenity of the Civic Plaza.

Perimeter traffic should not cut off the flow of people from perimeter buildings and sidewalks into the Civic Plaza. In many Western countries, the "traffic calming" movement has increased the livability of Civic Plazas and Downtown Parks by making the adjacent roadways deliberately narrower, forcing traffic to slow down, and providing "bump-outs" or "neck-downs" at pedestrian crosswalks located in the middle of the block as well as at each end of the block to increase pedestrian access and safety. In addition, cities may close perimeter streets to vehicular traffic on Sundays so that pedestrians can stroll in the roadway itself.

图 5.12
大教堂广场，米兰
Piazza del Duomo, Milan

这个城市广场很幸运地被周边尺度一致的建筑围合，最重要的是大教堂（右侧，正在被装修）与著名的埃马努埃莱二世长廊都由巨大的拱门进入。广场的铺装和室外家具都很优雅。机动车是禁行的。人们涌入广场聚集在临时大屏幕下观看世界杯。
This Civic Plaza is fortunate to be enclosed by architecture of consistent scale, most importantly the cathedral (right side, being repaired) and the famous Galleria Vittorio Emmanuele II retail mall entered through the huge archway. Paving and furnishings are elegant. Vehicles are restricted. Crowds flock here to watch World Cup broadcasts on a temporary large screen. (Credit: Dodo)

图 5.13
市中心公园，丹佛
Civic Center Park, Denver

市中心公园的设计由于缺乏对周边地区的设计掌控力而面临很多的挑战，不把进入空间的主景挡住是重要的，但是在这个案例中强有力的边界种植抵消了一部分周围建筑的无序感，这种无序感减弱了公园空间的几何感。
Civic Center design faces many challenges from the lack of control over the design of adjacent parcels. It is important not to block the main vistas into the space, but here stronger edge planting could offset the perimeter architectural forms whose randomness undermines the strong geometry within the space. (Credit: Vertigo 700)

当周边建筑的高度很一致时，城市广场像是一个户外围合空间。相反的，当周围建筑有大的空隙，比如空场地、停车场或者宽车道时，城市广场就失去了大部分的吸引力。周围高的建筑不仅使空间看起来变小了，并且会有大面积的阴影区。如果允许北京天安门广场、上海的外滩、法国巴黎的协和广场或者罗马的纳沃那广场周边有高楼，那它们的体量感将大大减少（图 5.11）。如果一个城市真正在意它的城市广场，并把它作为城市的重要形象标志，那么就不会屈服于商业利益而允许过高的建筑存在于城市广场周围。城市总体规划应该限制高度。

理想的周边建筑应该设有零售店面、咖啡店，同时，人们还可以在地上进行公共艺术，并且广场应是通透的，也就是说，允许大量的步行交通进进出出。好的城市广场周边应该有至少一个街区深的行人活动区域。例如米兰的大教堂广场通过一个巨大的拱形门廊与一个 200 米的商业零售拱廊——埃马努埃莱二世长廊直接相连，上面有巨大的天窗，并且，昼夜营业（图 5.12）。如果街道旁的建筑是"死的"或者"不透气的"，则会大大减少城市广场的活力与魅力（图 5.13）。

6．周边运输不应打破城市广场的宁静

去城市广场的人们不应该被周边的交通所阻隔。在很多西方国家，"交通减速"运动增加了城市广场和市区公园的活力，"交通减速"是通过故意缩窄临街道路，使交通减速，在街区的中间或者末端提供"瓶颈式的"或"凸出式的"人行横道，增强步行的可达性和安全性。不仅如此，城市还应在周日的时候限制广场

伦敦的特拉法加广场（图5.1）最近就在移除它一边的交通线路之后极大的获益，这个做法使它与文化地标性建筑——国家美术馆之间没有缝隙。同时使广场的使用率增加了250%。纽约的时代广场（图5.2）也从移除与百老汇相邻五个街区的行车道中受益，这一举措使之变为一个步行广场。上海的外滩通过减少侵入型的交通和提供更便利的人行通道而获利，人们在行走的同时还可以欣赏黄浦江中的游船。数世纪以来罗马坎皮多里奥广场和澳门的参议院广场都允许汽车行驶，但这最终还是被禁止了，只有这样，建筑周边和人行空间才会紧密相连。当然，独特又宏伟的铺装设计也因此获得了更多的展现机会。

7．从周边街道看城市广场的视野不应该被阻隔

当城市广场的周边不是建筑而是街道的时候，广场的边界应该被围合起来。大多使用较低的护栏或者有铁链的护栏，所谓"道牙的魅力"，是因为这是广场给访问者的第一印象。这是没有重来的机会的。不管白天还是晚上，城市广场都要时刻迎接人们，因此广场的周围也应该保持通透性，人们的视线可以穿过广场，并且可以在不是车行交通的任何地方进入广场。适用于残疾人使用的人行道牙应在一定间距内重复设置。大门并不适用于城市广场，除非这个广场在建造之初就有大门。因为广场四周并没有具体的围墙，而且在夜晚广场也不会关闭。为了防止流浪汉的借宿，监视和强制措施必不可少，当然要很小心地处理，不能以暴力解决。（图5.14-图5.15）。

8．城市广场应包含食物供应以及座椅

如果周围没有提供户外座椅的咖啡店，那么城市广场应该在不妨碍其他活动的前提下，在合适的空间尺度上设置固定的长椅或者可坐的矮墙，食品贩卖亭或者室内/室外咖啡店都应该提供室外桌椅，但是要制定规范去控制桌椅的质量，0.5公顷左右的城市广场适合放置一个食品贩卖亭。澳门参议院广场（0.5公顷）的食品亭就隐蔽地放在了树下。同时也为这个铺装为主的广场一侧提供了树荫下的座椅区域（图5.16）。如果城市广场没有食品供应区，可移动的座椅也是很合适的，纽约时代广场现在就提供可移动的座椅。

Trafalgar Square in London (5.1) has vastly benefited from the recent removal of vehicular traffic on one of its sides, allowing the space to flow seamlessly to the National Gallery, a major cultural landmark; that change increased usage of the Civic Plaza by 250%. Times Square in New York City (5.2) has benefited from the recent removal of traffic lanes from 5 city blocks along Broadway and their transformation into a pedestrian plaza. The Bund in Shanghai has benefited from reduced vehicular intrusion and better pedestrian accessibility, while pedestrians enjoy overlooking another mode of transportation, the boat traffic on the Huangpu River. Vehicular traffic, long allowed within the centuries-old Piazza del Campidoglio and Piazza Navona in Rome and Senate Square in Macau, was finally banned, allowing the architectural perimeter to meet the pedestrian zone seamlessly, and showcase the magnificent paving design.

7. The view into the Civic Plaza from perimeter streets is not blocked.

Where the edge of the Civic Plaza abuts a street rather than a building, the treatment of the perimeter edge between crosswalks should be enclosed, if at all, only by a low railing or post-and-chain fence. So-called "curb appeal" is important because it offers visitors their first impression. There is no second chance to make a first impression. Visual access from the perimeter is essential. The Civic Plaza should welcome all people at all times, day and night, and as such the perimeter should remain transparent so that people passing by can see right across the space and, except where the space is edged by vehicular traffic, enter at any point. Curb cuts to accommodate the disabled and wheelchair users should be provided at regular intervals, typically no more than a block apart. Except for instances where the space historically had them, gates are inappropriate. Because there is no perimeter enclosure, there is no closing at night. Preventing vagrancy, loitering and sleeping overnight requires surveillance and enforcement, but it should be done

城市广场规模
Civic Plaza Scale

表5.1
Table 5.1

位置 Location	米 Dimensions in meters	英尺 Dimensions in feet	说明 Notes		资料来源 Source
西方 West	80 × 125	265 × 410	中等	average	1
欧洲 Europe	50 × 150	190 × 495	中等	average	2
罗马 Rome	33 × 100	110 × 320	中等	average	2
罗马人民广场 Piazza del Popolo, Rome	100 × 130	340 × 430	缺乏魅力	uninviting	3
那不勒斯普雷比席特广场 Piazza del Plebiscito, Naples	150 × 150	500 × 500	缺乏魅力	uninviting	3
佛罗伦萨圣彼得教堂广场 Piazza San Pietro, Florence	210 × 240	690 × 790	宗教使用	religious use	1
大连星海广场 Xinghai Square, Dalian	610 × 1067	2000 × 3500	亚洲最大	largest in Asia	3

资料来源 (Sources)：(1) Robert F Gatje, *Great Public Squares, an Architect's Selection* (New York, W. W. Norton: 2010); (2) Robert Tullis AIA, lecture at BuildBoston, Boston, MA, November 2010; (3) Google Maps.

图 5.14
联合广场，旧金山
Union Square, San Francisco

这个倾斜的场地中心是一块平整的铺装区域，嵌套在几何形草坡条带中，整个广场比典型的城市广场有更多的植被。可坐的墙能够同时满足很多人就坐，并且在空的时候不会显得荒凉。
Within the heart of the sloping site is a flat paved area, nested within geometric lawn strips and more planting than is found in the typical Civic Plaza. The sitting walls can accommodate many sitters at one time and never seem deserted if empty. (Credit: Thomas M. Paine)

图 5.15
波茨坦广场的索尼中心，柏林
Sony Center at Potsdamer Platz, Berlin

包含在一个私人开发项目中的广场受益于控制周边的空间，从而有一个富有想象力的统一设计，就如照片中这个被水景装饰的室内广场。这个广场也展示了一个被毁宫殿的立面。
Plazas contained within a private development project may benefit from control over the perimeter of the space and result in imaginative and unified design, as in this indoor space animated with water. The plaza also showcases the façade of a destroyed palace. (Credit: Stefan-Xp)

sensitively, certainly free of brutality. (5.14-5.15)

8. The Civic Plaza includes food services and chairs.

Stationary benches and sitting walls are to be expected, but if the perimeter private sector does not provide any cafés with outdoor movable seating, then the Civic Plaza itself should do so on a scale that is appropriate to the size of the space and that does not preclude other activities. The food kiosk or indoor/outdoor café should include outdoor tables and chairs. This concession should be carefully regulated to assure quality. No Civic Plaza over 0.5 hectares is too small for at least a food kiosk. In Senate Square in Macau (0.5 ha.) the food kiosk lies unobtrusively under trees that provide a shaded sitting area to one side of the dominant paved area. (5.16) Even without food services within the Civic Plaza, movable chairs may be justified. Times Square in New York City now provides movable chairs.

9. Architectural elements within the Civic Plaza are unobtrusive.

The main focus of the space should not be a large structure. Whatever structure is required—information kiosk, small café, restrooms, subway entrance—should not dominate the space as if the outdoor space were simply the setting for the building. That structure should be of the highest design quality, worthy of such an important location, should be sustainably designed, but should remain subservient to the outdoor space. (5.17)

10. Sensitively designed lighting encourages lively nighttime activity.

Particularly in hot and humid parts of the world, the night is a terrible thing to waste. With soft lighting, the Civic Plaza could provide an ideal

图 5.16
参议院广场，澳门
Senate Square, Macau

即使在这个小的空间里也留有地方给咖啡店以及座椅，以便于人们在树荫下小憩以及享用零食。
Even this small space has room for a café and chairs so people can stay for a while and enjoy a snack in the shade of trees. (Credit: Thomas M. Paine)

9．在城市广场中的建筑元素应不显突兀

大型的建筑结构不是焦点，不能让城市广场成为这些建筑的附属。不管是信息询问亭、小咖啡店、洗手间还是地铁入口都不应该占主导地位，它们在位置选择上应该为广场空间让路，但是要保证这些的设计高质量，使其值得占据这个重要的位置，并且做到可持续（图 5.17）。

10．易感灯光设计为夜晚活动添彩

夜晚不能被浪费，特别是在世界上拥有炎热和潮湿天气的地区，有着柔和灯光的城市广场是休闲放松和跳舞的理想场所。一个管理良好的夜间市场，可以在固定时间段内给游客提供本地工

图 5.17
都市阳伞，塞维尔亚
Metropol Parasol, Seville

建筑通常不应该支配城市户外空间，也不能试着用大胆的陈述来无视当地的文化。但这个例子可能会诱导人们不遵守这个建议。巨大的木制小品赢得了赞赏者并且吸引了人群到户外的中央市场。五年以后它会怎样还需要继续观察。
Architecture generally should not dominate urban outdoor space and not try to make a bold statement that ignores local culture. This example may tempt others to disobey that advice. The huge wood structure has won admirers and attracted crowds to the outdoor central market. How well it is maintained after five years remains to be seen. (Credit: Sander Westerveld)

艺品和小吃，但是不能让它像普通市场一样：车道边整晚都过于热闹，而且每个摊位的照明强度都要受到监管和调节。

两个音乐公园：水上篝火公园与吉米·亨德里克斯公园

罗德岛州普罗维登斯市的城市广场和绿道坐落在一个火车站旁边，俯瞰着以前毫无生机的河流，直到巴纳比·埃文斯这位艺术提倡者去设计了一个水上空间。许多个夏季夜晚都有这项活动，水上篝火表演吸引了成千上万的人去见证水、火与音乐在傍晚的壮丽交融。漂浮的火柴堆被无声木船所点燃，随着缓慢多变的音乐触动着听众。人们沉迷于篝火，拥抱、跳舞或者表达爱意。这种情绪相当有感染力。埃文斯把他称作共同体。他甚至雇用演员在观众中表演模拟情景剧，这种形式深深吸引了临近的看客。水上篝火赢得了人们的心，并且无限地扩大成为了普罗维登斯的城市品牌。

为了纪念被称为史上最棒的摇滚吉他手的本土音乐传奇人物，西雅图准备建造 1 公顷的吉米·亨德里克斯公园。亨德里克斯以他的"墙的声音"而闻名，在"墙的声音"，听众在可以想象的三维空间和运动中幻想自己精神世界中的风景。村濑协会设计了 10 米长的混凝土"墙的声音"，这个设计将由喇叭播放音乐，并且给游客提供可插电的乐器。豆荚形的两个座椅区域的设计灵

venue for leisurely relaxation and dancing. A well regulated, limited night market offering local handicraft goods and snacks to tourists is compatible but cannot replace the more extensive, messy excitement of a night market on a street closed to night-time traffic, ideally located nearby. Any lighting that the market vendors provide to supplement the lighting on poles should be closely monitored and regulated.

Two Parks with Music: Waterfire and Jimi Hendrix Park

The Civic Plaza and Greenway that Providence RI created on the site of a railyard overlooking a daylighted river remained lifeless until the city partnered with Barnaby Evans, an arts advocate, to program the space. An event held on many summer evenings, Waterfire attracts thousands of participants to witness the magical interactivity of water, fire, and music at dusk and into the early evening—floating pyres are lit from noiseless wooden boats, as music from a wide variety of musical traditions soothes the audience. Recorded music is actually preferred to live performances to avoid the usual passivity of concertgoers listening to live musicians and allow people more freedom for spontaneity, movement, and social interaction. Hypnotized by the fire islets, people hug, dance, and otherwise openly display affection. The mood is contagious. Evans calls it communitas. He even hires actors to stage mock scenes among the onlookers—incidents of park theater that captivate nearby onlookers. Waterfire has won the hearts of the people and added immeasurably to the "brand" of the city of Providence.

Seattle is poised to create a 1-hectare Downtown Park near a museum to celebrate the musical legacy of native son Jimi Hendrix, considered by many the best rock and roll guitarist who ever lived. Hendrix was famous for his "wall of sound" in which listeners imagined a mental landscape with palpable three-dimensionality and movement. The proposed design by Murase Associates features a 10-meter concrete "Wall of Sound" that will have speakers for playing music and places for visitors to plug in an instrument. Two seating areas that look like pods are based on a drawing by Hendrix. A 50-meter long wall made of perforated steel will have cutouts including one in the shape of Hendrix playing his guitar. Visitors can walk through the cutouts. There is a stage for musical performances. Paulownia trees, which produce purple flowers, will honor Hendrix' hit song **Purple Haze**. The park provides a multisensory immersion experience, in its way recalling "the Jimi Hendrix Experience" and memorable three dimensionality of hits like **All Along the Watchtower**.

■ Downtown Park

As important and visible as the Civic Plaza, the Downtown Park likewise has vast potential to make a difference in people's lives, in the social sustainability of the urban community, and in the success of the community "brand". The Downtown Park too lies in the heart of the city, near diverse uses (retail, commercial, transit). While some examples are called a public garden, the word "park" likewise connotes the strong presence of Nature - trees, shrubs, perhaps lawn and water. Hectare for hectare, the Downtown Park cannot contain as many people at one time as the Civic Plaza because at least a third of the space is given over to planting or a water feature. Although it may not accommodate as many users as the Civic Plaza, users may have more reasons to linger longer, because there are more places to relax, alone as well as in groups, and there is more dappled shade under trees, rather than the dull flat shade cast by high-rise buildings. The Downtown Park will attract as many out-of-town visitors as the Civic Plaza. Therefore its role in setting the "tone at the top" and influencing public opinion about "the brand" of the city as a whole is, like that of the Civic Plaza, enormous.

1. The Downtown Park is a lush oasis.

The design of the Downtown Park should impart the experience of tranquility and multisensory delight, an experience to be shared with other visitors in a spirit of mutual serenity. What makes a Downtown Park truly memorable and powerful is when it exceeds one's expectations of what is possible in a public space, aesthetically and programmatically, in a dense downtown location, and offers so much, especially in the way of planting and water features, that the visitor wants to return again and again. Simplistic design is unlikely to

感来源于亨德里克斯的绘画。一个50米长的墙壁上面都是打孔的不锈钢形成的一些图案，其中一个是亨德里克斯弹吉他。人们可以在那些图案中间穿梭。公园中还有提供音乐表演的舞台。公园中种植着紫色花朵的泡桐树，是为了纪念亨德里克斯的歌曲"紫雾"。公园提供了多重感官的体验，以此来纪念"亨德里克斯体验乐队"和值得纪念的三维弹奏"沿着望塔"。

■ 市区公园

市区公园和城市广场一样，对人们生活的影响是巨大的，特别是城市社区可持续性和社区自己的品牌效应方面，尤其的重要和明显。市区公园同样和城市广场一样，位于市区的中心，周边功能多样化（零售、商业、交通）。有些案例被叫作公共花园，"公园"的字眼意味着有着强烈的自然气息——乔木、灌木甚至草坪以及水景。它至少1/3的面积是种植区或者水景，所以市区公园不能像城市广场那样一次性容纳那么多的人，但是使用者很可能在这里逗留更久，因为里面有更多的个体或群体休闲空间，树荫也更多。市区公园会像城市广场一样吸引外地游客前往。因此，它在"决定城市上限"和打造城市品牌方面的影响，同城市广场一样，是巨大的。

1. 市区公园是茂密绿洲

市区公园应该提供给游客宁静和愉悦等多重感觉，游客应平静地分享彼此的视觉感受。怎样的市区公园是出挑的并且让人记住的呢？就是当它超出了人们对于公共空间期待的时候。想象在城市中有那么一公园，在美学上和空间结构上对植物和水景的安排都那么的精彩，怎么会不让人流连忘返！既简约又精彩不是那么的容易，漂亮的亲民的设计需要不断地强调每一个细节，无论种植还是结构，使用优质的、耐久的材料，做到这一点就会让一个场所与众不同，在城市的中心，一个潜心设计的场所将直通人心，就想城市的心一样。

种植可能使我们回忆起传统园林里面对于乔木、灌木、草坪、地被、多年生和一年生的花境的运用。那是种在城市荒漠上种植出绿洲的保险方法。盆栽的植物可以做墙体绿化或者"绿植雕塑"。如果形式风格很简单，假如只是种植成排的树或者在铺装广场上的树池里种树，那么绿洲效果就大大减少了，尽管可能因为有现实问题（人们需要在铺装上行走）弱化了这个绿洲的效果。

建在地下停车库混凝土顶板上的市区公园，应该有1.5米左右深的土壤保证植物茂密地生长，同时也让地形不再成为约束（图5.18）。

2. 因特色的活动或者关注广泛的公众效应而著名，市区公园在社区的文化生活中扮演重要角色

公园设计应该体现出一个建筑焦点，比如水景、表演舞台或者雕塑等等，21世纪科技允许人们以新的方式展示自己，比如安装创新的艺术装置和媒体设施。这些装置最好应用在硬质景观区域，而不是软质景观区域。文化寓意通过这些设施传播，或者体现在市区公园设计中的形式和颜色上，这些寓意应该让使用者明了，而不仅仅是设计者自己知道，这些信息既不应该很不易亲近也不应该很做作（图5.19 - 图5.20）。

图 5.18
诺曼·B·利文撒尔公园,波士顿
Norman B. Leventhal Park, Boston

这个公园只有 0.7 公顷并且建在一个地下停车场上面,通过细心的种植和区分不同的子空间,比如喷泉区、草坪区和藤架区,使人觉得空间要比实际大得多。这个绿洲是市区办公室员工的最爱(另见图 1.112,图 4.7,图 4.10,图 4.12)。
Only 0.7 hectares in size and built over an underground garage, this Downtown Park manages to feel much bigger through careful planting and distinct subspaces like the fountain area, lawn and trellis. This oasis is a favorite with downtown office workers (See also 1.112, 4.7, 4.10, 4.12). (Credit: Thomas M. Paine)

图 5.19
公共花园,波士顿
Public Garden, Boston

这个标志性的传统市区公园是一个郁郁葱葱的绿洲,它被植物和中心水景所主导。中心水景的设计让人看起来像个自然的水池,当然还配有著名的天鹅船(另见图 4.16)。
This iconic traditional Downtown Park is a lush oasis dominated by planting and a central water feature designed to look like a natural pond and yet accommodate the Swan Boats. (See also 4.16). (Credit: Thomas M. Paine)

图 5.20
千禧公园,芝加哥
Millennium Park, Chicago

这个相对较新的市区公园变成了城市当前的地标,它用混合硬质铺装城市广场,与另一个市区公园相接的多个花园与空间以及和湖前的绿道取代了以前的铁路场院,人们坐在巨大的草坪上欣赏免费的午餐时间音乐会。
This relatively new Downtown Park became an instant icon for the city, replacing a rail yard with a mixture of hardscape Civic Plaza and garden subspaces linked to an adjacent Downtown Park and the lakefront Greenway beyond. People sit on the large lawn area for free lunchtime concerts. (See also 1.116). (Credit: Thomas M. Paine)

accomplish that. Beautiful, user-friendly design executed durably in the best materials, whether planted or constructed, down to the last detail, will instill pride of place. A place beautifully designed speaks to the heart, from the heart of the city.

The planting may recall traditional gardens in the use of canopy trees, shrubs, lawn, ground covers, perennial flower beds, and annuals. That is the safe approach to create a lush landscape effect, like an oasis in the city desert. Potted plants can be suspended on wire frames forming green walls or shaped into free-standing sculptural forms. If the style is minimalist, for example, simple rows of trees set in a bed of ground covers or in tree grates set in paving, then the oasis effect is diminished, although there may be practical reasons to fall short of this important goal.

Downtown Parks that are built on top of a concrete deck over underground parking should allow for the weight of sufficient soil to approach a depth of 1.5 meters to allow for planting to be lush and for landform design to be unconstrained. (5.18)

2. Known for its signature activity or focus of wide public appeal, the Downtown Park plays a strong role in the cultural life of the community.

The design should showcase a focal landscape feature rather than architectural one. Examples include a water feature, performance pavilion or sculpture, but twenty-first century technology allows communities to showcase themselves in new ways, such as installing innovative art forms and media. Such installations are best confined to hardscape areas rather than the softscape areas. The cultural meaning conveyed by such installations, or embedded in the forms, patterns and colors that go into the design of the Downtown Park, should be understandable to users, not just designers. The message should be neither aloof nor condescending. (5.19-5.20)

图 5.21
城市广场,绍兴
City Plaza, Shaoxing

市区公园西南角的壁画叙述了当地的历史,广场的设置如花园一般,并且靠近一处唐朝的宝塔以及一个新的歌剧院。
Murals in the southwest corner of the Downtown Park provide a narrative of local history in a garden-like setting near a Tang dynasty pagoda and a new opera house. (Credit: Thomas M. Paine)

图 5.22
奥林匹克雕塑公园,西雅图
Olympic Sculpture Park, Seattle

以前是工业场址的地方现在有着文化纪念活动以及提供了水边的休闲。废弃的场地越来越多地被转变成新型城市开放空间。这个巨大的斜坡由西雅图艺术展览馆管理。海岸线给濒危的鲑鱼提供了一个天然的栖息地。
The former industrial site now celebrates culture and offers recreation on the waterfront. Brownfields sites are increasingly being transformed into innovative urban open space. This dramatic sloping site is managed by the Seattle Art Museum. The shoreline recreates a natural habitat attractive to threatened salmon. (Credit: M. O. Stevens)

图 5.23
古当莱尼公园,伊斯坦布尔
Gülhane Park, Istanbul

开放于一个世纪以及前,这个市区公园得益于它的可视通透性边界。
Opened a century ago, this Downtown Park benefits from a visually permeable edge. (Credit: Gryffindor)

As in the Civic Plaza, public art may celebrate local heroes and talents or historic or mythic events or make some unique artistic statement. The swirling design of the Toronto Music Garden, dedicated by renowned cellist Yo-Yo Ma on the city's riverfront, is inspired by the dance suites for cello of composer Johann Sebastian Bach. The Downtown Park may host festivals, performances, outdoor art exhibitions, but in a cultural rather than commercial spirit. Some of these occur annually and may become well known far outside the community. Like the Civic Plaza, the Downtown Park should also

和城市广场一样,公共艺术应该包括纪念英雄、历史事件或者神话故事,或者传达某一独特的艺术观点。位于城市滨河的多伦多音乐花园的漩涡设计,由著名的大提琴家马友友先生投资,灵感来源于大提琴家约翰·塞巴斯蒂安·巴赫的舞蹈组曲。市区公园可以举办节日庆典、表演、户外艺术展览,这些活动应该注重从文化层面开展而不是商业形式。有些活动每年都会发生并且可能变得远近闻名。和城市广场一样,市区公园应该也包括了每周、甚至每天的公共活动,也应该由当地政府代表和私人团体一起管理,进行头脑风暴,去商讨是否需要发放商贩的经营执照、是否允许表演艺术家和团体演出,并进一步的推广(图 5.21 - 图 5.22)。

3. 边缘要质朴,视觉上要通透

用有吸引力的围栏或者矮墙而不是阻拦视线的高墙去围合空间,设计时要保证视线通透和出入口的数量。在外侧设计连续的残疾斜坡通道,同时,缓冲植被应该起到屏蔽外围交通车辆的作用。如果要在围合的地方施工,那一定要快。特别指出的是:围墙设计要有吸引力,这相当的重要(图 5.23)。

4. 景观、室外家具应该很好地维护并保持整洁

市区公园的维护往往比其他类型的公园更高,因为软质景观越多维护量就越大,整个场地应该坚持每周进行维护,并且照顾到每个角落,比如场地边缘的草地很容易被人们踩踏得露出了土壤,这可以用设置低矮围栏(大概低于 20cm)的办法来补救,但是别用视觉隔断空间,也不能禁止人们走进场地。软质景观草

图 5.24
波士顿公共公园
Boston Public Garden

这个波士顿标志性的传统市区公园有着高水平的维护，其中包括步道周围的低矮篱笆，这里不绝对禁止，每日不计其数的游客进入草坪。
The high level of maintenance of Boston's iconic traditional Downtown Park includes low selective path fencing that discourages but does not forbid access to lawn areas by its thousands of daily visitors. (Credit: Daderot)

图 5.25
芳草地公园，旧金山
Yerba Buena Gardens, San Francisco

这个当代市区公园的每个角落都有高水平的维护，包括这个在清水池中具有高度观赏性的植物岛，维护使得水池一直保持清新。
The high level of maintenance of every part of this contemporary Downtown Park, including this highly ornamental planting island in a basin of clean water, keeps it looking fresh. (Credit: Thomas M. Paine)

include weekly or even daily public events, managed through a partnership of local government and a private support group, to issue licenses to vendors and permits to performing artists and interest groups, brainstorm new ideas, and reach out to the public. (5.21-5.22)

3. The edges are pristine and visually permeable.

If there is an enclosure, it should be an attractive fence or low wall allowing views in and out, with access through plenty of gates, rather than a high and forbidding wall blocking views, and with only a few gates. Outside the fence is a continuous perimeter sidewalk with curb cuts at crosswalks. At the same time, buffer planting should screen perimeter vehicular traffic. Damage to the enclosure is fixed promptly. Again, "curb appeal" is of paramount importance. (5.23)

4. The landscape, furnishings, and facilities are well maintained and clean.

The level of maintenance may exceed what is required in other parks types in order to achieve the same standard. Intensely used softscape areas do require more maintenance effort than hardscape

图 5.26
古城公园，上海
Gu Cheng Park, Shanghai

临近豫园和外滩的这个公园有着良好的草坪保养，也许可以尝试着让人们进入，这是目前所不允许的。
A proper level of lawn maintenance in this Downtown Park near Yu Yuan and the Bund could permit access by the crowds of visitors to the areas from which they are excluded. (Credit: Thomas M. Paine)

areas and less intensely used softscape areas. Maintenance should stay consistent week after week, throughout the space. No areas should be chronically under-maintained. Where the edges of paths are bare earth due to people trampling the grass or ground cover, low fencing (as low as 20 cm) in short lengths may help maintain a pristine path edge without visually partitioning the space or forbidding people from walking on the grass altogether. Softscape lawn areas should be fertilized, fed, aerated, top-seeded, and sodded according to a regular seasonal schedule. Trees should be pruned (no trees should be left chronically unpruned), fed, and treated for pests. Damaged furniture should be promptly repaired, and lighting should be kept in working order. Snow removal damage to path edges should be corrected the next spring. Ornamental fountains should be operational during the months when there is no risk of freezing. Litter should be removed before it overflows the receptacles. The restrooms should be clean and smell clean. (5.24-5.26)

5. Maintenance equipment and material are stored off-site.
If the area is small compared to the level of visitation, then the

坪区域应该按照季节规律进行合理的施肥、浇灌和通风，乔木应该定期修剪、除虫，损坏的室外家具要及时地修理；保证照明灯具运行良好；道路两旁的积雪应该在春季前清理完毕；结冰的季节别开启观赏喷泉。垃圾不要堆积，卫生间也要保持干净、味道清新（图5.24 - 图5.26）。

5．维护设施和材料要储存在场地之外
如果相对于使用者的来访量，空间较小，土地就变得更加珍贵，可谓寸土寸金。小于30公顷的市区公园，就不应该放置任何维护设备与材料。特别是使用频率很低的大型机器，就更不应该来占据空间了。比如，只在下雪季节使用的除雪机，应该放在场外的库房，使用时用拖车拖过来。大于30公顷的市区公园，经过隐蔽又巧妙的设计，是可以存放维护设备和材料的，这样既可以保证充足的活动空间，又不会让游客感到唐突。

6．提供可移动的座椅
活动座椅向人们传达了文明友好的信息，甚至可以被称为是一个城市拥有世界级市区公园的标志。它使人们更愿意驻足停留在一片区域，并把这片区域当作一个与他人交流沟通的场所，就

图 5.27 - 图 5.28
静安公园,上海
Jing'an Park, Shanghai

这个成功的市区公园吸引着人群但并没有牺牲质量。由于公园不提供座椅,一些人自己带着座椅。其他人则倾心于草坪坡地,或在池塘边休息。
This successful Downtown Park accommodates crowds without sacrificing quality. Some people bring their own movable chairs, since the park does not provide them. Others love to use the sloping lawn or relax by the pond. (Credits: Thomas M. Paine)

像自家的起居室或花园一样。巴黎、纽约和斯德哥尔摩的许多公园里都提供可移动座椅,卢森堡公园(图 1.56)的座椅成了公园的标志,纽约和波士顿也效仿了此法。

怕被盗窃或者破坏常常成为公园不提供可移动座椅的理由,然而,在纽约中央公园的大都市博物馆,情况是这样的:博物馆前面台阶上提供了 200 个可移动的座椅,并且 24 小时都放在那里。博物馆发现,增补被偷座椅花的钱比每晚去储存它们花的钱少。[3] 最近,刚建完的圣路易斯的城市花园也设置了可移动的座椅(图 5.27 - 图 5.28)。

7.市区公园开设咖啡店或小吃店

人们喜欢在露天下吃东西,公园应该设置这样的咖啡店、小吃店以吸引更多的人停下脚步。咖啡店可以提供可移动的室外桌椅,甚至大部分的座椅都应该在室外。市区公园绝对能容下一个咖啡店。即使在像波士顿的邮局广场公园那样的小市区公园(只有 0.7 公顷)也有空间打造一个轻松惬意的咖啡店。

ground is precious, all of it. In a Downtown Park with an area well under 30 hectares, none of its ground should be sacrificed to storage of maintenance equipment or materials. In particular, large machines that are not used weekly throughout the year should not take up space in the Downtown Park. For example, in locations with severe winter snowfall, unless snow removal is by shoveling, snow blowers should be brought in by truck from a system-wide equipment storage facility located offsite. On the other hand, if the Downtown Park covers more than 30 hectares, it is more likely that, with careful design, expropriation of ground for maintenance equipment and materials storage would remain unobtrusive to the vast majority of users, and still leave adequate space for all program activities.

6. There are movable chairs.

Movable chairs convey a civilized, welcoming atmosphere inviting people to linger in space that functions as the people's community outdoor living room or garden. No other improvement so dramatically signals a city's claim to world-class parks status than this one grand gesture. Paris, New York, London, and Stockholm provide movable seats in many of their parks, and some—like the little chairs in the Luxembourg Gardens (1.56)—have become park trademarks, copied in places like New York and Boston.

Vandalism and theft are often given as reasons for not providing movable chairs. However, this runs contrary to the experience at New York's Metropolitan Museum of Art, located in Central Park. It provides 200 movable chairs along its front steps, and leaves them out 24 hours a day, seven days a week. The Museum has found that it is less costly to replace stolen chairs than to pay for storage each night. [3] Recently completed Citygarden in St. Louis also includes movable chairs. (5.27-5.28)

7. The Downtown Park includes a café.

People love to eat in open space. Cafés provide snacks that encourage rapid turnover to accommodate as many people as possible. The café should include movable tables and chairs outdoors in addition to limited indoor seating. Most of the seating should be outdoors. No Civic Plaza is too small for a café. Even a Downtown Park as small Norman B. Leventhal Park in Boston (0.7 ha.) fits one in effortlessly and elegantly.

8. Active recreational facilities are secondary and subordinate.

Activity areas like a children's playground or *tai chi* area are appropriate. Unless the Downtown Park is not too crowded, allocating space to most types of active recreational activity such as field sports is inappropriate. For example, allocating space for flying model airplanes, camping, or planting community gardens should not be considered. Other examples can be found in the Open Space Activity Table (Table 4.1).

On the other hand, during the winter in temperate or cool climate regions, special programs such as winter carnivals or festivals should be introduced, like Ottawa's cleverly named Winterlude. Detroit's Campus Martius Park is a prime example of how an outdoor skating rink can help bring a great city setting to life. Seasonal holiday night lighting, music, and ice sculpture such as in Harbin's world famous winter festival or St. Paul's winter carnival in Rice Park add a magical feeling. Small outdoor tents can be set up with outdoor space heaters to offer hot drinks such as cocoa and cider as well as soups to warm people up.

9. There are no overhead wires or other intrusions of infrastructure.

The entire space should be open to the clear sky, and defined by elements that are meant to be seen rather than ones that are preferably left out of sight.

10. There is accountability of management.

The public should be able to communicate to park personnel and their supervisors with concerns and suggestions, and expect to receive a thoughtful response. Private uses should be strictly regulated. While occasional use of public space for private functions occurs, it should not become the norm and it should always financially benefit the space by supporting upkeep or underwriting the cost of improvements. Commercial promotion by the private entity should be strictly controlled.

Recent Downtown Park Success Stories

In Xiamen, Snowy Egret City Park is spectacularly located on Egret Island in the heart of the city. Its 16 hectares include a plaza where crowds feed the pigeons imported from the Netherlands and take photographs of the 13-meter tall snowy egret goddess statue perched on a rock in the lake. The statue is the symbol of Egret Island and of Xiamen. Park improvements include a Hong Kong Reunification monument, Chinese zodiac columns, a musical fountain plaza which is popular in the summertime, art and calligraphy museums, galleries, craft gift shops featuring local crafts and plants, restaurants, clubs and sports facilities. A creative lightshow inspired by the lanterns of local night fishing boats returning to YunDang Harbor attracts many visitors. Seen against a backdrop of modern high-rise buildings across the lake, the shimmering reflection is beautiful. Called the "living room" of Xiamen, its most important park delivers the iconic takeway for the city's out-of-town visitors contributing to the Xiamen brand. (5.29-5.30)

In Beijing, SOHO New Town is organized around four 2.5-hectare Downtown Parks themed as follows: ① history & culture, ② art & performing arts, ③ technology & information processing, and ④ nature & science. The private provision of urban public open space is always welcome but this program is particularly exciting and ambitious. The potential to communicate these themes by touching people on an emotional level is limited only by the resources of developer and local authorities, and the imagination of the design team.

In Miami, Soundscape Park (West 8), a retreat of lawn, palms and vine-clad trellises, brings classical music, so often considered remote and inaccessible, into the public realm, free of charge. Simulcasts of New World Symphony Performances taking place inside Frank Gehry's adjacent concert hall are projected onto the exterior façade with

8．活动的娱乐设施应该退居其次

用于儿童活动和练太极的场地可以在市区公园出现，但不能妄想安置所有的活动场地，比如体育场、用于模型飞机和野营的场地是不合适的，否则空间会很拥挤，具体的例子请参看表4.1。

另一方面，在寒冷的冬季，应该多考虑有地域特色的活动。像渥太华有冰雪节，底特律大学的战神广场在冬季就是个户外滑冰场，这给城市带来蓬勃的生气；哈尔滨也有享誉世界的冰雪节；圣保罗市莫斯公园有冬季嘉年华。季节性的灯光、音乐和冰雕给人们带来魔幻的感觉。可为游客提供可以保暖的小帐篷，还可以再来点苹果汁和暖暖的汤。

9．人的头顶不要出现电线或者其他的建筑构造

整个场地上空应该是开放的，在场地中的所有元素都不应该被遮挡或者被忽略。

10．有必要的管理

公众可以向公园管理者、监管人提出建议和意见，也有权利得到合理的回应。私人使用应该被严格控制。虽然偶尔的会将公共空间给私人使用，但是不能变成常规，而且应该从提供场地这点上得到资金的受益。私人商业的用途应该被严格控制。

近期市区公园的成功案例

厦门白鹭洲公园坐落在城市中心的鹭岛上。16公顷的面积，包括一个人们喂养荷兰鸽的广场，湖中岩石上竖立着13米高的白鹭女神雕像，人们可以在这里拍照留念。这个雕像寓意着鹭岛以及厦门。公园还建有香港回归纪念碑、中国生肖柱、书画院、展馆以及贩卖当地手工艺品和盆栽的商品店、餐馆、酒吧、体育设施，夏天有非常受欢迎的音乐喷泉广场。还有不得不提的创意灯光秀，很多游客慕名而来，它的设计灵感来源于当地每晚回港时岸边挂起的灯笼，灯光秀开始时，背景是湖对岸拔地而起的高楼，灯光倒映在水中，随波摇曳。白鹭洲公园被称为厦门的客厅，它也帮助厦门打造了一个独特的城市品牌（图5.29 - 图5.30）。

北京SOHO新村周边2.5公顷的市区公园的主题有：①历史与文化，②艺术与表达，③科技与信息，④自然与科学。这些针对城市公共开放空间的私人条款通常很受欢迎，并且在这个项目上尤其有趣和刺激。由此产生的潜移默化的交流会使人们在情感上有感触，一些特定因素制约着这些主题在情感方面的表现形式，如开发商与当地政府提供的资源或者设计团队的想象力。城市鼓励私人建造公共空间，但这需要很多激励机制和详尽的计划，同时依靠开发商和当地政府资源，再配合优秀的设计团队，才能深入人心地表现公共空间的主题。

迈阿密的声景公园有一个后退的草坪、棕榈以及藤架，公园中免费播放古典音乐，音乐声音缥缈。有次播放了旁边的弗兰克·盖里设计的音乐厅里的新世界交响乐团的演出。余音绕梁的声音从建筑外立面投射过来，从外面缓坡草坪上提升的管道中传来的奇妙的多重体验达到了给不同使用者带来交流的目的。

位于圣路易斯的城市公园（Nelson Byrd Woltz）是一个由某基金会资助建设的共有公园。在1.2公顷的公园中，包含一个雕塑园，内有23座雕塑，欢迎人们触摸甚至爬上雕塑。蜿蜒的座凳矮墙和瀑布墙形式活泼，影像墙则由4家博物馆共同协作提供的不停变幻的影像，特殊的LED灯每3分钟变化一次。咖啡

图 5.29 - 图 5.30
白鹭州公园，厦门
Snowy Egret Park, Xiamen

在市中心白鹭洲上的是被人民称作为"客厅"的市区公园，公园中灯光秀的灵感来自当地渔船上的灯笼，可以跨过在水面上的城市天际线看到，另外公园中还有工艺品店和画廊等文化景点。
In the heart of the city on Egret Island is the Downtown Park called the peoples' "living room," with a light show inspired by the lanterns of local fishing boats seen against the city skyline across the water, and cultural attractions including craft shops and galleries. (Credits: Weiming Ran)

图 5.31
城市花园，圣路易斯
Citygarden, St. Louis

这个最近建好的市区公园有着足够多的变化和亲密感来鼓励人们多次游览，虽然它的周边建筑是写字楼而不是能够吸引游客的商业销售。人们喜欢公园里可以移动的椅子，并且公园富有想象地利用了光和水，如照片中的喷雾广场所示。
This recently completed Downtown Park contains enough variety and intimacy to encourage many repeat visits, despite its less than ideal perimeter of office buildings devoid of retail uses to attract visitors. People love the movable chairs. The park uses lighting and water imaginatively, as in the spray plaza. (Credit: Tyler Burrus)

"immersive sound" from elevated tubes edging the gently sloping sitting lawn for a magical multisensory experience, achieving communitas among diverse perimeter users of this "maximalist" park.

In St. Louis, Citygarden (Nelson Byrd Woltz), a city-owned park whose construction was funded by a foundation, combines in 1.2 hectares a downtown sculpture garden that invites people to touch or even climb on the 23 sculptures, with the playful forms of a meandering seat wall and the sweep of a waterfall wall, the narrative possibilities of a video wall whose ever changing content is managed jointly by four local museums, and special LED lighting that changes every three minutes. A café seats 80 people inside and out. As if all that were not enough, a shallow reflecting pool and a spray plaza choreographed to 10 musical selections attract crowds like a magnet, despite the less than ideal office building perimeter lacking in retail activity or "permeability." There are gardenesque touches like lawn and stonedust paths, sustainable touches like a rain garden. The excellence of design resides in the balance of rectilinear geometry and curving forms, irregular planting and subtly varied topography, creating an engaging environment that invites exploration and many repeat visits. (5.31)

厅有室内和室外座椅，可供80人在这里享受。如果这些还不够，尽管周边不甚理想的办公楼缺乏一些零售活动或"通透性"，然而倒影水池和伴有10种不同音乐的喷淋广场像磁铁一样吸引着人群。这里有传统花园的形式，比如草坪和石铺路，也有可持续的设计，比如雨水花园。这个设计的非凡之处在于它有着几何线条和曲线形式的平衡，有着不规则种植和微地形的平衡，它创造一个有吸引力的环境，邀请人们来一次次的探索（图5.31）。

在休斯敦，绿色探索花园占地5公顷，包括了0.4公顷的湖，0.8公顷的草坪，小树林，表演舞台，4个公共艺术的场地，游戏场地，弹滚球的场地，2个跑狗场地（一个给大狗、一个给小狗），网状的喷泉，湿地公园以及雕塑式的地形，这些全部建在了一个容纳630辆车的地下停车场上面。

在波士顿，城市中心公园由两个相邻但独立的公园组成，波士顿公园和公共花园。在公共花园中，人们划着著名的天鹅船，喂着鸭子。然而波士顿公园的青蛙池塘则提供人们滑冰和涉水的活动。现在，两个公园也提供了类似剧院效果的特定场地，人们对于漫步其中增加了更深入的体验。为公众使用的设计计划包括了露天桌子，雨伞，可移动的座椅，卖食物的手推车，卖书的手推车和一个在午饭时间提供音乐的专业电子乐队。

■ 大型公园

建设大型公园的目的就是为城市中的人们提供接触自然、享受自然的机会。那么让人们接触自然、享受自然就成为大型公园设计最应优先考虑的因素。那些绵延数公顷的公园尽可能地给人们提供一个享受自然的空间，无论是那些喜欢独自漫步的沉思者，还是那些热衷结伴出游的年轻人，他们都无法抗拒这样的公园所散发出的魅力。如果有充足的场地，那么相应的设施也是必需的，尤其是针对各种不同的主题活动，各项设施不应成为制约主题活动或影响活动进程的因素。在这方面，很多公园仍然没有提起足够的重视，预算的缺乏和设计上的缺陷仍然屡见不鲜。当然，真正可以适应各种活动的场地在所有场地中只占很小的比例。

在所有大型公园中，对于场地，很大一部分要为自然保留，并使其免受汽车交通、网球场和娱乐场地结构构造的影响，甚至在大型公园中不应该有高尔夫球场。有些大型公园包含了像城市广场那么大面积的广场，但是只有当人们需要这么大空间时才这么做。再一次强调，当在铺装上花销很大，但是没有达到预期的很熙攘的效果时，没有什么能比这个更不幸了。

大型公园是典型的免费娱乐场所，不像游乐场以及主题公园那样，那是商业公司通过让人们穿过有趣的有镜子的房屋或者惊喜或者吓唬人们，让人们花钱买刺激。大型公园核心的娱乐应该与游乐场的娱乐有所不同。理想状态是，和吵闹的游乐场相比，人们在大型公园中能够获得更为健康的体验，所获得的幸福感也更深刻。这一启发式的想法应用在了莫斯科最近改造的109公顷的高尔基广场上，在改造以前，这里的气氛过于混乱和疯狂以至于常被描述成像精神病院一样。过山车是游乐场最受欢迎的项目之一，但是侵入的形式和过山车产生的噪声，就大大地降低了大型公园的特性。虽然如此，也还是有些例外的，一些大型公园里面也会有旋转木马以及迷你小火车。

在大型公园中考虑到动物园、水族馆以及博物馆是好的。动物园最好选址在别处。他们需要足够多的空间，以至于很容易就从大型公园中分离出。武汉最近把中山公园里的动物园分离出来了。动物园通常要收费，其他外面的公共场所就对所有人是免费的。这当然会引起道德伦理方面的话题，是否将动物关在动物园里，人们就不愿意再去花钱去参观动物园了。动物表演（例如海豚）和动物主题公园当然要放到公共公园之外。这和西方的马戏团有着共同点，马戏团是商业娱乐，从不在公园中举行。博物馆和水族馆需要稍微小一点的空间，也更容易在大型公园里面落户。室内的空间在城区里会变得更加易被体验，因为在那里不会占用人们沉浸在自然中的时间和宝贵的户外娱乐空间。

1．大型公园很少分布在城市中心

城市几乎不会在中心提供大片未经开发的场地。只要社会可持续发展是靠的高密度中心，城市广场以及市区公园就要足够大。人们愿意花更多的时间去中心外围的大型公园去体验。的确，旅程本身就有诱惑力。越来越多的大型公园会占据以前裸露的土地，这需要完整的对自然水系的生态恢复，还包括用新的表层土去取代受污染的土壤并重新引入本土植物。雅典希腊规划了400公顷的海莱尼都会公园，取代了之前的国际机场，计划成为欧洲最大的公园，设计具有可持续性，80%的水都将被一侧收集。

In Houston, Discovery Green (Hargreaves) in five hectares manages to include a 0.4 ha lake, 0.8 ha lawn, grove, performance stage, four public art pieces, playground, putting and bocce areas, two dog runs (for large and small dogs), gateway fountain, wetland gardens, sculptural landforms, all on deck over a 630-car underground parking garage.

In Boston, one Downtown Park consists of the Boston Common and Public Garden, two adjacent but distinct spaces. The Public Garden's famous swan boat rides let people feed the ducks, while Boston Common's Frog Pond offers skating or wading. Now the two spaces are also offering site-specific soundtracks of theatrical productions whose plot takes place in them, allowing visitors an immersion experience adding meaning to their walk around the park. The Common provides patio tables, umbrellas, movable chairs, food truck, book carts, and a professional-grade electric keyboard offering lunchtime music.

■ Large Park

The Large Park is the principal destination for immersion in Nature in the middle of the city. That benefit should take precedence over all design decisions. Parks extending over many hectares offer people the broadest range of possibilities for immersion in Nature in solitude or in the company of friends or fellow citizens. If there is sufficient area, facilities for every programmatic activity that a locale may need can be accommodated, although in many Large Parks the lack of budget or design vision may preclude some of these facilities. In others, the actual usable area suitable for many of the more demanding activities may be only a fraction of the whole.

In all Large Parks, a significant portion of the site should be reserved for immersion in Nature, free of vehicular traffic, free of structured facilities like tennis courts and playing fields for active recreation, free even of a golf course. Some Large Parks may include a plaza on the scale of a Civic Plaza, but should do so only if the crowd levels require it. Again, nothing is quite so unfortunate as a large expanse of paving lying empty when it was anticipated that it would be crowded.

The Large Park is typically free of the activities found in an amusement park or theme park, which are commercial enterprises that require people to pay fees for thrills usually in the form of rides or walking through fun houses with mirrors and possibly elements designed to surprise or frighten people for their amusement. Amusement park entertainment is distinct from the recreation that should be the core of any activity programming in the Large Park. Ideally people go to Large Parks much more often than to amusement parks, and the benefits to health and well-being of regular recreation in relative quiet are much more profound than what is gained at a noisy amusement park. This enlightened thinking has guided the recent restoration of 109-hectare Gorky Park in Moscow, reclaiming it from an unregulated carnival atmosphere described as a madhouse and magnet for unlicensed and decadent businesses. The intrusive form and noise generated by a roller coaster, a favorite ride in many amusement parks, would greatly detract from the character of the

Large Park. Nevertheless there are minor exceptions: some Large Parks include carousels or miniature train rides.

The temptation to consider including zoos, aquariums and museums in a Large Park is great. Many Large Parks created decades ago did decide to include these facilities, and they are hard to move. Zoos require so much space that they could easily dominate the Large Park as a whole; they are preferably located elsewhere. The cluster of cages and animal shelters, like a small city, could easily detract from the landscape character of the park. Wuhan relocated its zoo from Zhongshan Park to a separate site. Zoos which charge a fee belong outside a public park open free to all. In addition, there is of course the ethical issue of whether it is right to subject animals in a zoo to the very stresses that humans seek to avoid when they come to that park. Animal performances (dolphins for example) and animal theme parks certainly belong outside a public park. They have much in common with Western circuses, which are commercial entertainment and are never located in parks. Museums and aquariums require less space and can more easily be accommodated in a Large Park but as indoor experiences make much more sense in a city block where they do not detract from the experience of immersion in Nature outdoors or take precious space away from outdoor activities.

1. The Large Park is rarely centrally located.

Cities do not usually offer large undeveloped sites downtown, nor should they. As long as social sustainability depends on high density downtown, Civic Plazas and Downtown Parks are large enough. People are willing to spend more time to arrive at a more peripheral location to experience the Large Park. Indeed, the journey itself is part of the allure. Increasingly, Large Parks will occupy brown fields sites and require total ecological restoration of the natural drainage system, including new topsoil in place of contaminated soils, and reintroduced native plantings. The 400-hectare Hellenikon Metropolitan Park in Athens, Greece, planned to be the largest in Europe, is replacing the former international airport and will be sustainable, with 80% of its water collected onsite.

2. The Large Park balances ecological preservation, sustainable design and human accessibility.

The Large Park should contribute significantly to the ecosystem even as it provides a host of facilities for people. Access to ecologically sensitive areas such as wetlands should rely on a trail system such as a raised boardwalk that does not interfere with the water and lies gently on the land. During the winter in temperate or cool climate zones, active recreation such as skating or sledding or cross-country skiing should be encouraged, even celebrated.

3. The Large Park is the showcase in the urban park system.

The standards of design and maintenance should both be high for a space that is a showcase attracting many visitors. The reputation

2．大型公园平衡了生态保护、可持续性设计以及人的可达性

即便为人提供了大量的设备，大型公园也应该为生态系统做出巨大的贡献。在接近生态敏感地区，例如湿地，公园应该依附一个轻放在土地上并且不与水接触的道路系统，例如木栈道。当冬天气温低的时候或者在寒带地区，滑冰、雪橇或者滑雪越野赛就应该被鼓励或者庆祝。

3．大型公园是在城市公园系统中对大型空间的展示

大型公园的设计和维护应比作为吸引游客的展示空间具有更高的标准。如果维护水平很低，包括到达区、广场区、游戏场地、高尔夫球场、网球和篮球场、咖啡馆、洗手间、道路系统、家具、引导标志和照明，那再高水平的设计也会被破坏。因为如此大的空间，无论好的或者坏的影响，品牌美誉度都会受到很大影响。

4．设计可以大胆，但是对于市区公园合适的设计密度，对于大型公园就不见得合适了

作为城市居民经过的最大的自然区域，城市大型公园应该传达自然的设计理念。提供和自然亲近的机会是一种负责的值得信任的做法，设计师永远都不要屈服和妥协于在大面积上寻求不自然的几何图形。在赤裸的土地上创造自然是个有意义的设计目标。

尽管如此，如果一个大型公园具有高水平的设计并有细心的专人看管，那么安置一个雕塑区就成了一个有意义的特色。雕塑公园要鼓励人们前来散步其中，不要只在路上溜达，要去体验其中的片段，甚至触碰雕塑（除非雕塑的表面是易碎的或者锋利的）。大型公园中应该有多种景观设施。

植物园或者树木园是另一种值得称赞的想法，但同时也需要与其他活动匹配。典型的植物园主要集中在乔木上，每一种都标明植物学名称，通常按照种类去分，用于科普教育目的，通常绝不成排地种植。不规则的自然的群组对于公众更有趣。这种想法在植物园中还延伸到了灌木、藤本、地被、一年生和多年生植物上，包括花卉的多种颜色与样式。除了自然的群落外，有很强几何设计感的空间如规则的花园代表了不同的形式与时代的特征。植物园和树木园需要园艺专业人士专门的维护（图5.32 - 图5.39）。

图 5.32
阿诺德树木园，波士顿
Arnold Arboretum, Boston

从世界各地引进的自然乔木、灌木和地被群组布满了整个山坡和山谷，其间还有小溪和池塘做点缀，这些都使这里变成了受人欢迎的目的地。这里可以骑自行车，同样也可以在小路漫步，或者到"丁香周日"来野餐。这个还包括温室和图书馆的大型公园属于波士顿公园系统，并由哈佛大学来管理。来自中国的植物均有中文的标签。
The natural groupings of trees, shrubs and ground covers from all over the world covering hills and valleys interspersed with brooks and ponds make this a popular destination. Bicycling is permitted, as is walking off the trails, and picnicking is permitted on Lilac Sunday. This Large Park within the Boston Park System is operated by Harvard University and includes greenhouses and library. Plants from China are labeled in Chinese. (Credit: John Phelan)

图 5.33
范渡森植物园，加拿大
VanDusen Botanical Garden, Vancouver, Canada

这个公园被设计成适合于任何年龄段和背景的人群，而不只是一个研究机构。它陈列了生态学的基本要素，并且包括了一个图书馆。被标记的植物按照之间的联系和地理起源来排列。此植物园提供有导游的旅行、演讲和实验室。
The garden is designed to be enjoyed by people of all ages and backgrounds rather than as a research facility. It showcases ecological principles, and includes a library. The plants are labeled and arranged to demonstrate botanical relationships or geographical origins. Guided tours, lectures, and workshops are offered. (Credit: Stan Shebs)

图 5.34 - 图 5.35
上海植物园
Shanghai Botanical Garden

人们来到这个远离城市的大型公园漫步，或者坐在草坪上。如果维护得很好的话，这个草坪就能承受长时间的使用。中国更多的公园应该向人们开放草坪。多种风格修剪的灌木和亭台展示出植物多样的作用。
Families come to this Large Park far from downtown for a stroll or to sit on the lawn, which can tolerate the heavy use if there is proper maintenance. More parks in China could follow this example in opening up lawns to the use of people. Topiary shapes and pavilions in no one dominant style showcase a variety of floral effects. (Credits: Thomas M. Paine)

for having a high standard of design of facilities—arrival areas, plaza areas, playing fields, playgrounds, tennis and basketball courts, cafés and restrooms, trail systems, furnishings, signage, and lighting—is undermined if the standard of maintenance is low. As the largest space in the system, the impact, good or bad, on the "brand" reputation is dramatic.

4. Design can be bold, but the design intensity appropriate to the Downtown Park is inappropriate across most of the Large Park.

Being most likely the largest natural area to which urban dwellers have access, the Large Park should convey a design character inspired by Nature. Providing access to Nature is a responsibility and a trust that a designer should never compromise or violate by seeking to impose an unnatural geometric order over a vast area. Creating the illusion of unspoiled Nature on a brown fields site is itself a worthy design objective.

Nevertheless, in some Large Parks, programming may include a sculpture collection, a worthy feature if handled at a high design standard with thoughtful curatorial control. The sculpture park should encourage people to wander around the pieces, not stay on a path, to experience the pieces "in the round", even touch them (unless the surface is delicate or sharp) but not climb on them. It should include a variety of landscape settings.

A botanical garden or arboretum is another worthy feature, and compatible with most other activities. The typical arboretum focuses mainly on trees, each labeled with its botanical name and usually arranged by species, for educational purposes, although by no means

图 5.36
莱茵河休闲公园，波恩，德国
Rheinaue Leisure Park, Bonn, Germany

位于莱茵河两岸的"绿色心脏"几乎与波恩市中心一样大。这个山地包含了一个日本公园、盲人公园、玫瑰公园和一个湖，并且最初是在 1979 年为国家园博会建造的。
The "green heart" of Bonn on both banks of the Rhine River is almost as large as downtown Bonn itself. The hilly site includes a Japanese Garden, Garden for the Blind, Rose Garden and a lake, and was originally created for the National Horticultural Show in 1979. (Credit: Leit)

necessarily arranged in rows, since irregular, natural groupings are more interesting to the general public. Botanical gardens extend that idea to include shrubs, vines, ground covers, perennials and annuals, and take advantage of the greater variety of forms and flowering color to include, besides natural groupings, spaces designed with strong geometry like formal gardens evoking different styles and eras. Arboretums and botanical gardens require specialized maintenance led by professionals qualified in the fields of horticulture and botany. (5.32-5.39)

5. The Large Park celebrates local heroes, events and achievements, showcases local talent in the arts, and interprets sustainable services being provided by the space.

Whatever constraints the limited area of a downtown location may impose on programming Civic Plazas and Downtown Parks, by definition the Large Park should include space enough to include commemoration or environmental education. The result does not need to be "large" just because it is in a Large Park or even be the primary focus. (5.40-5.44)

6. The Large Park encourages contemplation and reverie.

Ample space should be set aside where people, alone or with a friend, can find serenity. The Civic Plaza or even Downtown Park may not have enough space to allow full immersion in a natural world free of the clutter of manmade things where the serenity of finding spiritual meaning is truly possible. The Large Park should provide precious service as a safe haven supporting many human needs all the way up to self-actualization as described in Chapter 2.

5．大型公园纪念当地的英雄、事件以及成就，是当地艺术的展示地，并且说明空间所能提供的服务

无论什么因素约束了城市广场以及市区公园的区域，大型公园都应该有足够的空间去进行庆典或者进行环境的教育（图5.40 - 图5.44）。

6．大型公园鼓励人们沉思与冥想

我们要给予独自的人或者和三五好友一起的人足够的平静的空间。城市广场或者甚至市区公园可能不能提供逃避现代城市的"桃花源"让人们在其中去沉思冥想。城市中理想的天堂应该提供宝贵的服务去满足人们的需求去达到正如第二章所说的自我实现。

7．内部交通系统要减到最少，游览线路要成环

在大型公园里开放交通极具诱惑力，但这会导致其他对人更有益的公共开放空间活动的减少，然后会降低大型公园使用水平。在西方，开放的园内道路系统现在看来是个错误，某些公园在周

图5.37
南部公园，索非亚，保加利亚
South Park, Sofia, Bulgaria

在这个位于国家首都的公园中的大型的建筑设施并没有破坏树木、草地和水的效果。这里有游乐场地和咖啡馆，还有禁止机动车通行的步道。其他地方几何形状的种植与照片中英式园林的自然形成对比。
The showplace of this capital city does not undermine the soothing effects of trees, grass and water with a clutter of large built facilities. There are playgrounds and cafés, and paths free of vehicles. Geometric plantings elsewhere contrast with the informal so-called English style shown here. (Credit: © Plamen Agov • studiolemontree.com)

图5.38 - 图5.39
西湖，杭州
West Lake, Hangzhou

中国以这个大型公园为傲，也许这是世界上最早的大型公园。沿着古迹漫步湖边或者泛舟其上，人们活在中国的山水画中，享受着博大的中国文化。在其他许多国家会要求在湖边装设栏杆，但是这里并没有，这种简约的保持增加了景色的美感。
China is justly proud of this Large Park, perhaps the earliest anywhere in the world. Walking its perimeter walkway and boating its waters, passing storied sites, people are immersed in a living Chinese scroll painting and in touch with what is timeless in Chinese culture. The lack of railings, which would be required in many other countries, enhances the beauty by maintaining simplicity. (Credits: Thomas M. Paine)

图 5.40 - 图 5.41
中山公园,上海
Zhongshan Park, Shanghai

很多公园为了纪念孙中山先生而命名为中山公园,这里就是其中之一。中山公园建在原是英国移居人的住所场地,公园提供划船以及几个给孩子的游乐项目。在周日,人们齐聚这里高歌一曲,的确很有团体性。
One of many parks in China named in honor of Dr. Sun Zhongshan, this Large Park on the site of a former expatriate Englishman's estate offers boating and includes several rides for children. On Sundays people gather for a traditional sing-along, truly an experience of *communitas*. (Credits: Thomas M. Paine)

日对机动车禁止,反而对行人、骑自行车的人、玩轮滑的人,以及散步者有益。(图5.45)停车场应该在场地的外围,范围也不应太大,尽管提供大型停车场的诱惑很大。那些在当地交通系统中是必要的道路,如果在周日或者特殊日子禁行,这样的建议可以被采纳。如旧金山的金门大桥公园,周日道路的禁行措施让使用者的数量增加了216%。⁴

成环的步行系统提供了一个户外的有氧体验。因为公园的安全大部分取决于可视性,成环的步行系统应该在其他使用者的视线里。道路应该很好地铺装,为了跑步的人和散步的人应该尽量避免锐角。

8. 照明要适当地使用,要节能以及不能造成光污染

开放到晚上(例如到23:00)的公园,要在路上和活动场地、运动场地提供照明。照明要有屏蔽以防止光打到非目的照明的地

7. Internal vehicular circulation system is minimal, and recreational circulation includes a loop.

There is a huge temptation to open up Large Parks to vehicular traffic. A road system that is too intrusive or too extensive will displace or discourage other public open space activities that are more beneficial to people, reduce the level of non-vehicular park usage and detract from the quiet character of the park. Internal road systems open to vehicles in Large Parks in the West are now seen as a mistake and in some cases are being closed to vehicles and turned over to pedestrians, cyclists, rollerblades (inline skaters) and strollers, at least on Sundays. (5.45) Parking should be peripheral and minimal, despite the temptation to provide abundant parking. For those roadways that are felt to be essential to the local urban transportation system, road closure on Sundays or another regular schedule that the people can depend on should be adopted if at all possible. In San Francisco's Golden Gate Park, Sunday road closure results in a 216% increase in the number of users compared to Saturday usage. ⁴

Walking loops provide an extended aerobic experience in a natural setting. Because park safety depends largely on visibility, walking loops should be within the sightline of other park users. They should be paved and laid out free of sharp corners to allow use by runners and joggers.

8. Lighting is used in moderation, is energy-efficient and does not contribute to light pollution.

Large Parks which are open into the evening (for example, until 11 PM) require lighting along paths as well as along the perimeter of activity areas and playing fields. The lighting should be shielded to prevent light-scatter into non-targeted areas, especially the sky, to protect the night-time character of the park. Providing safety and security after dark for visitors with path and area lighting does not preclude an unobtrusive, energy-efficient approach, for example using solar-powered LED lights. (5.46-5.47)

9. Facilities do not crowd out the natural environment and they are shared.

No matter how extensive the Large Park may be, the principle of sharing not only makes efficient use of scarce resources but also builds a sense of community. Adding too many playing fields in place of a natural meadow deprives people who do not use the playing fields of the chance to enjoy unstructured Nature. Many Large Parks face the tension between providing facilities and providing a natural environment free of facilities. Sports fields are shared according to a simple and fair reservation system. Large Parks should also include a subarea designated as a Dog Park (see Neighborhood Park below for more information).

图 5.42
比拉容·马尔公园，墨尔本
Birrarung Marr Parkland, Melbourne

比拉容和马尔（雾之河与河岸）这两个名字都是澳洲当地原住民的语言，这个大型公园中以当地动物为主题的雕塑也是用来纪念澳大利亚的原住民。
Both the name Birrarung Marr ("river of mists" and "river bank"), from the Woiwurrung language, and the sculpture in indigenous animalistic motifs in this Large Park honor the Wurundjeri people, the original inhabitants of the area. (Credit: Donaldytong)

图 5.43
西里西亚动物花园，西里西亚中心公园，波兰
Silesian Zoological Garden, Silesian Central Park, Poland

位于这个大型公园内动物园中心的恐龙谷里有 16 只由一支波兰探险队在戈壁滩上发现的恐龙化石的等比复原模型，反映出这里的文化和自然历史。
Poland Dinosaur Valley in the middle of the Zoo within the Large Park includes full-scale reproductions of 16 dinosaurs found by a Polish expedition to the Gobi Desert, celebrating cultural and natural history. (Credit: LUCPOL)

图 5.44
西贝柳斯公园，赫尔辛基
Sibelius Park, Helsinki

在芬兰首都这个大型公园内的纪念碑纪念着伟大的芬兰作曲家让·西贝柳斯。他的音乐在芬兰人民心中有不可动摇的地位。埃拉·希尔图宁在 1967 年设计了这个公园，尤其在下雪的冬日里使人联想起教堂的风琴或者凝固的音乐。
This monument in a Large Park in the capital of Finland honors the great Finnish composer Jean Sibelius, whose music played an important role in the formation of the Finnish national identity. It was designed by Eila Hiltunen in 1967 and is suggestive of a church organ, or frozen music, especially fitting on a snowy winter's day. (Credit: Henryk Sadura)

10. Perimeter enclosure is of a high quality and appropriate material.

To limit the use of the park to daytime and early evening hours, walls or fences with gates are unavoidable. The materials should be durable and require only low maintenance, and the style should suggest local vernacular design or a rural rather than urban style. (5.48)

方，尤其是天空，这样可以保护公园夜晚的特色。在天黑后在道路和某些区域给游客提供安全和安保并不影响到节能这一点，因为可以使用太阳能的 LED 灯（图 5.46 - 图 5.47）。

9．设施和器材不应该让自然环境变得拥挤，而应该适度分享

无论公园有多大，高效利用资源的同时还应该提供简单公平的预定制度来保证"分享"。在天然的草地上加上太多的游乐场，

图 5.45
中央公园，纽约
Central Park, New York

中央公园中交通流线系统的设计经验是宝贵的。大型公园应有内环路，禁止机动车通行，把不同形式的交通分流，应使用曲线做道路分段并在设计桥时模仿自然的形式、颜色和材料。
The lessons of circulation design from Central Park are timeless: Large Parks deserve internal loops, banishment of cars, separation of modes of movement, curving alignment and bridge design mimicking natural forms, colors and materials. (Credit: Jet Lowe; Library of Congress, National Park Service, Historic American Engineering Record. Survey number HAER NY-194-7)

供人们享受自然的空间就会变少，很多公园也面临着这个选择：提供更多的人工设施还是更多的自然空间。大型公园应该也包括了一个宠物公园的分区（请参阅社区公园部分）。

10．外围的围合应该使用高质量的合适的材料

为了限制从白天到傍晚的公园的使用，墙和门是不可避免的。材料应该是耐久的并且低成本维护的，设计语言应该是当地的或者当地乡村的而不是其他城市的样式（图 5.48）。

滑板运动场

滑板、轮滑、极限单车（BMX）给青少年提供了愉悦的体验。从已有的公园中汲取经验，这些活动必须分离出有预制混凝土的硬质专用场地，不然非常容易损坏公园的铺装。1976 年在加州兴起，在美国、英国和澳大利亚建造的滑板公园中，一般包括 1/2 的管道、1/4 的管道、完整的管道、弧线碗状场地、台阶、

Skateparks

Skateboarding and rollerblading (inline skating) and race bike (BMX) riding provide an exhilarating experience for young people from pre-teen years. The wear and tear on existing park furnishings has necessitated the provision of separate hardscape spaces with precast concrete designed especially for skateboarders, rollerblades (inline skaters), and BMX riders to engage in their moves. Introduced in California in 1976, and found in the US, UK and Australia, a skatepark may contain half-pipes, quarter pipes, full pipes, bowls, stairs, ramps, pyramids, walls, ledges, pools and many other obstacles intended for wheel contact. Some skateparks are indoors. "Street plaza parks" are the preferred type, because they mimic the urban streetscape. Bowl parks emulate pools and are typically 1 to 4 meters deep. Many Skateparks provide walls where youth are permitted to create graffiti. (5.49-5.50)

图 5.46
伊比拉普埃拉公园,圣保罗市,巴西
Ibirapuera Park, Sao Paolo

在南半球的夏季,这个大型公园点亮了灯来庆祝圣诞夜。
In this Large Park night lighting celebrates Christmas Night in the summer of the southern hemisphere. (Credit: Silvio Tanaka)

图 5.47
奥尔巴尼湖公园,北岸市,新西兰
Albany Lakes Park, North Shore City, New Zealand

过多的灯光浪费能源并且降低了景观的质量,尤其是当没有树反而都是灯柱的时候。
Excessive lighting wastes energy and detracts from the landscape quality, especially when there are no trees and the lighting poles themselves are the only forest. (Credit: "Uploader")

■ Greenways

Greenways are open space corridors. The reality of urbanization means that the form of much urban open space is squeezed into a linear form like a riverwalk or harborwalk, a buffer along the edge of a development project superblock, a strip reserved from development connecting larger parcels of open space or shared with grey infrastructure (that is, non-green infrastructure). In the West, established Greenways follow rivers (like the Riverwalk in San Antonio, Texas) or line the waterfront replacing dockyards with signature amenities for local residents and tourists alike. Likewise Greenways are sometimes created under an elevated highway, or along a canal or railroad right-of-way. In Europe and Asia, some

斜坡、锥体形墙、平台、水池以及其他的适用障碍。有些滑板场是在室内的,但街道广场是更好的形式,因为他模仿了城市的街景。碗状的公园模仿了水池,典型的为 1-4 米深。很多滑板公园都提供给年轻人涂鸦的墙面(图 5.49 - 图 5.50)。

■ 绿道

绿道是开放的空间走廊。都市文化的真实性意味着大多数城市开放空间的形式要被挤成线条形,就像沿着开发项目街区的缓冲带,或沿着河流、港口,由大型开放空间演变下来的条状地带,或是共享灰色基础建设(非绿化的基础设施)。在西方,最近建成的绿道,在滨水区用当地居民和旅游者喜爱的有特征的活动取代了造船厂。同样绿道有时也在高架桥的下面建设,或者沿

图 5.48
海德公园，伦敦
Hyde Park, London

并不只是一些老的围栏会成就一些大型公园。这个富有艺术气息的栏围区域既抓住了艺术又抓住了自然，并且还抓住了这个充满百年老树、草坪和湖水的大型公园的精髓。
Not just any old fence will do for some Large Parks. The artistic flourish of this section of fence captures both art and nature, and the spirit of this much loved Large Park with its centuries-old trees, meadows and lake. (Credit: Supermac1961)

着运河或铁路线路。在欧洲和中国，绿道有时利用了古代城墙防御系统，用给散步和慢跑者的风景优美的环形绿带替代了历史悠久的护城河。南京建在明朝的城墙，在20世纪60年代被缩短了，在1980年变成了文化遗产，在2002年提供了一个环形的绿道。在21千米的遗址中，一些在大门附近的片段得以保留。其他绿道被当作林荫大道或者风景区干道，通过景观廊道提供了多模式的车行活动。中国的第一个景观绿道也是在南京，中山路，使用悬铃木为行道树，直通中山陵。几乎但不是所有的绿道强调连接其他的主要的公园系统的线性交通（图5.51-图5.52）。

绿道有着可以将开放空间很好延伸进社区的可能性，这是其他几种空间形式通常因为远离社区而无法做到的。一些国家建立的道路系统可以延伸到很远的距离，连接城市和乡下。在伦敦，泰晤士河的两侧可以从城市的一端步行到另外一端，超过了64公里长。此外，泰晤士河的河边步道其实是泰晤士河国家路径的

Greenways take advantage of ancient fortifications in the form of city walls, sometimes replacing the historic moat with a scenic loop for walkers and joggers. Nanjing's Ming Dynasty city walls, truncated in the 1960s, went on to became a cultural relic in the 1980s and a proposed ring greenway in 2002; of the 21 kilometers that remain, several segments near gates have been restored so far. Other Greenways have been created as boulevards or parkways, providing for multimodal vehicular movement through a landscape corridor. One of the first in China is also in Nanjing, Zhongshan Road, a beautiful planetree canopy corridor leading to the mausoleum of Dr. Sun Yat-sen. Most but not all Greenways emphasize linear movement to connect with other major elements of the urban public open space system. (5.51-5.52)

Greenways have the potential to extend open space equity access across the community, into neighborhoods that otherwise would consider open space to be inconveniently far away. Some countries establish trails that can extend for great distances, connecting the city to the countryside. In London, both sides of the Thames River are open for walking from one end of the city to the other—over 64 kilometers (40 miles). Moreover, the Thames riverfront is actually part of Thames Path National Trail running the entire length of England's most famous river, 294 kilometers (184 miles), and is one of over twelve National Trails extending over 4,000 kilometers (2,500 miles). England is the world leader in providing public footpaths across the countryside, in public rights-of-way through private land. (5.53-5.55)

A benefit often overlooked is that Greenways can help offset the threat of global warming to biodiversity by allowing easy movement through an otherwise hardscape world for wildlife, even the migration of plant species. Some Greenways protect an ecological resource so fragile or so narrow that human access must be strictly limited. In areas with steep topography, because north-facing slopes are much cooler than slopes facing in other directions, protecting north-facing slopes will give plant species the best chance of survival in our age of rising temperatures.

图 5.49 - 图 5.50
轮滑公园
Skate Parks

为了避免滑板、轮滑（单排轮滑）和竞速自行车对其他公园使用者的安全威胁和对公园设施的损坏，城市可以提供单独的封闭的轮滑公园。所有地面都为可在上面骑行的硬质景观，并且应经受住特定轮胎或装备的磨损。
To avoid the problem that skateboarding, rollerblading (inline skating) and race bike riding put other park users at risk and cause damage to park furnishings, cities can provide separate enclosed skate parks, hardscape spaces in which all surfaces are available to be ridden on and are designed to withstand the excessive wear and tear. (Credits: Landscape Structures Inc. and John McAllister)

图 5.51 - 图 5.52
中国的传统绿道
Traditional Greenways in China

今天滨水区的绿道令人回想起古老的中国传统，就像这些云南和浙江的传统村落。在岩头，这个被称为水镇的中国沿海城市，一个直接通向各户家庭的遮阴漫步大道至今还在使用。连续设置的长凳让人们可以逗留并且欣赏景色、雨和阳光。这种有魅力的建筑形式在西方基本是闻所未闻的。
Waterfront greenways today hark back to ancient Chinese traditions like these villages in Yunnan and Zhejiang. Still in use in Yantou, a so-called water town in coastal China, is a covered community promenade onto which individual houses open directly. A continuous bench allows people to linger and enjoy the view, rain or shine. This charming building type is almost unknown in the West. (Credits: Joachim Czogalla and Thomas M. Paine)

1. Gaps that interrupt the continuity of Greenways are eliminated.

Sometimes the connection between interrupted sections of a planned Greenway requires structures to allow people to pass through a site that is intensely developed and lacks room for an open space corridor at ground level. The gaps severely reduce the potential for the Greenway to provide accessibility throughout the city. Boston waited three decades to complete the dream of providing continuous waterfront access extending for many kilometers along the three main rivers that flow into Boston Harbor, crossing creeks, and passing the most densely developed downtown residential and office waterfront development. Many brownfield areas had to be crossed, and many new structures created.

2. Except on downtown or waterfront Greenways, buffer plantings create the illusion of greater depth than exists and screen unsightly uses that detract from the effect of a lush landscape, and landscape effects humanize otherwise stark conditions.

Strengthening the effectiveness of a narrow link may require strong design, for example creating bold landforms and introducing dense planting. Nevertheless, native plantings and habitats should be retained, protected unobtrusively, and left undisturbed during construction. Narrow sites should be designed so as to suggest spatial expanse and direct the eye away from ugly adjacent development. The principle of borrowed landscape invented in China may apply to many situations. Greenways may benefit from other landscape effects such as the placement of specially selected boulders, clearly adding social

一部分，这条步道贯穿这条 294 千米长、全英国最著名的河道，并且是总长超过 4000 千米的 12 条国家路径之一。英国是世界上提供穿过乡村的步行途径的领跑者，擅长在私有用地上使用公众权利（图 5.53 - 图 5.55）。

一个经常被忽略的好处就是绿道可以帮助缓解全球的生物多样性的危机，通过为野生动物迁徙建立一个简单的移动廊道，甚至是为植物的迁徙所建。一些绿道项目的生态资源是如此脆弱和紧张以至于人们的进入要被严格的控制。在陡峭的地形上，由于北面的斜坡会比其他朝向的坡面温度更低，在北面山坡保护植物是我们减缓全球变暖的一个手段。

1. 消除使绿道连续性被破坏的缝隙

城市中，人们要从一个区域到达另一个区域需要一个连接的结构，比如绿道，而这种节点往往早已做好了其他规划并开发得很成熟，却唯独缺乏公共空间。这种"缝隙"严重地降低了作为绿道提供穿越城市的可行性的潜力。在波士顿，花了 30 年才实现提供连续的滨水空间的梦想。这个滨水空间沿着 3 条主要的流向波士顿港口、流经最高密度的中心商业和办公区域的河段，延长达数千米。很多棕色地带不得不被穿过，很多新的结构也建造出来。

2. 除了在城市中心和绿道的滨水区域，缓冲种植会给人带来种植深度更深的错觉，同时，屏风式的设施减弱了茂盛的种植带来的景观效果，景观也改变了荒地

在一个狭窄的节点强调景观效果需要一个强有力的设计，例如创造一个大胆的地形以及种上浓密的植物。然而，本土的植物

图 5.53
湖畔小路,芝加哥
Lakefront Trail, Chicago

沿着密西根湖,这个绿道给骑自行车者、漫步者、慢跑者和轮滑者提供了便利。
Along Lake Michigan, the Greenway includes lanes for cyclists, walkers, runners and inline skaters. (Credit: http://www.flickr.com/photos/katjung/199062645/)

图 5.54
兰布拉斯大道,帕尔马
Las Ramblas, Palma

有宽敞的受装饰的人行道,有繁茂的树冠遮阴,有户外咖啡店和花丛,这个市中心的城市街道起到了绿道的作用,而不像是城市边缘的公园道路。这个在西班牙东部的马略卡岛的例子使人们想起巴塞罗那兰布拉斯大道本身。
With wide decorative sidewalks, a robust shade tree canopy overhead, outdoor cafés and flowers, a downtown urban street serves as a Greenway, not unlike a parkway on the edge of the city. This example on the Spanish island of Mallorca recalls Las Ramblas in Barcelona itself. (Credit: Rafael Ortega Díaz)

图 5.55
西风公园,阿姆斯特丹
Westerpark, Amsterdam

步道沿着水岸后退了一定距离。这样既允许人们在水边放松,又可以保证人们不去打扰水体内部。
Keeping the trails back from the water's edge permits people to relax on the edge of the water without interference. (Credit: David van der Mark)

value by appealing to our deep-seated love of Nature, which includes geology. (5.56-5.57)

3. The Greenway that includes significant cultural or historic sites showcases them rather than ignores them or replaces them.

The human scale and sustainable use of local materials in a farmhouse, hamlet cluster, small mill or kiln provide reminders of the former way of life, and add positive cultural and visual character to the landscape. As rapidly urbanizing countries start to lose their historic character, recognition of the importance of preserving vernacular style surviving from simpler times becomes a powerful force. Examples of vernacular style may even include ancient farmhouses or village clusters. In waterfront Greenways, design may appropriately evoke the historic nautical heritage in the forms of furnishings such as railings, benches and lighting as well as materials (wood, steel, and canvas have been used in many Western projects).

This approach is consistent with a related phenomenon, the historic walking trail. In Western cities, many guidebooks for international tourists include excellent maps that suggest the best walking route to tour the historic sites. Many cities promoting tourism, whether domestic or international, distribute on-line or printed maps that do the same. Some cities go further. They erect markers that briefly explain the historic or cultural significance of a site, or even mark the route on the pavement. The Freedom Trail in Boston links culturally significant sites with a continuous line of bricks embedded in the concrete sidewalks in the heart of the city. Greenways linking cultural sites can offer the same. A walk through culture and a walk in Nature both accommodate fundamental human needs. (5.58)

4. In downtown locations the Greenway tames brute infrastructure.

Most narrow corridors of open space in outlying parts of the community do not accommodate heavy usage. On the other hand, Greenways that are located downtown should take on some of the characteristics of the Civic Plaza or Downtown Park. At Sydney's Darling Harbor, a Greenway connecting with the waterfront park tames the underside of an elevated superhighway by a seamless plaza paving creating a civic space that provides grand shelter during rain. A decorative channel of water runs straight underneath, even a row of stately palm trees passes under the highway without interruption, and a children's merry-go-round is no less merry there. (5.59-5.61)

Consider the Bund, which is paved to accommodate some of the most intense use of any urban public open space in the world with as such functions more as a linear Civic Plaza than a Greenway. Green is not the trademark color of the Bund, which best celebrates the iconic historical backdrop of commercial banks and the Peace Hotel with hardscape and without trees to mask the architectural glory to be seen from the Pudong side of the Huangpu River. The Yan'an Elevated Road Downtown Greenspace in Shanghai compensated for the divisive effect of a new elevated superhighway by creating a parallel

和栖息地应该在建造中不被打扰地保留和保护。狭长的场地应该被这样设计，延伸空间并且引导视线远离丑的邻近的地方。中国的借景就可以很好地应用在这种情况下。绿道可以从别的景观中获益，例如布置精心选择的岩石，通过我们深爱的自然来体现自我的社会价值，同时也有地质学的意义（图5.56 - 图5.57）。

3．包含了文化或者历史的绿道地段需要被展示，而不是被忽略或者替换掉

人类尺度下可在农场、小村庄、小磨坊或者窑厂中持续地使用的当地材料，可以提醒人们以前的生活方式，同时给景观增加积极的文化和视觉特点。现在迅速崛起的都市化的国家开始慢慢地丢失他们的历史传统特点，要承认保存本土形式的重要性。在滨水的绿道中，设计可以涉及一些有关航海历史的遗产，可以用在户外家具的形状上，比如栏杆、座椅和灯具，或者材料上（木、钢铁和帆布就被大量西方的项目所使用）。

这种方法与一种现象相似，历史走廊。在西方的城市里，很多的为国际旅行者准备的地图小手册中都包括有建议的徒步路线来了解当地的历史。很多城市为了促进旅游业，无论是国内还是国外的市场，同时准备了在线的和印刷的地图。有些城市先行一步，特意树立起标记，及时地解释了当地的历史和文化，或者甚至在铺装上标出路径。波士顿的自由路径与重要的文化场地相连，用连续的砖块镶嵌在混凝土的人行步道上，甚至在铺装上还有路径的标记（图5.58）。

4．在城市中心的绿道使得生硬的构造物变得人性化

很多狭长的户外开放空间的廊道不能应对过大的使用量。在城市中心的绿道应该参考一下城市广场和市区公园的处理手法。在悉尼的情人港，绿道连接着滨水公园，通过一个天衣无缝的广场铺装创造了一个提供遮雨的市民空间从而弱化了一个高架桥的下层。一条有装饰的运河径直流到下层，即使是在高速路下面的一排连续的棕榈树以及儿童的旋转木马，都不足以像这种方式一样让这个地方很活跃（图5.59 - 图5.61）。

说到外滩，被铺装覆盖的外滩承载着在全世界的开放空间里最高的使用密度，这种功能使得外滩更像是一个长条形的城市广场而不是一个绿道。绿色不是外滩的品牌颜色，设计用没有树的硬质铺装最好地强调了标志一样的具有历史背景的银行和和平饭店，同时让在黄浦江对岸的浦东的人们能看到外滩的建筑。上海的延安高架绿色区域通过设计了一个平行的包括水景、草地和有想象的缓坡的单独节点和茂密种植的绿道，缓解了新架起的超级高速公路的分裂影响。这一定会被波士顿肯尼迪绿色通道嫉妒的，这个绿道把它不大好看的架起的高速路藏入一个隧道里，里面有草坪和喷泉，但是树很少，几乎没有能静心的地方。波士顿这个绿道的地面的层次很分明，但是需要通过更多大胆的种植和规划来转变空间（图5.62 - 图5.63）。

在罗德岛的普罗维登斯，正如前面所讲，很长的普罗维登斯河最近拥有了新的绿道和水景公园，那里是水火表演的舞台。韩国的首尔也是，它移除了丑陋的12个小巷长的清溪川高速公路，重新安排一个绿道使污染严重的清溪川重见天日。清溪运河部分由在波士顿白手起家的景观设计师金美净设计了个水景，它由从韩国的

图 5.56
奥姆斯特德公园,波士顿
Olmsted Park, Boston

我们称之为"翡翠项链"的绿道包括了在窄小的地方密植植物来阻挡城市楼群以及机动车流——这种手法与借景相反。
The Greenway called the Emerald Necklace includes dense planting in narrow spots to screen out urban forms and traffic movement—the opposite of borrowed landscape. (Credit: bikeable)

图 5.57
湖边步道,芝加哥
Lakefront Trail, Chicago

步道和机动车道很近,并且画出条纹线来指引自行车流。慢跑者和步行者更喜欢道路两旁没有铺装的区域。树有强烈的围合感,并且封住了天际线。
Trails are closed to vehicles and are painted with stripes to direct the flow of bicycle traffic. Joggers and walkers prefer the unpaved edges of the roadway. The trees provide a strong sense of enclosure that blocks out the skyline. (Credit: Alan Scott Walker)

图 5.58
中式绿道
Chinese Greenway

传统的地砖为这个绿道形成了吸引人的铺地表面。绿道同样有传统的灯饰和座凳以及维护良好的种植。
Traditional bricks form an attractive surface for this Greenway trail complemented by traditional forms for lighting and benches, and well maintained planting. (Credit: Keng Po Leung)

Greenway offering the immersion experience of water features, lawns, and secluded spots with imaginative grading and lush planting that would be the envy of Boston's Rose Fitzgerald Kennedy Greenway, which replaced its own ugly elevated superhighway now hidden in a tunnel with lawns and fountains but few trees and little immersion. It needs to transform space to place by more ambitious planting and programming. (5.62-5.63)

In Providence, RI, the long buried Providence River was "daylighted" as the new Providence River Greenway and Waterplace Park, home to Waterfire, discussed earlier. Seoul, Korea did both, removing the ugly 12-lane elevated Cheonggye Expressway, daylighting the heavily polluted Cheongyye River below, and replacing it with a Greenway. In its ChonGae Canal section, Boston-based landscape architect Mikyoung Kim designed a fountain that children can play on and in, framed by stones from all the provinces of Korea, expressing the idea of eventual reunification. As in Halprin's fountains that invite play, no safety railings mar the elemental power of stones meeting water. In Kaohsiung Port Station Park in Taiwan, a railroad trainyard is being transformed into a cultural and commercial mecca. In Santa Fe Railyard Park and Plaza, a public-private partnership has transformed a 5-hectare site which hosts the twice-weekly Santa Fe Farmers Market and provides additional space for arts, festivals, and day-to-day life. It includes a pedestrian and bicycle path parallel to the rail tracks, the first of several pathways that will link the park and plaza to districts throughout the city and beyond. (5.64-5.65)

High Line, New York City

The 22-block long rail trestle elevated 10 meters above street level can be said to invent a new urban public open space type, a park that celebrates its weedy and rusty origins with its signature striated concrete plank walkway and provides contemplative strolling uninterrupted by city traffic and overlooking dynamic views of the cityscape from a height offering serene detachment. The elevated park is reached by elevators and metal staircases.

The $115 million which the city has spent on the park generated $2 billion in private investment surrounding the park. Some of the new apartments, art galleries, restaurants and boutiques which High Line overlooks were designed by celebrity architects like Jean Nouvel, Annabelle Selldorf and Neil Denari. In addition to the 8,000 construction jobs which those projects required, the redevelopment has added about 12,000 permanent jobs in the area, according to Mayor Michael Bloomberg. In one abutting building the price of apartments has doubled since the park opened, to about $2,000 a square foot. Since opening in June 2009, the High Line's first segment, running between Gansevoort and 20th Streets, has drawn nearly two million visitors a year. The momentum has continued: the third segment ends at proposed 10-hectare Hudson Yards, the "Rockefeller Center of the 21st century". The park attracted strong publicity and financial support from the private Friends of the High Line.

图 5.59 - 图 5.61
情人港，悉尼
Darling Harbor, Sydney

悉尼将这个被废弃的水边废地转变成了一个极受欢迎的充满商业和文化活动的休闲去处。沿着水边是连续的步道和大量的座凳。绿道走廊的种植和优美的灯光装置延续了高速桥下的连贯性，这同时也启发了一个水平水景的设计。人们欣赏着关于生态设计的教育展。高速桥下的空间就如同是宏伟的柱廊。

Sydney transformed a derelict brownfields waterfront site into this popular leisure destination with retail and cultural activities. Along the waterfront is a continuous walkway with ample seating steps. The Greenway corridor planting and elegant lighting fixtures continue uninterrupted underneath the highway bridge, which inspired the design of a horizontal water feature. People enjoy an outdoor educational exhibition of ecological design. The space under the highway bridge is treated like a grand portico. (Credits: Thomas M. Paine)

图 5.62
北端花园，罗斯·肯尼迪绿道，波士顿
North End Park, Rose Kennedy Greenway, Boston

图 5.63
西南道廊公园，波士顿
Southwest Corridor Park, Boston

早先的高速大桥把城市肌理切割成两部分，现在人们使用隧道，取而代之在地面层上则建为贯穿城市中心的绿道。这个绿道赞美城市生活，并且强调低矮种植从而展现城市天际线的景色。其他形式中使用的材料十分讲究，体现了绿色设施的到来。
The former highway bridge that used to cut the urban fabric in two was recently replaced with a tunnel, covered with a Greenway running through the heart of the city that celebrates city life and emphasizes low planting and open vistas of the city skyline. The use of elegant materials in spare forms celebrates the arrival of green infrastructure. (Credit: Thomas M. Paine)

州政府计划并设计了州内交通系统中的一段。场地中心的建筑在改造之前被清除，高速路被废弃。取而代之的是对公共交通和公共开放空间的强调。这个绿道具有很长的并且不被间断的步行道以及骑车道。
The State government planned and designed a segment of the interstate highway system and cleared the site of buildings before it changed course, abandoned the highway and instead emphasized public transportation and public open space accommodating long uninterrupted walks and bicycle rides on this Greenway corridor.　(Credit: Grk1011)

图 5.64 - 图 5.65
戴尔池塘，弗吉尼亚大学，弗吉尼亚
Dell Pond, University of Virginia, Charlottesville, Virginia

这个绿道，把一个之前的涵洞水溪开放化从而形成湿地和池塘，以此帮助城市破损的生态系统得以恢复。小路时而沿着湖边，时而跨过水面，是一个饶有趣味的步道。
This Greenway, created by daylighting a formerly culverted stream to create a wetland and pond, helps heal the damaged ecology of the city. A trail alongside and crossing the water provides a lovely walk. (Credits: Thomas M. Paine)

Bloomingdale Trail, Chicago

Chicago's nearly 5 km (3 mi.) long elevated Bloomingdale Trail, twice the length of New York's High Line, features 8 access points from adjacent pocket parks, and for half the length provides separate paths for pedestrians and bikers/roller-blades.

5. In popular scenic areas Greenway roadways are closed to vehicular traffic on Sundays.

The necessity of accommodating vehicular traffic is reduced on Sundays when many workers have the day off, and people have leisure time to visit urban public open space. When it is free of vehicles, the Greenway roadway is open to a variety of activities including walking, jogging, rollerblading (inline skating), bicycling, walking baby strollers, playing street hockey. In Cambridge, Massachusetts, Memorial Drive, a stretch of beautiful parkway lined with huge sycamore trees and facing the Charles River, has been closed to vehicles each Sunday from late April to mid-November for twenty years, to accommodate crowds of people at leisure and not in their cars. The 2.5 kilometer-long stretch is now called Riverbend Park. New York City has recently adopted a program called Summer Streets in which certain blocks of streets are closed to vehicles on August weekends; now crowds of pedestrians and cyclists can gather for a leisurely walk right down the middle of a quiet block of Park Avenue.

The related livable streets movement is springing up in a number of cities in Europe and the Americas. For example, in its Ciclovia Program, the Parks Commission of Bogota, Colombia, offers car-free Sundays on hundreds of kilometers of roads blocked off for walking, biking, rollerblading, picnicking, and aerobic workouts. Temporary food kiosks provide refreshment. Bikes are available for rent. Over a million residents participate, and their safety is assured by hundreds of patrolling uniformed "Bikewatch" volunteers. The Ciclovia Program has helped unite the people and instill a sense of community. (See www.streetfilms.org/ciclovia/).

■ Neighborhood Parks

Neighborhood Parks are smaller than the other open space types, typically under 2 hectares in size. They may be like islands, not well connected to the major components in the overall urban public open space system. They are usually located in residential neighborhoods, although "pocket parks" are sometimes located in commercial areas, like famous privately-owned Paley Park in New York City, with its movable chairs and 6-meter high water-wall whose soothing "grey noise" blocks out the sounds of the city. (1.110) The typical Neighborhood Park in an outlying suburban area can add badly needed open space to a part of the community that is not near a Greenway or a Large Park. On the other hand, opportunities may arise to connect the park to the urban public open space system, with benefits to the mobility of people and wildlife alike, and even the diversity of native plant species. The atypical Neighborhood Park can

各个省里面的石头构造而成，象征着最后的团结统一，孩子们可以在上面和里面玩耍。哈普林的喷泉，邀请人们攀爬玩耍，没有安全围栏来玷污石头与水之间的碰撞。圣菲铁路广场公园和广场则用一种公众和个人的合伙模式将5公顷的场地变成了一周两次的圣菲农贸市场，并提供了额外的空间给艺术、节日庆典以及日常生活，它包括了一个人行以及自行车行的平行于铁路的线路。一些小路首次将环绕城市及其周边的广场连接起来（图5.64-图5.65）。

纽约城高铁

建在高于街道10米，穿越22个街区的铁轨高架桥上的高线公园可以说开辟了城市公共开放空间的新类型，公园以看似蔓延的杂草和生锈的铁轨以及条形混凝土加木板通道而出名，提供连续的不被城市交通所打断的散步场所，人们可以在高空静谧的园内俯瞰城市街景，可以通过滚梯和金属楼梯上到这个架高的公园。

这个花费1.15亿美元的公园，为周边地区带来了20亿美元的私人投资。诸如让·诺维尔，安娜贝尔·塞尔多夫，尼尔·德纳瑞等等的众多著名建筑师，纷纷着手设计公园沿线的新公寓、美术馆、餐厅和精品店铺。正如纽约市长布隆博格索说，这些项目，不仅仅创造了8000个建筑岗位，整体的深入开发更是为沿线带来了12000个长期的就业机会。一座毗邻建筑内的公寓价格，也翻倍上涨到了2000美元一平方英尺。自公元2009年6月对外开放以来，公园的第一阶段，即从甘瑟弗尔街道至20街之间的一段，已经接待了近200万的游客。而且这样的势头也一直在延续。项目的最后一期，即第三期，拟改造方圆10公顷的哈德逊铁路场站以及21世纪洛克菲勒中心。公园从私人组织"高线之友"获得了强有力的公众支持和资金支持。

芝加哥布鲁明戴尔小径公园

芝加哥布鲁明戴尔小径公园总长约5公里（3英里），相当于纽约高线公园总长的2倍，改造后的小径公园与周边小型公园相连的出入口共有8处，并且近一半的区域将人行道和自行车道分开。

5．在一些风景名胜区的绿道，周日对机动车禁行

在周日人们不用上班的时候减少机动车是有必要的，这样人们就有空闲的时间去探访城市公共开放空间。绿道的道路不受机动车的干扰，这就利于开放多样的活动，例如步走、慢跑、轮滑、骑车、推婴儿车或者玩街道曲棍球。在马萨诸塞州坎布里奇的纪念大道，一个延展2.5公里长的美丽的公园路，列植悬铃木，对面是查理士河，20年来每年4月底-11月中旬的每个周日都对机动车禁行，每个周日满满的都是行人。现在这个地方叫作河湾公园。新纽约城最近采用了一个项目叫作"夏的街道"，这个项目让一些特定的街区在八月的周末对机动车禁行；现在行人群和自行车群可以一起享受沿着派克大街的景色。

有关适于居住的街道运动在一些欧洲和美国的城市里涌现。例如，城市运动项目，由哥伦比亚的波哥大的公园委员会发起，在周日封锁道路，让人们进行散步、骑车、轮滑、野餐和有氧运动。临时的食物贩卖亭保证了食物的新鲜，并且还提供自行车的租赁活动。超过了百万的居民参加了这一项目，他们的安全由统一着装的志愿者保障。城市运动项目促进了公民的团结，同时也建立了社区感。

■ 社区公园

社区公园比其他形式的公园要小，大小不超过2公顷。它们很像是小岛，与城市里面其他的主要公共空间和设施的联系较少。这些社区公园通常是在居民区的旁边，当然也有一些微型的公园建在商业区里，比如纽约著名的"佩利"私人公园，其有着移动座椅和6米高的流水墙，通过这个流水墙和座椅"佩利"隔绝了城市的喧嚣，还给人们以片刻的安宁（图1.110）。典型的社区公园坐落于城市郊区，它们迎合了那些离绿道和大型公园比较远的社区对开放空间的需求。另一方面，得益于类似人群及野生动物的迁移，甚至天然植物种类的多样性，社区公园和周围的城市公共开放空间系统仍然有着相互关联的机会。人们也会因为某些特殊的原因建立一些非典型的社区公园，比如在一片高楼耸立的区域里取代一座被破坏的老建筑，或是重新利用部分的铁路，就像纽约绿色街道计划那样（看下面的介绍）。在某些情况下，当地规划者并不会按照总平面的方案去做，因为这样的设计方案不是总能够分块详细调查特定的场地再去建造社区公园。

纽约的绿色街道计划

在过去的15年里，纽约的城市公园部门和交通运输部门合作拆除了那些过度的道路铺装并建设了更多绿色空间。到目前为止，已经建立了2574个迷你公园，共计40公顷的新的绿色空间。在这里绿树成荫，花团锦簇，随处可见长椅及栅栏，那些幽静的小路和收集雨水的设施一起构成了纽约的新气息。减少的铺装同时改善了交通流量，并且加强了公共的安全性。不一样的是，非直角的街道模式比典型的直角网格街道、十字路口、隔盘带、转盘道和道路的尽头等处可以产生更多的可能性。虽然迷你公园不能取代大型的社区公园，但在城市公共开放空间里也是一种备受青睐的模式。

社区公园可以提供给附近居民一个远离喧嚣的环境，在这里你看不到大量停靠在路边的汽车，也听不到嘈杂的鸣笛声。这里为人们提供了一个安静祥和的环境，使人们有意愿与周围社区的邻居们友好相处。人们在这里各得其所，其乐融融。在社区公园，人们很少会因为距离太远而忽视了他人的存在，更多的时候，人们能够感受到他人的动作、表情甚至是情绪。举例来说，那些带孩子的妈妈会在游乐场会面，她们会谈论孩子，谈论家庭，陌生人的擦肩而过几乎不会发生在这里。当人们在社区公园中扮演积极的角色时，必须要明白理解和关心他人的重要性。因为这样不仅仅可以改善公园的气氛，更重要的是，他们共同建设了一个更为包容更持久的社区（图5.66 - 图5.67）。

1. 社区公园并不是过于精致复杂的设计

没有足够的空间来完成一个大的设计是常常出现的事情。一个好的设计强调实用性而不牺牲品质。种植可以简单地体现当地植物群落，而不是一个过分雕饰的种植设计。

2. 空间总是有限的，所以对空间的多重利用十分必要，需要管理者精心的安排

有些活动会共同使用铺装场地和未铺装的场地。所以，维护要跟得上草坪周边的破损部分。适当的活动要在拥有秋千、沙滩、立体方格铁架、滑梯、喷泉、座椅区、野餐区、户外健身器材以

图 5.66
汤姆平纪念公园，芝加哥
Ping Tom Memorial Park, Chicago

在中国城中保存有一个邻里公园，公园有一个塔式的凉亭、一个竹园以及这个游乐场地。这个公园纪念了一个当地的商人和城市的领导者，他建设了这个公园，满足了人们对于社区开放空间强烈的需求。
Within Chinatown the City maintains a Neighborhood Park including a pagoda style pavilion, a bamboo garden, and this playground. The park honors a local businessman and civic leader who put together the real estate deal to create this public park to provide badly needed open space for a formerly underserved neighborhood. (Credit: Torsodog)

arise opportunistically, replacing a demolished building in a densely built-up area, or replacing part of the roadway, as in New York City's Greenstreets Program (see below). In such a case the location may not follow a master plan recommendation, because such plans do not always investigate parcel-specific opportunities to create a Neighborhood Park.

New York City's Green Streets Program

For the past 15 years New York City's Parks Department in a partnership with the Transportation Department has been replacing excess paving with greenspace. The 2,574 mini parks built so far total nearly 40 hectares (100 acres) of new green space, many with trees, sitting benches, paths, litter receptacles, fences, many collecting stormwater. The reduced paving actually improves traffic flow and public safety. Irregular, non-rectangular street patterns yield more opportunities than a simple rectangular grid, at roadway intersections, in median strips, traffic circles, and at the end of dead-end roads. They cannot substitute for larger Neighborhood Parks, but are a most welcome addition to Urban Public Open Space.

Neighborhood Parks provide a benefit that private open space within superblock residential projects cannot, and that is a venue for people to build a sense of neighborliness within a larger community. Local residents can run into each other and feel they belong. In Neighborhood Parks, people are rarely too far away to notice one another, and can perceive each other's facial expressions. Mothers with children may meet other mothers here in a play area. When people take an active role in caring for their Neighborhood Park, they get to know, trust, and look out for one another. They are not just improving the park: they are building a more socially sustainable community. (5.66-5.67)

1. The Neighborhood Park is not overdesigned.

There is usually no room for a big design idea. Design emphasizes practical considerations without sacrificing quality. The planting may even reflect simple local plant communities rather than elaborate planting design.

2. Space is usually limited; therefore multiple use of spaces is a matter of policy and requires careful management and design, particularly children's play areas.

Paved and unpaved areas may be shared by several activities. Maintenance should keep up with the wear and tear on grass areas. Appropriate activities and features include play area with swings, sand, jungle gym and slide, spray fountain, sitting areas, picnic area, outdoor fitness systems (exercise equipment), and interpretive markers.

Play Areas should do much more than provide manufactured play equipment. They should encourage random play, that is, activity that is freely chosen, child-directed, and self-motivated. The ideal setting for free play includes natural elements such as trees and boulders. In Lovejoy Fountain Park in Portland OR (Halprin) no safety railings interfere with a child's exploration of natural elements and imaginative role-playing among the tall layered waterfalls. In New York City a former parking lot now features fountains activated by the motion of the elevated subway line nearby, a device that makes the once-unsettling noise and vibration seem fun and exciting. Through creativity, exploration, physical activity, friendship, and adventure, free play helps children grow to be healthy, happy, confident, inquisitive, and successful. Since play often involves physical activity and risk, play is closely related to the development and refinement of children's growth and fine motor skills as well as the ability to make decisions, solve problems, and regulate emotions. Children who play are less likely to be obese and develop obesity-related health problems such as diabetes and heart disease. Children whose only outdoor activity is organized sports in school or otherwise structured activity sacrifice valuable opportunities to develop self-directed exploration and risk-taking, especially climbing and overcoming a fear of heights.

Play helps children feel more connected to their communities. Community play spaces have a positive effect on social cohesion, foster positive attitudes toward racial and cultural diversity, and help reduce feelings of isolation or exclusion. Children who play build their confidence and learn the social skills that help them become happy, well-adjusted adults. Play is a factor in improving attention, attitudes, creativity, imagination, memory, and cognitive skills, all of which are critical for learning.

The Children's Discovery Garden is the name sometimes given to garden-like spaces with elements for children that encourage learning about Nature, plants and animals in their natural habitat, rocks and water, and might also include interactive elements and sculpture.

The attractiveness of letting a child take risks, implicit in the "adventure playground" popularized in Europe three or four decades

及示意标语牌的游乐场里进行。

游乐场地的作用不单单是提供娱乐设施。我们鼓励孩子们自由随性地玩耍，就是让游玩者出于童真和最本质的动机自由地选择想要玩的设施。理想的免费娱乐设施应该包含更多的自然元素比如树或是岩石等。波兰的快乐喷泉公园没有安全栅栏的打扰，孩子们自由地去自然中探索发现，在高大的瀑布中还有一些角色扮演的游戏。在纽约，一个以前的停车场由于振兴地铁站沿线计划被装饰上了喷泉，这样的喷泉让那些本来嘈杂的声音和扰人的震动也变得有趣和令人激动。在这里，通过创造性、探险、活动身体、友谊和冒险以及自由的玩耍可以帮助孩子们变得健康、快乐、自信、有好奇心并且有成就感。由于探险和身体活动是玩耍中的一部分，玩耍和孩子的成长以及运动能力的发展紧密相连，同样通过玩耍还可以锻炼孩子们的判断力、解决问题的能力和调节情绪的能力。那些经常玩耍的孩

图 5.67
口袋公园，纽约
Pocket Park, New York City

这个城市的绿色街道计划是将交叉口过多的铺装转变成三角的口袋公园。疏缓交通的同时也为居民提供了一块小小的自然。
The City's Greenstreets Program has converted excess paving into triangular pocket parks at intersections, calming the traffic while providing a small retreat for neighbors. (Credit: "participant/team Zefferus as part of the Commons:Wikis Take Manhattan project on October 4, 2008")

图 5.68
奥林匹克公园，布尔诺，捷克共和国
Olympia Park, Brno, Czech Republic

这个为青少年和成人准备的攀岩墙具有肢体挑战，它要求人们必须集中精力，并且具备通常在家中和在学校中不具有的身体强度和柔韧性。
The climbing wall for teens and adults poses physical challenges and risks that demand concentration, physical strength and dexterity beyond what can be acquired in the usual home or school setting. (Credit: Pavel Ševela / Wikimedia Commons)

图 5.69 - 图 5.70
亚历山大 W. 坎普游乐场地，坎布里奇，马萨诸塞
Alexander W. Kemp Playground, Cambridge, MA

这个公园设计的目的是为 2-12 岁的儿童提供具有身体挑战性，并且能激发探索力、想象力以及社交能力的游乐设施。这个设计包括小丘、山谷、沙滩、树板、树桩、植物、材料以及可供搭建的木块。还有一个可供轮椅使用者使用的旋转木马。
The design intent is to provide physical challenges as well as stimulate exploratory, imaginative and social play for 2-12 year olds. The design includes hills, valleys, sand, wooden branches and stumps, living plant material, and loose wooden building blocks. A "merry-go-round" provides wheelchair access. (Credits: Thomas M. Paine)

ago, which fell into disfavor later, as local officials became concerned with safety issues and fear of litigation and schools reduced recess time, is making a comeback in parks such as Imagination Playground (Rockwell Group) in New York City, which offers elements that children can use to build things and indulge their fondness for thrill-seeking, competing with less healthy indoor activities like video games and computers.

Nevertheless, near play areas, safety surfaces such as rubber, pea gravel, mulch, or sand are a huge improvement over hard paving like concrete or asphalt. Some playground equipment is custom designed using thin and slightly irregular rot-resistant tree trunks to support swings, slides and nets. (5.68-5.70)

3. The Neighborhood Park is too small to accommodate vehicular access or parking.

The whole premise is that the Neighborhood Park is close enough to home to walk to it, allowing users to get exercise on the way. On the other hand, several parking spaces for the disabled may be warranted.

4. The Neighborhood Park can benefit from the active participation of local residents in its policies, activities and maintenance.

It is reasonable to assume that the local residents can help the responsible central agency do a better job of accommodating local preferences. Local residents can be expected to feel a sense of proprietorship toward the space, and be motivated to assure its proper upkeep.

5. Community gardens are appropriate in some Neighborhood Parks.

Local families may join in a group of community gardens to raise their own vegetables and cut flowers. The community garden movement in the West over the last century has evolved from providing scarce food in wartime into a recreational activity for those who do not have a garden at home and want to feel connected to the land and to Nature. (5.71)

Community Garden Standards

An area of ground of at least a half hectare is divided into side-by-side plots for individual gardens. They can measure as small as 3 by 3 meters, as large as 6 by 6 meters. In the U.S., the average is 19 plots per 10,000 population, according to The Trust for Public Land. People are required to pay a small fee annually. In cases where demand exceeds supply, plots are allocated in a fair and open process. One of the challenges is how to prevent theft of crops.

See http://en.wikipedia.org/wiki/Community_gardening

Community gardens parallel the urban agriculture movement in some Western cities that replicates conditions in many Chinese cities: a small parcel of urban land is transformed into a small farm to raise locally grown produce for sale locally.

子不容易肥胖，自然也不容易得一些和肥胖有关的疾病，比如糖尿病或是心脏病。那些仅仅是在学校有组织的运动或者在其他方面被组织而活动的孩子们，就没有了良好的机会去发展例如攀岩和克服恐高等自我探索和冒险的能力。

游戏加强了孩子和社区之间的联系。社区的玩耍空间对社会凝聚力起着正面的影响，可以让孩子在面对种族和文化差异时变得更加乐观和积极，同时也可以有效地缓解孤独感和对外界的排斥心理。那些孩子们在游戏的时候建立起自信并懂得一些社会技巧。这些都使他们变得快乐，并且成熟得更快。总的来说游戏对于孩子的各方面的成长都有不可估量的积极作用。

有些类似公园的空间由于具备了鼓励孩子学习自然万物的环境而被称作孩子们的发现王国。在这里孩子们会学到很多知识，比如一些动物的习性，岩石和水的特点，还包括一些互动的元素和一些雕塑。

400年前的欧洲，大家都想让孩子们去冒险，于是那时的社会流行冒险游乐场一类的公园。但在其后，由于政府官方更加关心安全问题，希望减少诉讼的比例并增加学校的授课时间，所以在公园里面加入了更多的主题，比如纽约的想象游乐场。在这个游乐场里面，孩子们会得到一些积木似的东西去建造自己想要建造的东西并且让孩子们沉溺于寻找刺激中，与没那么健康的室内游戏例如电动游戏和电脑相抗争。

并且，游乐场周围，在坚硬的路面比如混凝土或者沥青上铺设一些柔软安全的表面覆盖物比如橡胶、小鹅卵石、覆盖物或是沙子，这样的举措无疑是一种巨大的进步。一些游乐场所的设施是特别订制的，使用了又薄又轻的非常规的防腐树干去支撑秋千、滑梯和网（图5.68 - 图5.70）。

3．由于社区公园普遍比较小，所以并不适合车辆的穿行和停泊

这么做的大前提是社区公园要离居民区足够近，让人们可以步行到达，并且允许人们可以在路上做运动。另外，园中应该设置有确保残疾人专用的空间。

4．社区公园附近居民的一些志愿活动有利于公园的政策、活动和维护

当地的居民能够帮助负责人去更好地维护社区公园是一种合理的设想。一旦公园成为他们日常活动的一部分，他们会更为关注他们身边的事情，自发地去维护和保养公园。

5．社区花园在一些社区公园中无疑会起到锦上添花的作用

一些公园附近的居民非常愿意在公园里拥有一块土地种植自己的蔬菜或是花草。从20世纪仅仅是一种为战时储备食物的无奈之举，变成现在的一项休闲活动，社区花园的活动为那些家中没有花园但是希望更多接触自然的人们提供了一个有效的途径（图5.71）。

社区花园的标准

一片至少有半公顷的空地被分成并排的几份分配给各自的居民作为他们自己的花园。他们可以获得9~36平方米见方的土地。

图 5.71
社区花园，布鲁克斯公园，旧金山
Community Garden, Brooks Park, San Francisco

在很多的西方城市，那些没有自己花园去种植蔬菜和花草的城市居民可以只付一小部分的钱去共享一部分的社区公园来满足他们的需求。
In many Western cities, urban residents who have no yard of their own to grow a few vegetables and flowers for their own use can share a portion of a Neighborhood Park to fill that need, for a small fee. (Credit: Kevion Krejci)

图 5.72
宠物公园标识
Dog Park Sign

这种可以让狗自由在其中奔跑的围合空间类型越来越流行，它的标识注明了狗的主人需要负责的注意事项。主人应该清理狗产生的排泄物。有一些公园为此提供免费的生物降解塑料袋。
For this increasingly popular type of enclosed space where dogs can run free, signage outlines the special responsibilities of dog owners, who are expected to clean up the waste generated by their pet. Some sites offer free biodegradable plastic bags for the purpose. (Credit: Henryk Sadura)

图 5.73
汤普斯金广场宠物公园，汤普斯金广场，纽约
Tompkins Square Dog Run, Tompkins Square Park, New York City

围合的遛狗空间挤满了主人和他们的宠物。这个空间最近被"高贵狗"杂志评选为美国最受欢迎的5个宠物公园之一。
The enclosed Dog Run space can be crowded with dog owners and their pets. This space was recently named by *Dog Fancy* magazine as one of the top five dog parks in the United States. (Credit: David Shankbone)

图 5.74
社区公园
Neighborhood Park

在一些社区公园中，宠物狗们可以获准无束缚地跑，做它们自己的肢体活动，例如在空中跳跃接住飞盘。
In some Neighborhood Parks, dogs are permitted to run free, and engage in physical challenges of their own, like jumping to catch a Frisbee in midair. (Credit: Ashleigh Nushawg)

6. The Neighborhood Park with sufficient area may include a subarea for a dog park.

This may fulfill a real need in neighborhoods where superblock projects do not permit dogs off leash in their private shared open space. The design and scale should be less elaborate than is appropriate for the Large Park. (5.72-5.74)

Dog Parks: Current Trends in U.S.

Dog parks are becoming very popular in the U.S. and globally. Currently there are up to 5.6 per 100,000 population in cities, according to The Trust for Public Land (2008). They range in size up to 20 hectares and are enclosed by a 1.2 to 1.8 meter high metal fence so that dogs can run free, not on a leash, with gates kept shut at all times. Double gates allow dogs to enter and leave safely. The best dog parks provide separate areas for large and small dogs. Turf is the preferred groundcover. Some dog parks offer old tennis balls for dogs to play catch and water fountains for dogs and people, even dog-washing stations. Others have swimming areas set aside exclusively for dogs, in lakes, ponds, rivers and ocean. Lights and benches may be included as well. Some dog parks even have canine memorials.

Owners are expected to clean up the waste left by their dogs using a so-called "pooper scooper" or a plastic bag, both typically provided in dispensers near a covered waste receptacle. A good example is on Chicago Lakeshore East.

This chapter has covered a lot of ground. Taken together, the urban public open space types that this book advocates and the activities that should be accommodated in excellent open space design are an essential antidote to the stress of living in the unnatural world of cities. If parks are distributed equitably so that all people have reasonable access, and if the design, construction quality, programming and maintenance of each transcends mediocrity to achieve excellence, then the public realm is doing what it should to meet the many needs of the community, to contribute to the well-being and happiness of the community, to inspire people and enrich peoples' lives.

注释
1. 凯西·马登．公共空间设计项目汇报，"广场中的生活：公共广场在城市空间营造中所扮演的角色"，构建波士顿会议，波士顿，2010年11月7日
2. 杰克·汉米尔顿访谈，2011年4月27日
3. http://www.waterfronttrail.org/pdfs/books/regeneration/chapter%206%20-%20pages%206-10.pdf, September 22,2005
4. 彼得·哈尼克，本·威利．《从健身区到医疗区，城市公园系统如何最佳的提升大众健康》（公共土地信托出版社，2011），14

NOTES
1. Kathy Madden. Project for Public Places, presentation, "Life in the Square: the role of the public plaza in urban place-making", Build Boston Convention, Boston, 17 November, 2010
2. Interview with Jake Hamilton, 27 April, 2011. http://www.timeout.com.hk/feature-stories/features/42240/lord-norman-foster.html
3. http://www.waterfronttrail.org/pdfs/books/regeneration/chapter%206%20-%20pages%206-10.pdf, September 22, 2005
4. Peter Harnik and Ben Welle, *From Fitness Zones to the Medical Mile, How Urban Park Systems Can Best Promote Health and Wellness* (The Trust for Public Land, 2011), 14

根据公共土地部门的报告，在美国，平均10 000人有19块自己的社区花园。人们需要每年交一小笔费用。为了预防供小于求的情况，这些土地的划分程序会尽可能的公平公开。唯一的挑战是如何预防小偷。

见 http://en.wikipedia.org/wiki/Community_gardening

和一些西方城市的城市农业进程相类似，社区花园在中国的城市也开始出现：一些城市里面小块的土地变成了小块的农田。人们在这些农田上种植并出售收获的产品。

6．如果街旁公园的面积足够大，那么在设计时可以专门为遛狗的人们和宠物狗们准备一片区域

宠物公园可以弥补一些不允许遛狗或是不允许解开狗链的地方。专门为遛狗准备的区域可以给那些有狗却又不知道到哪里遛狗的人们一块区域得以同他们的宠物共同活动并让他们的宠物自由地享受大自然。当然这样的设计和规划应该尽可能的简单（图5.72-图5.74）。

宠物公园在美国的趋势

宠物公园在美国和世界范围内都越来越流行了。根据美国权威调查（2008年），城市里每100000人口生活的区域里有5.6个宠物公园，他们的范围已经达到了20公顷并且有1.2-1.8米高的金属栅栏包围。这样宠物狗可以更为自由地在这些区域内活动，脱离开狗链的束缚。双向门使宠物狗可以安全地出入这些区域。好的宠物公园为大中型犬和小型犬提供不同的并且隔离的区域。草皮自然是首选的植被。一些公园为这些宠物狗提供一些可以玩耍的旧网球，还有宠物狗饮水机，甚至那里还有宠物狗洗浴中心。其他的一些狗公园有专门为宠物狗准备的专用游泳区。灯光和长椅也包括在内。有些狗公园甚至还有宠物狗纪念碑。

不过宠物狗的主人要在自己的宠物离开之后清理自己宠物留下的排泄物。当然，在垃圾箱旁边会提供专门的塑料袋来承装这些排泄物。芝加哥湖滨东公园就是一个很好的例子。

这一章已经包含了很多种场地的介绍。总结起来，本书提到并推荐的这些典型城市公共开放空间和为这些开放空间专门设计的活动可以有效地缓解由于生活节奏快且脱离自然对城市人产生的压力。如果公园可以分散到足以让城市中每个人都有机会接近它们感受它们的程度，并且这些公园的设计都能够做到超凡脱俗，那么公共领域才真正能够做到想民之所想，忧民之所忧。

齐心协力展望空间的未来发展
Taking Heart, a Vision for the Next Level of Excellence

Chapter Six
第六章

玉不琢不成器，人不学不成才。
Jade must be carved and polished before it becomes an ornament, man must be educated before he can achieve great things.
—— 中国谚语 (Chinese Proverb)

十年树木，百年树人。
If your vision is for a decade, plant trees. If your vision is for a lifetime, plant people.
—— 中国谚语 (Chinese Proverb)

我们不能用导致问题的思维去解决问题。
We cannot solve problems using the same kind of thinking we used when we created them.
—— 公认为是阿尔伯特·爱因斯坦的现代西方谚语
Modern Western proverb popularly attributed to Albert Einstein

路易斯·康说直到贝多芬给我们一个贝多芬交响曲，我们才知道我们还需要它。
Lou Kahn said we didn't know we needed a Beethoven symphony until Beethoven gave us one.
—— 丹尼斯·斯科特·布朗，2005 年 [1]
Denise Scott Brown, 2005 [1]

这是苹果的先天基因（根植于骨髓的信念）：光有技术是不够的。技术必须与艺术结合，与人文结合，才会让我们的心灵歌唱。
It's in Apple's DNA that technology alone is not enough - We believe it's technology married with liberal arts, married with the humanities, that yields us the result that makes our heart sing.
—— 史蒂夫·乔布斯，2011 年
Steve Jobs, 2011

在接下来的数十年中，城市公共开放空间会延续之前150年的做法，但是也会面对前所未有的挑战，让人们接触树、灌木、草地、水、开朗的天空、干净的空气，同时也要保护和净化自然系统，与被破坏的世界做斗争。随着城市移民和人口流动性不可避免的增加以及气候变化，自然系统的压力也在增加。在所有的混乱下，城市公共开放空间将会提供一个绿洲式的避风港，在那里人们参与尽可能多的活动去恢复思想和身体，重述他们对于社区的归属感，然后感觉焕然一新地回家。中国人民正不断地探索新趋势，接受新的闲暇的形式，发明其他的形式，许多全新的开放空间活动将降临中国。

大多数的城市公共开放空间会墨守成规，仿照之前的例子，合理建造，并不去尝试新东西。但是有些空间应该走得更远。当一个地方的领导者能够深谋远虑，能够了解接受一个好的大胆的设计对地方可持续性的重要性，能够愿意将富有远见的设计想法付诸行动并愿意看到整个设计的建成，这样做出来的城市公共开放空间才会到达一个新的高度。我们要知道我们不能用产生问题的思路来解决问题，那么这些大胆进步的领导者就要欢迎那些有深远意义的新思路，他们不会混淆于表面的大胆和深刻的内涵，我们知道他们会抛光城市规划和设计的粗糙边缘，会强调更完美的设计来面对土地和人的需求的挑战，最终他们会获得全世界的感激。

这一章讲述城市是如何真正地抓住机遇满足人们对于优秀城市公共开放空间的需求。我们分两个阶段带领读者走入一段未来旅程，第一阶段介绍一些创新项目以及在中国、其他亚洲地区和西方未被建成的一些城市项目，第二阶段会讲述8个趋势，这8个趋势在一起，将会带领我们超越目前所看到的世界。还未建成但是可以实现的项目是激发新鲜设计灵感的绝佳资源。不可建造并不能总是解释为什么一个有愿景的设计不能实现，决策人必须分享这一愿景。

作为中国第一家在西方设立全职办公室的私营综合设计公司，意格对在中国如何把创新设计和实际作业完美地结合有着独到的见解。在第一个十年中，意格不仅在商业区和住宅区设计私人开放空间，也设计城市公共开放空间。中国的设计机构缺少与海外的联系，而海外设计公司又缺少一个在中国强有力的本土联系，它们在做设计时会有自己的独特视角，但是还未能达到以下七个案例的深度。

■ **七个案例**
1. 城市广场：增添深厚文化韵味
风之图腾，福里斯特，珀斯，澳大利亚

意格波士顿办公室最近与来自保加利亚索菲亚市的知名雕塑家Todor Todorov合作，参加了改善澳大利亚珀斯市区广场——福里斯特广场的设计竞赛。包括提供先进可视化技术的上海办公室在内，这个真正的国际团队的设计热情远远超越仅仅做一些常规无趣并且盲目的设计，而追求在深层次的文化背景和自然涵义中寻找设计共鸣。

没有任何能产生比南方的星座与强风更强有力的自然共鸣。也没任何能产生比西澳原住民色彩鲜艳大胆、设计抽象的图腾更深入的文化共鸣。

风之图腾充分地利用略为倾斜的基地平面，将永恒的当地特

In upcoming decades, urban public open spaces will continue to do what they have been doing for 150 years, but with ever greater urgency—combat a disruptive world by building community through shared recreational and cultural experiences and giving people access to trees, shrubs, grass, water, open sky, and clean air, while also protecting and cleansing natural systems. As urban immigration and mobility inevitably increase, natural systems will come under increased pressure from encroachment, and climate change. In the middle of all the turmoil, urban public open space will provide an oasis of calm where people come to participate in the widest possible range of activities to restore the mind and body, reaffirm their sense of belonging to a community, and go home feeling refreshed. Many of these activities will be new to Chinese open space, as Chinese people continue to explore new trends, adopt new forms of leisure, and invent others.

Most urban public open space will do the safe thing and follow the examples of other places, and try to do so competently, without aspiring to something greater. But some places will go further. The truly ground-breaking urban public open space of the future will achieve even greater things when thoughtful and committed community leaders who understand its importance to sustaining the community enlist the best advice available to decide on a bold plan, turn the most visionary ideas into compelling designs for the components of the plan, and see the designs through to implementation. Realizing that we cannot solve problems using the same kind of thinking we used when we created them, such progressive community leaders will welcome profound new thinking. They will not confuse the superficially bold with the profoundly deep. They will indeed polish away the rough edges of design and urban planning as we know it, address more perfectly the many challenges to meeting the needs of the earth and of people, and earn the world's gratitude.

This chapter presents a vision for how cities can truly seize the opportunity to meet the people's need for excellence in urban public open space design. It takes the reader on a journey into the future in two stages, first by describing examples of innovative projects and un-built work for cities in China, elsewhere in Asia, and in the West, then by describing eight trends that, taken together, will take us beyond anything the world has seen to date. Unbuilt but eminently buildable projects are an excellent source for inspiring fresh design ideas. Being unbuildable does not always explain why a visionary design does not get built; the decision makers must share the vision.

As perhaps China's first private-sector multidisciplinary design firm to open a full office in the West, AGER may be uniquely positioned to understand the realities of how innovation and best practices come together in Chinese cities. Over its first decade, AGER has designed urban public open space as well as private open space in commercial and residential superblocks and knows intimately the challenges inherent in designing and constructing each. Design institutes lacking an overseas office and foreign design firms lacking a strong local presence in China may cover the same ground but with different perspectives. They would be unlikely to come up with the design thinking behind the following seven projects.

■ **Seven Projects**

1. Civic Plaza: Adding Deep Cultural Meaning
Wind Totems, Forrest Place, Perth, Australia

AGER's Boston office teamed up with Todor Todorov, a well known sculptor from Sofia, Bulgaria, to enter a competition for a makeover of Forrest Place, a Civic Plaza in Perth, Australia. The truly global team, also including our Shanghai office for state of the art visualizations, was fired by their passion for thinking beyond the usual insipidity and slavish replication into the realm of deeply resonant cultural and natural meaning.

No natural resonance could go deeper than the southern constellations and the strong winds. No cultural resonance could go deeper than that of the aboriginal people of Western Australia, known for their totems and boldly colored abstract design patterns.

Taking full advantage of the slightly sloping plane of the site, the spirited and dramatic grouping clearly marks the northern entry to Forrest Place. Expressive of their role as a wayfinding device, the four totems are arrayed like the Southern Cross, the constellation used for celestial navigation in the Southern Hemisphere, and a symbol of Australia's nationhood. Discs on the totems rotate kinetically, celebrating Perth's signature breeziness. In form, motifs and patterns the totems also celebrate Aboriginal culture, as do the paved circles on the ground plane out of which totems seem to grow.

The dominant totem is aligned on axis with the central archway of Perth Station, reinforcing the connectivity between Perth Station and Forrest Place. As pedestrians emerge from the station, they first see that one totem on the west side of the space, before taking in the three totems in asymmetrical balance on the opposite side. The central zone is left unobstructed for pedestrian flow, emergency vehicle access, and for visibility of stage performances from the north. The tallest totem is in scale with the height of the performance stage canopy, the shortest in scale with Aboriginal totems.

Wind Totems adds festiveness, dynamism, and gravitas to Forrest Place. Contemporary in feeling, they complement the historic facades enclosing the space. The design is sufficiently permeable so as not to obscure views through Forrest Place in either longitudinal direction. They inform drivers that Forrest Place is a car-free zone. The totems lie within a zone of pedestrian movement, on a ramp, near seating on the flat zone south of the site. No seating within the site distracts from the power of their presence. Integrating Wind Totems' setting to the rest of Forrest Place, the paving between the totems and outside the granite-paved circles in which they stand is identical to that of the rest of Forrest Place. The use of reflective paint for the disks enhances Wind Totems' magical night-time presence. Uplighting from within the paved circles at their bases is coordinated with the lighting decisions for the rest of Forrest Place.

The design integrates well with the urban fabric, while its fresh contemporary style and dynamism are broadly appealing among all

质融入这个标志性的公共艺术品当中来，它横跨珀斯车站，与城市第一公共广场相得益彰。风之图腾富有生机并且大胆的组合方式明确标志着福里斯特广场北面的入口。在南半球南十字星用于航海指向，同时也是澳大利亚的国家象征。四支图腾依照着南十字星的方位排列，充分地显示了其作为导航标志的角色。图腾上的圆盘随风转动，颂扬着珀斯市的特有的微风。在主题和样式上，图腾颂扬着原住民文化，地面上的圆形铺装也使得图腾柱好像是从地上长出来一样。

主图腾的轴心与珀斯车站的中央拱门对齐，加强了珀斯车站和福里斯特广场之间的连接。当人们走出车站，在见到东侧不对称平衡的三个图腾之前，他们将首先看到位于西侧的主图腾。中央部分没有任何构筑物，以保证人流、紧急车辆的通畅使用，并且保证了北侧的舞台表演不被遮挡。最高的图腾与表演舞台顶篷高度相当，最短的图腾与原住民图腾规模相当。

风之图腾为福里斯特广场注入了欢庆、活力与庄严。它充满了现代感，但也因为周围的历史建筑不乏历史性。这个设计具有相当的通透性所以在福里斯特广场的任何纵向皆不会造成视线遮蔽。并且这些图腾明确地告知驾驶人福里斯特广场是个禁行汽车的地方。图腾阵列在一个斜坡上，靠近位于南侧平坦的座椅区域，属于人行的范围内，图腾周围并不设置座椅，这使得图腾本身显得更加强有力。为了整合风之图腾以及福里斯特广场，除了图腾环形花岗岩铺装外，其余铺装皆与福里斯特广场其他部分相同。圆盘上反光涂料的使用增强了风之图腾夜晚时的魔幻感觉。从图腾底部环形铺装透出的照明是以福里斯特广场其余照明设备相互协调为基础做考虑。

这个设计与城市脉络极佳地结合。它的清新且充满活力的现代风格对所有年龄层与文化背景的人都拥有广泛的吸引力。其与众不同的布局从各个角度来看都是不同的风景。风之图腾是全年提供庆祝活动的场所，用以振兴福里斯特广场，弘扬珀斯特色（图 6.1 - 图 6.2）。

2. 市区公园：面对气候变化改善废弃地
高雄火车站，高雄，中国台湾

作为台湾第二大城市的滨水港口，高雄火车站之前是一个15公顷大的火车停留地。虽然场地仍然适合停放火车，但是大部分的场地已不再为铁路服务使用，但是作为一个大型城市文化艺术，整个火车站是值得以一个合适的形式保留下来。意格的波士顿公司接受了这项城市设计的挑战来完成这个城市开放空间设计。场地被均分为四个部分：酒店区、办公区、零售商业区以及一个市区公园。这个获奖的设计颂扬了场地的历史，设计了定向的生物降解，调节了由于气候变化而增长的极端天气状况。就像自然界的潮涨潮落，使用场地的人的数量也有兴衰起伏。游客喜爱这种把大自然、绿色建筑和历史遗留铁道结合在一起的体验。并且通过翻新传统风格的商店建筑，展示区域地标性的酒店，这个项目把周边的社区和建筑群衔接在了一起。社会和环境持续性真正地融合在一个和谐的设计中。

整个项目聚焦于创新性与绿色经济（绿色科技），它是把连续性和创新性相结合的催化剂。火车站和邻近的记载着蔗糖航运历史的仓库表达了一个甜蜜生活的主题，这些都激发了竞争型经

图 6.1 - 图 6.2
风之图腾，福里斯特广场，珀斯，澳大利亚
Wind Totems, Forrest Place, Perth, Australia

为了纪念当地的土著文化，这个设计呼应了珀斯著名的强风。喜庆的图腾将一个交叉口转换成一个无机动车通行的城市广场，吸引着人们不分日夜地前往。
Celebrating the design of local aboriginal culture, and designed to respond to Perth s famously strong winds, the festive totems transform a traffic intersection into a car-free Civic Plaza that is inviting to people day or night. (Credits: AGER)

济中各个层面的革新性与创新性。

廊道中改良的铁道景观设施用于引导水流的方向，同时渐进地适应变化的气候条件。景观、建筑以及基础设施之间的和谐关系决定了一个非常适宜的发展密度，并且无论在任何天气状况下都提供了一个受保护的交通环路。

为了修复受污染的土壤，地面种植池、突起的坡崖甚至是竖直墙面上都种有修复效果最好的植物材料。这种修复污染的方式比传统呆板的机械修复方式更加的创新、有效并且节约资源。粗略地估计，植被修复法所需的费用仅仅是焚化污染物费用的1/10，是土壤清洗法的 1/3。植物生物矩阵用来促进和增加自然生长的过程，完全没有附加的污染。修复形式和过程本身又可作为公园的一部分。

建筑设计上考虑自然环境的因素，比如台风和洪水。提高地面层高度并且在建筑第一行柱网后面设置分层隔断墙可以预防破坏，并且保护这里的居民。在储藏阁楼下的半开放商业空间设置可移动的隔断墙可以根据市场需求进行季节性的甚至单日性的调整。这一设计还有助于提高一楼开放空间效益，使得一楼可以获得充足的自然光，良好的通风性同样可以使室内空气清新，并且有助于保持夏季室内凉爽（图 6.3 - 图 6.5）。

3. 大型公园：展示可持续性的植物园
宁波植物园，宁波，浙江，中国

2010 年，意格与浙江农林大学园林与艺术学院合作设计了

age groups and cultural backgrounds. Its unusual layout imparts a different view from every angle. In helping to revitalize Forrest Place as a year round activity area, Wind Totems celebrates essential attributes of Perth's identity. (6.1-6.2)

2. Downtown Park: Transforming Brownfields while Confronting Climate Change
Kaohsiung Port Station, Kaohsiung, Taiwan

The port waterfront of Taiwan's second largest city formerly required railcar storage over a 15-hectare rail-yard. While the site still accommodates rail service, most of its area is no longer used for rail service, but is considered a culturally significant artifact worthy of preservation in some form. AGER's Boston office took on the urban design challenge of providing urban public open space within an economically feasible phased commercial development program. The site is equally divided between hotel, office and retail development and a Downtown Park. The award-winning design celebrates the history of the site, incorporates phased bioremediation, and accommodates increasingly extreme weather conditions brought about by climate change. As natural processes ebb and flow, people ebb and flow through the site. Visitors enjoy an immersion experience that combines Nature, green infrastructure, and historic rail infrastructure and stitches together surrounding neighborhoods with contextual architectural design in the form of updated traditional shop-houses, while showcasing the area with a landmark hotel. Social and environmental sustainability can indeed fuse in one harmonious design.

The development program is a catalyst for connectivity/creative synergy, focusing on the creative and green economy ("greentech"). The rail-yard and adjacent warehouse history of cane sugar shipping suggests a theme of sweetening life, inspiring innovation and creativity at all levels in the competitive new economy.

The modified "railscape" infrastructure of the corridor is designed to direct water flow and incrementally respond to changing weather conditions. The physical and contextual relationship between landscape, architecture, and infrastructure determines an appropriate level of development density and provides protected circulation throughout the site in a variety of weather conditions.

To remediate contaminated soils, plant species most effective for phytoremediation are planted on the ground, on raised berms and planters, and even vertical surfaces. Such a passive element approach is more innovative and cost-effective than aggressive structural and mechanical treatment options. Phytoremediation costs only about 1/10 the cost of incineration and 1/3 the cost of soil washing. Biological matrices are developed to stimulate and augment natural processes that create no additional contamination. The forms that accommodate these processes become park amenities.

The architecture is designed to accommodate typhoon flood conditions. Raising the ground floor and providing layered partition walls behind the first row of building columns prevent damage and provide protection for residents. Adjustable partition walls in the semi-public commercial open space below the lofts allow for a seasonal or even daily fluctuation in demand. Environmental benefits of this first floor open space strategy include ample natural light and flexible air flow for ventilation and cooling. (6.3-6.5)

3. Large Park: Showcasing a Sustainable Botanical Garden
Ningbo Botanical Garden, Ningbo, Zhejiang Province, China

In 2010 AGER collaborated with Landscape Architecture & Arts College of Zhejiang Agriculture & Forestry University on the design of a Large Park—322 hectares—almost the same size as New York's Central Park. Located 9.3 kilometers north of downtown Ningbo in Zhenhai New City, Ningbo Botanical Garden will figure prominently in the urban fabric, in the lives of Ningbo and Zhenhai residents, and in the urban ecosystem. Ningbo Botanical Garden will serve as both an ecological and educational arboretum and a large public green space and urban park, accommodating local residents and domestic and international visitors. It will be the first botanical garden surrounded by an urban park in China.

The plan preserves natural and cultural resources—existing streams, field patterns and historic village architecture. Water bodies artfully separate the urban park from the botanical garden. The botanical garden features indigenous aquatic plants as well as trees and shrubs, and preserves natural landforms in preference to artificial grading. The museum included in the rich program showcases natural architectural form. Conceptually, its design represents water as essential for life, the main theme of the park. The fluidity, reflectivity and transparency of the glass forms serve as a metaphor for water, which flows through the site and the building.

The existing historic villages to the west of the botanical garden proper are adapted as a restaurant area. The existing field pattern in the eastern part of the site is adapted as a nursery for both exhibition and sale of flowers. The plan makes wise use of land resources to promote agriculture and forestry and thereby reduce the financial burden to the government and the taxpayers incurred by this large project. The plan emphasizes the integration of landscape management and services.

Setting ambitious energy conservation and low-carbon-footprint goals, and incorporating best practices such as renewable energy utilization, multiple energy-saving traffic systems, and water purification by its wetlands, Ningbo Botanical Garden will play a significant role in enhancing the green infrastructure of Ningbo. (6.6)

4. Large Park: Maximizing the Benefits in Sustainable Design
Discovery Park, Changzhou, Jiangsu Province, China

Located in a city near Suzhou crisscrossed by streets of traditional houses and canals, Discovery Park is a new model for a destination park, planned, designed and constructed to be environmentally and financially sustainable. Filled with beauty, attractions, and activities,

图 6.3 - 图 6.5
高雄火车站, 高雄, 中国台湾
Kaohsiung Port Station, Kaohsiung, Tai

这个获奖的作品将15公顷场地的一半变成一个市区公园, 纪念当地的铁路历史, 同时该设计融合了分期生物治理措施, 使整个场地适应由于气候变化而带来的越来越严重的破坏性沿海条件。一个地标性酒店俯瞰一个雕塑公园, 植物修复工程与美丽的景色并存。
The award-winning design allocates half of the 15-hectare site to a Downtown Park that celebrates the rail yard history of the area, incorporates phased bioremediation, and accommodates increasingly disruptive coastal conditions brought about by climate change. A landmark hotel overlooks a sculpture park. Phytoremediation coexists with beauty. (Credits: AGER)

这个占地面积322公顷, 几乎和纽约中央公园一样大的大型公园。由于距离宁波镇海新城仅有9.3公里的距离, 宁波植物园必定会在宁波城市结构中占据显著的位置, 并且影响着周围城市居民的生活和整个城市的生态系统。宁波植物园将会是一个兼具生态性和教育性的植物园, 作为一个大型公共绿色空间和城市公园, 宁波植物园将会为当地居民、外来的游客以及国外的来访者展现出宁波别样的风采。这将是中国第一个被城市公园环绕的植物园。

这个项目设计保护了当地生态环境和文化资源——现存水系、土地肌理和历史性的乡村建筑。水体巧夺天工地将植物园和城市公园分开。植物园以本地水生植物为主（也包括乔木和灌木），

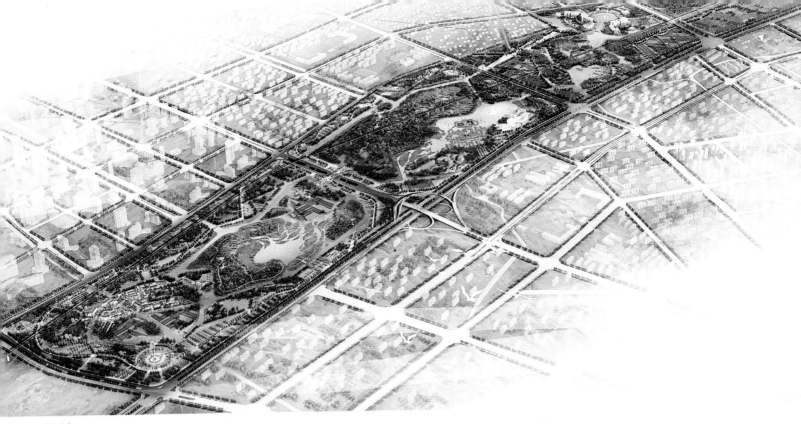

图 6.6
宁波植物园，宁波，浙江
Ningbo Botanical Garden, Ningbo, Zhejiang Province, China

这个和纽约中央公园面积一样的大型公园在一个巨大的尺度上结合了公园与植物园的功能，公园保护了文化资源，比如位于场地两端的当地原有村落，同时也达到了节省能源的目标。这个复杂的设计也同样作为一个高架铁路桥的缓冲带。
The Large Park which is the same size as Central Park in New York combines park and arboretum functions on a grand scale while preserving cultural resources like the village at the western extremity and achieving ambitious energy conservation goals. The intricate design manages to buffer an elevated rail line. (Credits: AGER)

并且在人工地形之外更加注重对自然地貌的保护。而活动类型丰富的博物馆则展示了自然建筑的形态。在理念上，整个公园设计以水是生命之源为主题。玻璃的易变性、反射性和透明性最能够体现水的特点，故而园区里的建筑充满了形态各异的玻璃形式。

现存的那些有历史意义的村庄坐落于这个植物园的西面，大部分的地方被改造成餐饮区域。而场地东面现有的田地则被改造成花卉的展销场所。这一规划巧妙地利用了土地资源，促进了当地林业农业的发展，并且减少了当地政府和纳税人的财政支出。这一规划强调了景区管理和服务的综合性。

有了保护周围环境、减少碳排放的决心，并且利用了各种最为先进的理念，如可再生能源、多重节能交通系统以及利用自身湿地进行水进化，宁波植物园必将有力地推动宁波环保基础设施的建设（图 6.6）。

4. 大型公园：利用可持续设计使收益最大化
探索公园，常州，江苏，中国

公园毗邻苏州这座充满古典建筑、河道交错的城市，旨在成为旅游目的地性公园的新典范。它的规划、设计和建造都是以环境可持续和经济可持续为出发点。公园内充满美景、趣味以及丰富的活动，邀请游客来体验发现之旅。公园重在营造休憩放松的多重感官体验，同时在体验和管理的过程中符合生态规律。长约5.5公里，面积超过 600 公顷的探索公园要大于旧金山的金门公园，而几乎是纽约中央公园的两倍。

the park invites visitors to an experience of discovery. They immerse themselves in a multisensory experience of rest and relaxation, managed and experienced in an environmentally responsible manner. Extending over 600 hectares and 5.5 kilometers, Discovery Park is larger than Golden Gate Park in San Francisco, and almost twice as large as Central Park in New York.

The plan for Discovery Park consists of eight "Biomes" connected by an "Eco-Spine". Each Biome is a portion of the park that is defined by a distinct dominant plant community and includes compatible cultural activities. The Eco-Spine is the contiguous body of water flowing through the length of the park that connects and gives life to all parts of the park. These are the Big Ideas that guide the design of the park as a whole.

Eight "Biomes" are roughly centered on the major north-south roads cutting across the park creating different visitor experiences within each. Contextually sensitive to existing site conditions and the proposed surrounding land use, the Biomes offer enough diversity of forest plant communities to ensure a year-round spectacle to attract visitors. The eight Biomes are organized east to west as follows: Community Farms, Wetlands, Lake Forest, Fall-color Forest, Conifer Forest, Flowering Forest, Bamboo Forest, and Agricultural Campus. They complement a diverse interplay of programmatic activities: ecological sustainability, cultural sustainability, forest regeneration, active recreation, fun educational and play experiences, and educational exhibits of world-class sustainable agriculture techniques.

The west end of Discovery Park includes community gardening among the existing villages retained on the site. The area includes individual garden plots for those living in the City, a daily farmers market, and an aquaculture area. It is assumed that residents living in the villages within the park will be employed to help run and maintain the community gardening facilities.

The Biomes are connected to the Eco-Spine through paths, boardwalks and bridges. Each is divided into parcels accommodating a variety of activities supported by parking and public transportation. The activities in each parcel are related to the character of the landscape and thus engage the visitor in a memorable communion with Nature. City dwellers will have the opportunity to experience the seasonal growing to dormancy cycle first-hand. In the beauty of the wetland cascade visitors can witness up close Nature at work restoring an ecological balance through vegetative and microbial processes for a vivid educational experience. Open-water activities include boating and fishing or spending romantic time with one's beloved. The fall-color forest area provides ample space for passive recreation, including picnics and barbecue parties, and experiencing the wonder of seasonal changes in Nature. The conifer forest is evergreen and lush providing opportunities for both solitude and group celebration with music, dance, and theater in a purely natural setting. The active sports zone provides facilities for both team and individual competition and fitness. A rich offering of art and cultural activities includes an array of dining venues, an education center, a music hall, an outdoor performance stage, a botanical garden, a sculpture park and a museum exhibiting the exquisite works of the Changzhou Painting school, where Nature can be contemplated on a whole new level. The Exploratory Play zone offers some of the most integrated family activities available anywhere, providing advanced learning experiences for children of all ages.

Multiple entries are provided along both the north and south edges of the park. This allows visitors to have direct access to the forest type and/or program of their choice. Parking is located at all entries and bike or electric vehicle rental are offered. The main path follows along the Eco-Spine and can be experienced on foot, bike, electric cart, or via the electric bus system. It also provides emergency vehicle and service access to all program locations. Secondary paths and a special tourist path provide access to all program locations and allow visitors to engage Nature on many levels.

The agriculture and aquaculture zones showcase sustainable farming practices that function in a closed-loop cycle where land and water are nourished to sustain life and the beauty of gardens is celebrated. The landscape is comprised of plots, beds, windrows, orchards, greenhouses, ponds and dykes. The Wetland Forest Zone is comprised of water-loving species in a series of small water bodies and streams that flow into the Open Water Zone exhibiting shoreline and littoral zone landscapes. The Fall-Color Forest Zone is a rolling

探索公园分为八个生物群区，由一条"生态之链"串联而成。每个生物群区作为公园的一部分，由一种主要的植物种群构成，并搭配适当的文化活动。水是万物之源，"生态之链"就是由一条贯穿场地的水体构成，它将场地各部分连接起来，并赋予其生命。生物群区和生态脊梁的概念决定了整个公园的序列和布局。

八个生物群区大致集中在公园南北向主园路的两侧，横穿整个公园，在每个生物群区创建不同的游览体验。为了呼应现有的周边场地条件和周围的土地利用，每个群区都有足够多样的植被群落以确保一年四季的非凡景色来吸引游客。八个生物群区由东至西依次是：农场、湿地、湖景森林、秋色林、针叶林、花树林、竹林与一个农业园区。除此之外还配有不同类型的主题互动活动：生态可持续活动、文化可持续活动、森林再生教育、康乐活动、趣味教学和游戏体验以及世界级的可持续农业技术教育展览。

社区农场位于探索公园的西端，由场地原有的现状村落改造而来。农场包括面向城市市民的个人菜园、日常农贸市场和一个水产养殖区，现有的村民可被雇佣来帮助社区农场的运营和维护。

探索园的生物群区在形态上被划分开，由"生态之链"连接，并结合路径、木栈道和桥。每个区块都有丰富的活动，配有停车场和公共交通站点。每个区块的功能布置都是跟景观特色以及森林特色紧密相连的，旨在给游客提供与大自然交流的独特体验。在农业体验区设有果蔬园，让人们能亲自参与并感受春耕秋收的农作体验。在湿地小瀑布美景里获取生态教育体验的人们可以亲眼见证植物生长与微生物分解的生态平衡。在水面活动区域人们可端坐垂钓、可泛舟湖面，亦可扣舷独啸。秋色林的田园背景恬淡清新，季相变化带来的颜色惊喜更是精彩纷呈；而且这里有宽阔的开放空间可供赏玩景色、放松呼吸新鲜空气，亦可席地野餐或开烧烤派对。针叶林是常绿的，四季郁郁葱葱，可沉思冥想亦可吹拉弹唱，用音乐和舞蹈在大自然中释放自己；剧场也是纯自然的设置，可供表演使用。活力运动区可作为正式或非正式比赛场地，也可以作为个人健身区域。艺术和文化的体验被安排在了下一个区域，被一系列的构筑空间烘托：用餐点、教育中心、音乐厅、露天舞台、植物园、雕塑园以及展示常州画派的艺术博物馆，向人们昭示着从一个全新的层面上对自然的审思和解读。探险区域给各个年龄段的孩子们提供了一个趣味空间，获取冒险和学习经历，并将各种家庭活动根据场地特色整合其中。

多个入口设置在公园的北端和南端，便于游客直接进入探索园的某一个生物群区或选择他们喜欢的活动项目。每个入口都设有停车位，也有提供自行车或电动车的租赁服务。探索园的主要道路沿着生态之链舒展开来，可供步行，自行车、电瓶车骑行或电动公车通过；同时有紧急消防通道和服务通道连通各个功能区块。森林公园次级道路和特殊的游园道路皆通往所有功能区域，确保游客能在各种层次上跟自然亲密接触。

农艺和水产养殖区域展示了可持续农业理念，封闭的环形场地里土地和水滋养着农作物，田园美景尽收眼底。该区域有田地、河床、垛堆、果园、温室、池塘和堤坝等景观。湿地森林区域里水生植物生长在一系列小水体和小河里，水流汇聚到一个开阔水域，形成岸线以及水岸景观。秋色林的景观随着起伏的缓坡和开放空间连绵展现在人们眼前。从果园里生长的果树到溪谷里的阔叶树林，延展出绝妙的开放空间形态。水顺着地形汇入溪谷，落

图 6.7 - 图 6.11
探索公园，常州，江苏省
Discovery Park, Changzhou, Jiangsu Province, China

这个大型公园比宁波植物园大一倍，由西到东有八个生物群区：农业观光园、竹林、花树林、针叶林、秋色林、湖景森林、湿地和社区农场。在景点中有一个艺术博物馆、雕塑公园、音乐厅、户外表演舞台和游泳池。公园的新颖之处在于它的尺度以及体验的多样性上并没有因为建筑群而混乱。
Twice as big as Ningbo Botanical Garden, this Large Park is organized into eight biomes, west to east: Agricultural Campus, Bamboo Forest, Flowering Forest, Conifer Forest, Fall-color Forest, Lake Forest, Wetlands, and Community Farms. Among the attractions are an art museum, sculpture park, music hall, outdoor performance stage, and swimming pool. What is innovative is the scale and and variety of experience possible in a park where variety is not overwhelmed by architectural clutter. (Credits: AGER)

英缤纷，让人恍若进入世外桃源。针叶林在高地，四季常绿，因为林间温度在寒冷的季节比外界暖，在热湿的季节比外界凉，这为林下植被创造了很好的生长条件。花树林的植被是组团搭配，自然随意而精致巧妙，为室内的活动提供了很好的户外空间。竹林区是趣味而且益智的，处处有意料之外的发现，每个转角都有新意和惊喜。人在与自然的亲密交流、互动中学到丰富知识（图6.7 - 图6.11）。

5. 绿道：提升当地的文化内涵
双湖公园，绍兴，浙江，中国

绍兴文化对于中国的影响是多方面深层次的，尤其是在城市公共空间方面的贡献更是不可忽视。中国20世纪初最为著名的作家鲁迅、久经考验的共产主义战士周恩来以及20世纪初许多诗人书法家都是来自绍兴。绍剧因为它的韵律和情调广为流传。上海的一个活动中心就坐落于绍兴街。游客们同样被绍兴城里那

landscape with hills and open fields. Some hardwood trees are planted in groves to showcase their magnificent wide-spreading forms. Small valleys lush with wildflowers offer intermittent streams. The Conifer Forest Zone planted on the upland provides year-round green color and an open understory where the temperatures are warmer in the cold months and cooler during hot humid weather. The Flowering Forest Zone is planted in both random clusters and organized rows adding a variety of outdoor rooms to complement the indoor program. The Bamboo Forest Zone is playful and intelligently designed to provide a surprise around every corner so visitors can be delighted by and learn from Nature through intimate interaction. (6.7-6.11)

5. Greenway: Adding Local Cultural Meaning
Twin Lake Parks, Shaoxing, Zhejiang Province, China

Shaoxing's cultural significance for China is many-layered and

worthy of celebration in its urban public open space. China's most celebrated 20th century writer, Lu Xun, and most sophisticated Communist, Zhou Enlai, both hailed from here, as did poets and calligraphers of earlier centuries. Shaoxing opera is known for its lyricism and musicality. A publishing center of Shanghai is on Shaoxing Street, surely no accident. Tourists also come for the intricate townscape of canals crisscrossing the city.

Within the newly expanded urban jurisdiction has been created a "Two Lake Development and Construction Office in the Textile City of China". For a proposed Greenway on the waterfront, AGER sent in a study team to research local history and material culture including architecture, bridges, boats, furniture, crafts, and opera stages. The interwoven water grid and road grid, warp and weft, make a fitting textile metaphor, or context, for this Textile Center. The metaphor suggests themes and imagery for designing everything from pavilions to markers to paving in a seamless, multilayered, meaning-rich design expressive of Shaoxing rather than something airlifted in from "out of context".

Most important to the context is the city's Civic Plaza, a really fine public space, anchored to the land by its proximity to a lone hill, like a citadel, surrounded by an ancient city wall. Dominating the space is an ancient Tang Dynasty pagoda, on the edge of a stone paved plaza with a map of historic Shaoxing spread over the plaza, the canals etched deep enough to hold water and make the city blocks into stepping stones and stone mural walls celebrating local heroes. On the edge of the Civic Plaza is a clone of the Sydney Opera House to house the local opera and fusion ballet.

The Greenway is environmentally and socially sustainable, including waterfront restaurants and retail in the festival marketplace tradition of Xintiandi in Shanghai and Faneuil Hall Marketplace in Boston. The space features as its signature amenity a traditional opera performance pavilion to showcase local stories, performances of all kinds, even Shaoxing opera. (6.12-6.13)

些错综复杂的城中河道所吸引。

新的城市管辖扩展催生出了"中国纺织城的两湖发展和建设中心"。为了更好地在滨水区建设绿道项目，意格组织了一个研究小组专门研究当地的历史文化，包括建筑、桥梁、船只、家具、手工和戏剧舞台。这里的水路交错纵横，或曲或直，就像纺织中的针线，左右穿梭，和这座纺织之城相得益彰。这样的比喻为那些设计者提供了更多设计的主题，可以更为深层次多角度地进行设计，来表达出绍兴这座城市丰富的内涵，这样比单纯地引入一些与绍兴毫无关系的内容更加符合绍兴的特点。

在这些城市的组成部分中最为重要的当属这个城市的城市广场，一个真正精致的公共空间，它坐落在一座孤山旁边，就像一座山顶城堡，四周围绕着古老的城墙。整个空间的主体是一座唐代的宝塔，宝塔位于一个石铺小广场的边缘，一幅绍兴古地图铺展了整个广场。深深的运河河道包围着城市，一个个街区就仿佛是水中踏步。在这片城市广场边缘地带有一个仿制的"悉尼歌剧院"，在这里举办了一些当地的戏剧和现代芭蕾。

绿道是一个环保而可持续的项目，比如上海新天地的节日市场和河畔餐饮以及波士顿的法纳尔厅市场。双湖公园深深地烙印着传统戏剧边沿的痕迹，表现出了当地的文化（图6.12 - 图6.13）。

6. 绿道：为废弃地带添加了新的生命
格瓦纳斯运河，布鲁克林，纽约，美国

把美好设计付诸实际能否成功很大程度上取决于是否有一个引人瞩目的主题并且加入完整的故事。意格为这个获奖的创新竞赛项目设定了一个富有想象力的标题，"国家的洗衣机：冲洗盆地、窗帘、床垫及枕头：格瓦纳斯运河的改革"。

这个设计考虑了当地工业遗产的过去、现在和将来，通过设计使这些遗产得以延续。工业化国家对减少不可再生资源的消耗和由于经济增长而被忽略的环境恶化负有重要责任。这个设计者的目的是将一个倒退的实用主义转变为健康的可持续发展的工业生态系统，这里的废物可以循环利用，资源被节约使用，改造的社区里使用了多种微生物治理的方法，所有培植方法都是根据场地的现有条件量身定做。

图 6.12 - 图 6.13
双河公园，绍兴，浙江
Twin Lake Parks, Shaoxing, Zhejiang Province, China

纪念非凡地方文化遗产的绿道不需要建得很宽，或者有很多绿色植物来与现代都市的喧嚣进行对比，来启发游客，来抚慰心灵。这个表演亭唤起了绍兴戏剧，也表演当地的故事或进行各种形式的表演。
A Greenway celebrating the extraordinary legacy of local culture does not need to be wide or even very green to provide a contrast from the chaos of the modern world, inspire visitors, and soothe the soul. The performance pavilion evokes Shaoxing opera and showcases local stories and performances of all kinds. (Credits: AGER)

"冲洗盆地"是指水净化的湿地，"窗帘"代表立式过滤器，"床垫"代表微生物母体媒介，而"枕头"指的是一条土壤净化道。当地居民通过游览和观察了解参与到这些修复的过程和持续的生态功能中，这样使他们与当地的联系更紧密。

设计的构想分为三个阶段。第一阶段里，污染最严重区域中的土壤和水体将会在实验中被仔细的检测。通过这些检测来确定最为创新并有效的生物治理办法。相较在每个分区都独立地安装昂贵的仪器来进行重复的检测，我们的设计采用了把所有分区联系起来共同合作检测的方法，即经济又可以保证对污染区的修复。水净化湿地位于每一个雨水和污水口，土壤驳船也被用来处理被污染河道的淤泥。这个计划的设想是让当地的利益相关者，比如商业集团和社区组织，都参与到这个公开的过程中来，作为工业区革新的一部分，为格瓦纳斯运河社区建立一个特有的身份。

在实验室进行了对场地生物治理的整体评估后，那些最有效的治理方法在第二阶段被强化实施。很快的，更多的场地将允许当地的公众进行正常、安全的进入和参观。这个设计旨在增加更多的行人交通、自行车交通和水上出租来连接布鲁克林区各个区域。继而各种商业和混合开发区就可以得到发展。

第三阶段中，土壤和水体的净化过程将被很完善的建立，并维护周围居民的健康。工业企业为当地居民提供现场的技术资源。当地居民将会以格瓦纳斯运河为荣，这个设计将作为工业废弃地可持续发展设计案例的经典。一旦被污染的运河可以焕发青春，那么居民和游客闲暇时间里就可以沿河散步，同时领略不同工业创新和城市生态系统的魅力。

在"冲洗盆地"和"窗帘"部分，未来的运河流域污染将会被水流中多样化的生物处理方式所限制。这其中就包括了生物浮岛，浮岛上暴露的植物根部吸收营养，为鱼类提供遮蔽和食物，同时亦作为有益微生物的定植培养基。同时地下水在进入运河之前就被植物修复梯田所净化。"窗帘"墙由一种建筑网格做成，这种特殊编造网格从当地工厂特殊定制而来。"窗帘"墙通过总基质和植物修复层过滤了场地内的所有排水径流。

6. Greenway: Regenerating Brownfields for a Sustainable Future
Gowanus Canal, Brooklyn, New York, USA

The success in bringing a visionary design to reality depends in no small part on defining a narrative in the form of a compelling title. AGER's imaginative title for this innovative and commendation-winning competition submission is Domestic Laundry: Flush Basin, Curtain, Mattress, Pillow: Regeneration of Gowanus Canal. (6.14)

The design takes into account the local industrial legacy, and makes connections that allow that legacy to continue. Industrialized nations share a critical responsibility to reduce consumption of non-renewable resources and degradation of the environment that undermine economic prosperity. The design intent is to transform retrogressive materialism into a healthy working industrial ecology where wastes are recycled, resources are conserved, and the community is regenerated using a variety of bioremediation and site cultivation methods rooted in the existing conditions of the site.

The *Flushing Basin* is a water-cleansing wetland, the *Curtain* is a vertical filter, the *Mattress is* a microbial matrix medium, and the *Pillow* is a soil-cleansing berm. Local residents gain an opportunity to observe, understand, and participate in the processes of site remediation and continued ecological functionality, and feel more closely connected to place.

The design is conceived in three phases. In Phase I, the soil and water in the most polluted site subareas are meticulously tested in a laboratory to determine the most promising and innovative means of site remediation. Rather than resorting to expensive methods that do no more than relocate the problem to another site, the design accepts that all site subareas are connected and incorporates effective yet still affordable methods for remediation of contaminated sites. Cleansing wetlands are located at all stormwater/sewer overflows and soil barges are used to treat contaminated canal sludge onsite. The plan

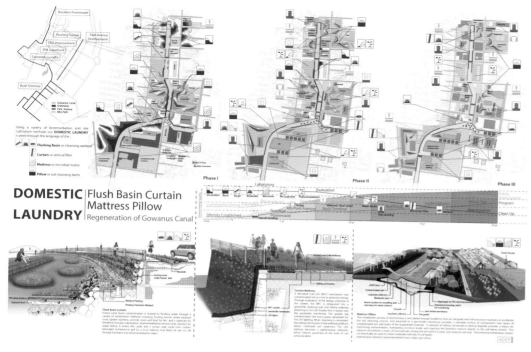

图 6.14
格瓦纳斯运河，布鲁克林，纽约
Gowanus Canal, Brooklyn, New York, USA

未来的设计应该体现过程和演变。在这个设计中体现了一个从已不能让人们接触的退化环境转变为可供人们散布、逗留与休闲的健康环境的过程。汇报展板上的细节解释了这块场地如何由三个阶段变为无害区域。在众多的创新应用中，LED灯光的能源来自于污染土壤供给的微生物燃料电池。
Design in the future will celebrate process and evolution, in this case from a degraded environment in which human contact must be avoided to a healed landscape suitable for strolling, lingering and relaxing. Explained in detail on this presentation panel is a three phase process to reach a non-toxic state. Among other innovations, LED lights are powered by microbial fuel cells fueled by contaminated soil. (Credit: AGER)

envisions an open process of engagement of local stakeholders, such as business and community groups, establishing an identity for the Gowanus Canal community as one of innovation and industry.

In Phase II, after initial evaluation by the laboratory of remediation methods tried on the site, the most effective methods are intensified. Soon public access and observation can be safely permitted at more locations. The design calls for the establishment of pedestrian, bike, and water taxi connections coordinated with ongoing efforts in the Brooklyn area. Commercial and mixed use development can then ensue.

In Phase III, the soil and water cleansing processes are well established and maintain neighborhood health. Industries established on-site offer technology resources to the region. Residents can now take pride in the Gowanus Canal as an epic example of how our sustainable future depends on brownfields cleanups. The once contaminated canal has become an amenity and residents and visitors stroll along the promenade in leisure or to view the diverse examples of industrial innovation and urban ecology.

In Flushing Basin and Curtain, future canal basin contamination is limited by flushing water through a variety of remediation methods including floating islands where exposed roots absorb nutrients, and provide cover and food for fish and a substrate for beneficial microbe colonization. Phytoremediation terraces cleanse sewer overflow water before it enters the canal. A curtain wall made from an architectural grill specially fabricated by a local industry filters all site run-off through a biomatrix and phytoremediation layers.

In Curtain and Mattress, a Microbial Fuel Cell (MFC) mechanism uses contaminated soil as a fuel to generate energy. Through evaluation of the design potential of the system, MFC is integrated

在"窗帘"和"床垫"部分，微生物燃料电池（MFC）机制利用受污染的土壤作为燃料来产生能源。通过对这个系统设计潜能的评估，MFC与带有地质纤维膜的挡土墙和滑动步道结合在一起，而镶嵌在地质纤维膜上的LED灯则会被MFC点亮。污染越严重，LED灯越亮。当净化的工作完成之后，步道中的滑动机制还将继续进行其他的土壤治理工作。净化过程中，LED灯则作为所处环境内污染的检测装置，实时地显示土壤污染的级别。

在"床垫"和"枕头"部分，农田的净化进程通过设计的地形一目了然，这些地形通过设计加入了加速土壤净化进程的必要基础设施。依附在地质纤维膜上的草皮为污染土壤层和生物增强材料层提供可移动表面遮盖。和垂直式干井相连的管网可以帮助管理者检测污染物情况、改善湿度等级和调节净化土壤中的生物母体种类。这个管网同时可用于土壤的机械混合直到土壤变得干净并且可以被转移。这个强化的生物治理系统最终可以以一种更柔和的方式用于基于植物修复和城市农业的土壤健康养护中（图6.14）。

7. 绿道：展现生态和人本理念
千岛湖珍珠广场和中轴溪景观，浙江，中国

千岛湖位于千岛湖国家森林公园内，面积573平方公里，为杭州附近的区域提供水源。作为一个大型公园，它提供了非凡的自然体验。中轴溪绿道和千岛湖公园完工于2013年，它满足了淳安这个建在半岛上新城的休闲需求。中轴溪为一条人工河，它的水源为附近卫星转播园区的冷却水。溪流沿着长3.5公里、宽150米的廊道穿城而过，一直流淌到俯瞰整个千岛湖的文化综合体中心——珍珠广场。无论水库的水位如何波动，意格设计的300米无边瀑布都会使珍珠广场与水库融为一体。全场地的人行坡道真正做到了无障碍设计，千岛湖公园无论从安全感和舒适感还是个人的平静感都为游客提供了一个与未破坏大自然的情感连接，而这一切都建立在对公共用水水源零干扰的基础上。中

图 6.15
千岛湖珍珠广场
Qiandao Lake Pearl Plaza

木栈道引领人们从左侧的珍珠广场经过湿地来到水台地的中心,在这里无边瀑布的边缘与千岛湖融为一体。
The boardwalk takes visitors from Pearl Plaza at the left rear past wetlands to the heart of the water terrace whose surface melds into the surface of Qiandao Lake beyond. (Credit: AGER)

图 6.16
千岛湖珍珠广场
Qiandao Lake Pearl Plaza

珍珠广场无边瀑布下的坝墙为晚间沿着千岛湖岸游走的人们创造了一个五彩缤纷的瀑布体验。
The weir wall below the infinity edge of the Pearl Plaza water terrace provides a colorful waterfall experience as visitors walk along the Qiandao Lake waterfront in the evening. (Credit: AGER)

轴溪绿道把绿道可持续设计提升到了新的卓越高度:真正的人车分流,真正的连续无间断,真正的无污染。所有的跨溪车行交通都通过溪上方的车行桥连接,从而避免的了对溪水和景观的打断。同时桥下方的无障碍人行坡道则给所有使用者提供了进入园区的便利。(图 6.15 - 图 6.20)

■ 八个新兴趋势

我们现在转到八个新兴的趋势,若同时满足这八种趋势,那么这样的设计将会远远超过现在世界上所有的任何一个案例。

into a geotextile retaining wall and sliding walkway powering the LED display that is woven into the geotextile membrane. The greater the contamination, the more power is generated for the LED lighting. When cleansing is completed, the sliding mechanism of the walkway platform allows continued soil treatment. The LED lighting becomes a performance indicator, which informs passerby of the level of soil contamination.

In Mattress and Pillow, the remediation process of land farming is articulated through landforms that are designed with the infrastructure necessary to accelerate the soil cleansing process. Sod attached to a geo-textile membrane provides a movable surface to contain layers of contaminated soil and layers of bio-augmented material. A network of tubing connected to vertical drywells provides a means for monitoring contamination, maintaining moisture levels, and injecting the biomatrix mixture specific to the soil being treated. This network also affords a means of mechanically mixing the soil until it is clean and ready for removal. This intensive remediation system can eventually be used for more moderate methods of soil health maintenance related to phytoremediation and urban agriculture. (6.14)

7. Greenway: clean, green, and totally accessible to all
Qiandao Lake Pearl Plaza and Central Creek
Zhejiang Province, China

The water supply reservoir for nearby Hangzhou, 573 sq. km.

图 6.17
千岛湖珍珠广场
Qiandao Lake Pearl Plaza

这个喷泉让年轻人在水花飞溅中受到心灵的震撼。在夜晚从千岛湖对岸的山坡望去,包括镭射动画的声光水特效动画使得广场的硬质景观成为一个真正多感官的体验。
A water jet fountain invites the young at heart to splash to their heart's content. At night, sound, light and fountain special effects including laser columns animate the hardscape in a truly multi-sensory experience, seen against the quiet waters of Qiandao Lake and its mountain backdrop.(Credit: AGER)

图 6.18
中轴溪
Central Creek

3.5 公里长的中轴溪绿道一直无间断地延伸到珍珠半岛新城的中心，这要归功于跨溪的车行桥，它使得车行交通远离了沉浸在自然中的行人和骑车的人。
Central Creek Greenway flows uninterrupted for 3.5 kilometers through the heart of Pearl Peninsula new town, thanks to overpasses, offering pedestrians and cyclists immersion in nature and time away from traffic.(Credit: AGER)

图 6.20
千岛湖珍珠广场
Qiandao Lake Pearl Plaza

木栈道随着珍珠广场的曲线延伸出来，镭射灯光柱、湿地台地和中心被水环绕的大曲线表演台都为珍珠广场增添生气。无边瀑布的尺度使整个千岛湖巧妙地成为舞台的"借景"。
The boardwalk curves from Pearl Plaza animated with its laser columns, past wetland terraces, and angles toward the large deck surrounded by water that serves as a performance stage to entertain visitors on Pearl Plaza. The scale of the infinity edge of the water terrace transforms Qiandao Lake into truly "borrowed landscape". (Credit: AGER)

图 6.19
千岛湖珍珠广场
Qiandao Lake Pearl Plaza

人们在最喜爱的广场音乐喷泉上嬉戏。
Immersion in the fountain brings joy to visitors of all ages.(Credit: AGER)

1. 技术和可持续发展

城市公共开放空间会融入更多的先进技术和可持续发展的理念。我们可以自信地说，在将来，可持续设计将是新的基准线——我们已经有的基准线并将一直持续下去——当然，对更有效果和更高效率的追求依然会继续，并且技术将在多方位提高可持续发展的浪潮中发挥更重要的作用。

在我们这个生态混乱的时代，"仿生学"引领了技术的新潮流，但是同时也带来了新的复杂性。这种设计手法的过程、进化以及以不平衡性为主导，颠覆了任何关于平衡性的言论。除此之外还有"仿生态学"，这种生物形态的设计手法自信地，甚至是过于自信地预测了生态功能，低估了易感的美学和叙事性。这些对于设计中几何学的影响是十分深远的。如今，简单的几何学的设计是十分大胆的，但这样大胆的设计依然赢得设计奖项，为设计者带来荣誉。举例来说，如今设计中流行铺设颜色交替的带状的硬质景观，甚至是在数条平行的石头线路中间铺设交错的软质景观。但今后，这种使用明显几何线条的人为设计习惯可能会逐渐消失。

几何形设计永远不会违背原始自然，就像奥姆斯特德的作品所呈现的那样；奥姆斯特德模仿自然，理想化自然，但是没有曲解自然。下一代设计者们会更加地依赖技术，技术不仅可以帮助设计精妙的生态系统，而且还能够实现那些比现如今所有公园更复杂的几何设计。即使视觉设计与可持续功能没有任何直接联系，将来的富有技术的设计者将会依靠技术来做出远超现有形式的复杂形式和图案，而这些复杂的形式和图案将是对可持续设计的补充，使之充分地被体现出来。

技术可以在多方面加强可持续性，比如在人造生态系统中检测物种组成，管理水质和水循环，检测土壤的组成和密实度，通过使用节能照明管理能源消耗，控制遮阴，利用太阳能控制灌溉系统，甚至是帮助受损材料复原。举例来说，半透明的光电玻璃面板能够同时吸收和释放适度的太阳光，还有正在研究过程中的生物混凝土，其中包含了细菌微生物，可以使出现裂缝的水泥自然地修复。

结合了手持设备的 GIS 软件和谷歌地球软件已经承诺释放多种详细的场地信息。本质上，这些开源计划软件将会给城市规划者和利益相关者提供更多的数据以及参与社区建设的机会。使用者可以读取所在位置的数据，添加到公共记录中，提交查询并接受私人的回答，甚至可以志愿协助设计者和规划者完成工作。

从长远的角度出发，技术同样可以帮助人们长期监控并精确地找出可持续设计的最优办法，这就是所谓的"生命周期分析"。但是这种叫法相比于公共开放空间更适合于建筑，因为开放空间的生命可能是无限的。长期监控可以防止由于人为蓄意或者无意偏移运行及维护标准所造成的性能下降。目前一些可持续设计项目仅仅关注在单一时间点上的生态效果，如果没有长期监控，这样的项目会让人担心它的节能效果或者节水效果会不会一直保持下去。

总而言之，技术会在设计过程、设计产品——城市公共开放空间本身——产品的运转和维护这三个部分中起到持续性的作用。

Qiandao Lake, within Qiandao Lake National Forest Park, offers a spectacular setting for a Large Park. Completed in 2013, Central Creek Greenway and Qiandao Lake Park accommodate the recreational needs of nearby Chunan, and soon-to-be-constructed Pearl Peninsula new town. A designed waterway fed by the cooling water discharge from a nearby television transmission facility, Central Creek flows in a continuous 3.5-km long, 150-meter wide Greenway through the city, to Pearl Plaza, focal point of the Cultural Complex civic center overlooking Qiandao Lake. Whatever the fluctuating reservoir water level may be, AGER's 300-meter-long infinity edge design melds the water terrace of Pearl Plaza with the reservoir. Made truly accessible to all by pedestrian ramps, Qiandao Lake Park provides visitors an emotional connection to unspoiled nature, from a feeling of safety and comfort to joy and personal serenity, all without compromising reservoir water quality. Central Creek Greenway advances sustainable greenway design to the next level of excellence: truly vehicular free, truly continuous, truly pollution-free. Overpasses carry cross-vehicular traffic over the uninterrupted landscape while pedestrian ramps in each block provide convenient accessibility to all. (6.15-6.20)

■ **Eight Emerging Trends**

We now turn to eight emerging trends that, taken together, promise to take design beyond anything the world has seen to date.

1. Technology and Sustainability

Urban public open space will combine more intensive adherence to sustainability and more extensive high technology. What we can confidently say about the future is that sustainable design is the new baseline—it is here to stay—but the search for ever greater effectiveness and efficiency will continue, and technology will play an ever greater role in enhancing sustainability in ever more ways.

In our time of ecological turmoil, "biomimicry" is on the ascendancy, but with a new complexity. This design approach now celebrates process, evolution and disequilibrium as dominant over any state of equilibrium. Otherwise known as "ecomimicry", such biomorphic design nowadays comes with a confident, perhaps overconfident, prognosis of ecological functionality and disdain of accessible aesthetics or accessible narrative. What this will mean for the geometry of design will be profound. Today, simplistic geometric design is bold, still wins design awards, and brings glory to the designer. For example, a current fad is to pave the hardscape in strips of alternating color, even slice across softscape in parallel lines of stone. Tomorrow, this habit of celebrating human manipulation through obvious, geometric design may lose ground.

Geometric design could never be mistaken for Nature preserved, as Olmsted's work has come to be; Olmsted mimicked Nature and idealized Nature, but did not denature Nature. The next generation of designers will increasingly rely on technology both to improve the success of such designed ecological systems, and to enable design to be geometrically complex beyond anything that can feasibly be fabricated for most parks today. Even when the visual design forms have nothing to do with sustainable functionality, tomorrow's skilled fabricators will rely on technology to lay out and install complex forms and patterns beyond anything commonly installed now, and those forms will complement and express sustainable processes.

Technology will enhance sustainability in such ways as monitoring species composition in the man-made ecological community, managing water flow and quality, monitoring soil composition and compaction, managing energy consumption for so-called energy-efficient lighting, managing shade, managing irrigation by solar power, and even helping damaged materials to heal. For example, semitransparent photovoltaic glazing panels can both collect energy and provide comfortable levels of sunlight, and research is underway to develop a "bioconcrete" containing bacteria and nutrients to allow cracks to seal naturally.

GIS and Google Earth combined with mobile devices promise to unleash a dense array of information about sites, their context, and how they are used. In essence, open-source planning software will provide citizen planners and stakeholders with ever better data and allow them to become ever more involved in the public spaces in their community. Users may have access to site data, add to the public record, submit inquiries, receive personal responses, and even volunteer to assist planners and designers in their work.

Technology will also enable accurate and continuous monitoring of sustainability best practices over the long term. This is sometimes called "life-cycle analysis", but that term is more appropriate for a building than for an open space whose life may be indefinite. Continuous monitoring will help prevent a decline in performance as people either willfully or out of ignorance depart from the expected standards of maintenance and operation. The current focus on a single "snapshot" to win certification as a sustainable project does raise concerns about how well the project will continue to perform in energy efficiency and water conservation if monitoring does not continue.

In summary, technology will deliver continued improvement in three areas: the process of design, the products of design—urban public open spaces themselves—and their operation and maintenance.

2. New Materials, New Artforms

In addition to the basic materials of landscape construction—plants, paving, walls, fences, and outdoor site furnishings—new materials will respond to the challenges and opportunities of the age of social sustainability and connectivity. New materials will be not only more durable and energy-efficient, but also richer in meaning and more able to engage people on an intellectual and emotional level. Recent inventions and introductions point the way.

Recent Inventions and Materials Introductions
- Light-reactive concrete and glass tiles for floors and walls
- Photographic bas relief imaging in concrete and ceramic tiles

2. 新材料、新艺术形式

除了基本的景观建造材料——植物、铺装、景墙、围栏和室外家具——新的材料将会应对社会可持续发展和连通所带来的挑战。新的材料不仅将会更耐久、更节能，而且还会有更丰富的意义，以及更能在材料本身的优点和情感上打动人们。最近的发明和介绍指引了这一方向。

新发明以及新材料简介
- 为地板和墙设置的光感混凝土与玻璃砖
- 用于混凝土、瓷砖的浮雕成像摄影
- 与环境呼应的建筑表面
- 嵌于纺织品、液体、3D 物品或打印图案中的多层塑料或树脂板，用于营造特殊视觉效果
- 光感高分子瓷砖
- 日间 3D 全息影像投影全透明表面
- 3D 真实深度屏幕
- 荧光材料的编织纤维
- 互动墙（纺织品或者瓷砖颜色可以改变）以及地板内镶嵌液体
- 像素金属墙壁图像（伪全息"光学瓷砖"）
- 合成四季冰面

设计师的挑战就是寻找一个把握住时机的艺术的声音，这样，公共空间就会充满动人的画面而不空洞乏味。高调的商业主义在某些地方是适用的，并且有时也是可以带来美感的。但是通常会降低游览者的体验。全息影像技术还并没有展示它在户外设计应用中的全部潜力，这种技术应该被更深入的发展，并且增加移动中的影像动能（图 6.21 - 图 6.25）。

光的盛宴

自 1999 年以后的每一年，坐落在法国中部的里昂市都会举行令人激动的光的庆典，在那里投影在墙上和雕塑上的灯光和影像把城市中主要的广场、街道和建筑都转变为展示前沿照明艺术的艺术品。这个庆典吸引了全世界著名的灯光设计师、建筑师和艺术家，并且还有超过 400 万的法国和世界各地的游客。这个为期四天的庆典要追溯到 1852 年，那时候里昂的人们拿着灯笼沿街游走，路过一扇扇点着蜡烛的窗户。

图 6.21 - 图 6.22
LED 墙
LED walls

LED 是一种有叙事功能的节能技术，促进了环境和社会的可持续。LED 墙通常结合大型透明影像。预制的玻璃墙可以结合 LED 墙来减少透明度。
LED is an energy efficient technology with narrative possibilities—promoting environmental and social sustainability. LED walls typically combine large scale imagery with transparency. Molded glass walls can be combined with LED walls to reduce the transparency. (Credits: Thomas M. Paine)

图 6.23
投影表演
Projection Shows

在法国里昂的光之盛宴上，一些有重要文化价值的图像在夜晚投影在像雾幕一样的非实物上，可以营造出奇幻的感觉，并培养与当地情感上的连接。这个每年12月举办的盛宴展示了最新的灯光技术。
In its Festival of Lights, Lyon, France, projecting culturally significant images onto something as insubstantial as mist at night creates a sense of wonder and fosters an emotional connection to place. The Festival held annually in December showcases the latest lighting technology. (Credit: © Andrei Iancu, Dreamstime.com)

图 6.24
灯光表演
Light Shows

在新加坡的码头沙滩上，其实还包括香港、澳门和许多其他城市，激光表演与烟花表演相竞争，与后者相比，激光表演没有噪声。
At Marina Sands, Singapore, indeed in Hong Kong, Macau and many other cities, laser light shows rival the excitement of fireworks, but without the noise. (Credit: Padsaworn Wannakarn)

3. 社会媒体互动

一些列在第二点中的材料的互动性可以说是"放大真实性"这一趋势的一部分，"放大真实性"意思是结合一体化的技术和通常用来建造户外空间的材料来把虚拟和现实融于一体。比如"QR"（快速反应）二维代码被用在了立面和人行道上面，那么移动设备就可以扫描或是"读取"这个代码并且连接到网络上获取相关的信息，不管是世界的还是地方文化的都可以（图6.26）。

同时，无线网络科技将会增强社会的可持续性，它的微波辐射可造成的潜在危害被解决后会更加受青睐。最为有趣的融合趋势之一便是社会媒体与公共领域的融合。其他一些令人激动的融合也正在发展中。

- Environmentally responsive architectural surfaces
- Laminated plastic or resin panels embedded with textiles, liquids, 3D objects or printed patterns for special visual effects
- Light reactive polymer tiles
- Transparent surfaces for projected images and 3D holograms, suitable in daylight
- 3D (true depth) screens
- Fabric woven with luminous phosphors
- Interactive walls (color changing textile or ceramic) and liquid embedded floor panels
- Pixellated metal wall images (pseudo-holographic "Optical Tiles")
- Synthetic all-season skating ice

[Sources include Blaine Brownell, Transmaterial, a catalogue of materials that redefine our physical environment (New York: Princeton Architectural Press, 2006)]

The challenge for designers is to find an artistic voice worthy of the opportunity, so that public space is crowned with deeply moving imagery rather than something inane. Introducing an obtrusive degree of commercialism may be considered suitable for some places, and may at times rise to artistic excellence, but usually can be depended on only to lower the experience for visitors. Holograms have not yet reached their full potential in outdoor space and should be developed further, including an increased illusion of movement. (6.21-6.25)

Festival of Lights

Every year since 1999, the city of Lyon, located in central France, hosts a breathtaking festival in which lighting and video projection on walls and sculpture transforms the main squares, streets and buildings into an illuminated urban showcase of cutting-edge lighting technology. The festival attracts famous lighting designers, architects and artists from all over the world, along with more than 4 million visitors, both French and foreign. The four-day festival dates back to 1852, when the people of Lyon walked down the streets holding lanterns and passed windows lit with candles.

http://www.weltlighting.com/3d-video-mapping-projections-at-festival-of-lights-in-lyon/#ixzz1rC69M0NM

3. Social Media and Interactivity

The interactivity of some of the examples of materials listed above are part of trend called "augmented reality," that is, the merging of virtuality and physicality that comes with the integration of technology and the materials which are used to construct outdoor spaces. If the "QR" (quick response) code is designed in a façade or pavement, a mobile device can scan it or "read" it and access a web site with relevant information about the site, whether mundane or cultural. (6.26)

At the same time, technology in the form of free Wi-Fi will strengthen social sustainability, preferably after its potential harm through microwave radiation is addressed. One of the most intriguing emerging trends is the place of social media in the public realm. Several examples of this exciting emerging trend are underway.

Canada Tourism Twitter Wall

The Canadian Tourism Commission (CTC) has installed 2.5 meter × 3 meter touchscreens (interactive twitter walls) in Times Square and Chelsea in New York City, Michigan Avenue in Chicago and The Grove in Los Angeles. They display twitter streams with photos and tweets about Canadian destinations, and passers-by can interact with the touchscreens, and pull up photos and information about Canadian destinations related to a tweet on the screen, or one in which they are interested. (6.27)

Saint Louis Media Commons

In Grand Center, an arts and entertainment district in Saint Louis, Missouri, the "Intersection of Art and Life", a proposed outdoor meeting place, unites engaged visitors and video, audio and social media technology in a landscaped hardscape space enclosed by projection walls. The "media mecca" is an unusual collaboration of Nine Network of Public Media including St. Louis Public Radio and the University of Missouri and local art museums, which see the value of synergy in jointly providing a "participatory" audience with the most innovative media in an outdoor space open to all, with seating and tables. Commercialism is avoided.

The Official Freedom Trail App

The popularity of applications ("apps") for hand-held devices like cellphones, iPads and tablets has spread to site appreciation. Visitors to Boston's famous Freedom Trail can now download an app to follow the route on their hand-held device using GPS updated maps, read historic information about nearby sites, and go online to access websites that provide even more site information. The app is modestly priced; the money supports the sites. Texting is sure to follow.

Johannesburg's Life Centre

On one of the city's most iconic high-rise buildings, Nike sponsors a digital interactive communications experience on an LED screen half a football pitch in size that allows sports fans around the world to submit a 57-character personal message to over 50 of Nike's athletes from around the world through Facebook.com/nikefootball, Twitter, Facebook, Mxit (South Africa) and QQ (China). Up to 100 headlines are then selected each night and transformed into digital player animations that light up the Johannesburg skyline, also captured in a photo that is sent to the submitter. The images do not sell Nike products directly. A degree of control assures that displays are not inappropriate. The idea is suggestive of possibilities in urban public open space through carefully regulated commercial sponsorship.

Crowdsourcing and Participatory Design

In the West there is a long tradition of public meetings held by governmental officials who manage proposed urban development projects, so that individuals may make suggestions about the design or redesign of a specific urban public open space. Such participatory design is sometimes called "community-based design" or "design by democracy". Led by meeting facilitators, local citizens gather in a room and collaborate

加拿大游客推特墙

加拿大旅游局（CTC）在美国纽约市时代广场与切尔西区、芝加哥密歇根大道、洛杉矶葛洛夫购物中心安装了2.5米×3米的触摸屏（推特互动墙），并在显示屏上滚动播放加拿大旅游胜地的照片及推文。路过的游客可以通过拖动屏幕上的图片或点击推文，或者搜索他们感兴趣的地方，来获取加拿大旅游相关讯息。（图6.27）

圣路易斯媒体公园

格兰特中心是密苏里州圣路易斯市的艺术娱乐区，被称为"艺术与生活的交汇"。一个新型户外会面中心计划将建在这里，它将把参与的游客、影像、音效以及社会媒体科技都集中在这个被投影墙环绕的硬质景观当中。这个"媒体圣地"不寻常地将九个网络和公共媒介结合在一起，其中包括圣路易斯公共广播、密苏里大学以及当地的艺术博物馆。这些合作者都看到了在具有最创新媒体的公共开放空间整合合作来培养共同听众的价值。场地将提供户外桌椅。商业主义在这里将被禁止。

波士顿自由之路官方应用程序

适用于掌上设备，比如手机、iPad以及平板电脑的各种应用程序已经普及到景点游览方面。要去游览波士顿著名的自由之路的游览者现在可以下载这个官方应用程序到他们的掌上设备，通过GPS更新地图，查看周边历史古迹信息，以及登录相关网站获取更多信息的方式来了解这一游览路线。这个应用程序有一定的合理收费，得来的钱都用于场地的维护。游客可以通过这个应用程序发表评论和看到其他人的评论。

约翰内斯堡生活中心

在这个城市最有标志性的高楼上，耐克公司赞助了一块足有半个橄榄球场大小的LED电子互动屏幕。这个用于交流体验的巨幕允许全世界的体育迷们通过耐克足球脸书、推特、脸书、Mxit（南非）和QQ（中国）向耐克旗下50多名运动员发送不超于57字的个人留言。每晚最多将有100条头条被选中，之后转变为运动员的电子动画照亮约翰内斯堡的天际线。这些动画将被拍照并送给发送者留念。这些图像并不会直接销售耐克的产品。会有一定的控制来保证这些图像演示的合理性。这个活动的想法是提出在公共开放空间管理控制商业赞助的可能性。

众包与参与性设计

在西方，公共会议是一个延续了很长时间的传统。通常由管理城市发展提案项目的政府官员来组织会议，公民们可以对每一个城市公共开放空间的设计和重建提出建议。这种参与性设计有时被称作"基于社区的设计"或者"民主设计"。在会议举办者的组织下，当地居民聚集在一起，合作讨论出新的概念或者备选方案。参与性设计是一种关注于过程的手段，并不是一种设计风格。对于"过程"的一个强烈推荐是公共空间项目。（www.pps.org）

一个新的融合趋势是更加注重对于包括社会媒体在内的所有媒体的应用，来接触全社区所有的人，并且邀请他们为组织者提出最最相关的新想法。这种"众包"方式意味着越过局限的专业人员通过公共联系来接触更广泛的社区群"众"。这种方式的目标是为当地文化、情感、精神以及实际需求创建一种更合适的、

图 6.25
感光镶嵌墙幕
Photomosaic

有高分辨率的 LED 墙可以展示壮丽的影像。这种巨大的影像的每一个像素都是由一副完整的影像缩小而成，而且可以被再缩小至一个像素，从而成千上万的缩小影像又组成了下一幅巨型影像。苹果电脑目前提供这种图像作为屏保。
LED walls of sufficiently high resolution will permit magnificent displays of imagery in which every pixel is itself composed of an image in miniature, with the ability for each image to shrink down to one pixel among the thousands of pixels making up the next big image, as Macbook computers now offer as a screensaver. (Credit: Nevit Dilmen)

更易回应的环境模式。人们可以通过社交网和网络参与到整个方案过程中。

在这个早已被史蒂夫·乔布斯预见的后个人计算机时代，无论是在信息空间还是在现实空间，也许甚至连拥有掌上设备的好处都将会被公共空间免费的电子通道概念的好处所替代。

高科技售货亭在将来的某一天会以电子信息公示板的形式出现。人们可以根据特定的地点来下载"应用程序"，并且发布他们的电子标记或者给这个售货亭做"标签"。就像社区的公告栏一样，一条条的电子信息也会成为持续更新的、有启发意义的信息大杂烩，并且鼓励社区的参与。特别是，售货亭可以为不同年龄的人群提供有教育意义的现场互动演示，并且鼓励人们写下自己的意见，提交疑问或者对改善管理提出建议。这些信息中心将是"可刷新的"，就是说，它会更新内容和与访问者的互动来吸引回访。被更新的内容演示可以与体育活动的大屏幕显示相仿，但是与那些由视觉智能软件随机产生的艺术布景效果将是天壤之别。它的理想效果将是大胆的、欢乐的，没有陈腐，拒绝商业主义，尽管谨慎但是充满言论自由。

一些人谴责这种科技趋势为破坏自然体验的入侵物，他们可能过于悲观了。不错，自然需要保持它的有利地位，而不是被科技掩埋。但同样可以确定的是，人们会继续拥有沉浸在自然体验之中的自由，他们想在自然中待多久就待多久，直到选择走出自然与外界相连。事实上，感受自然——无论是品尝自然的美丽还是从自然中寻找灵感——然后立刻把这美好时刻与远方好友分享是一种珍贵的礼物，这是过去几代人无法想象的。收到相关图片和影像所带来的美好仅次于亲身沉浸在其中，当然这种力量也会让人感觉更好。

4. 设计超级连通性

那些顶尖设计实例所带来的灵感将远远大于仅仅简单模仿其他成功者的外观所带来的灵感。好的设计是深思熟虑的，是谨慎的。有太多启发设计灵感的好资源至今依然被忽视，导致我们满足于对其他设计师作品的模仿。这些灵感就在我们周围，在自然形态里，植物就不用说了，还有昆虫、鸟和蝴蝶、细胞组织、几何分形、地理，甚至是星象。设计灵感还在类似的艺术形式中，比如珠宝、吹制玻璃、纺织品、书法、屏幕保护、平面艺术、艺术广告、舞蹈和音乐，例如电路板的工业设计和科技同样带给我们灵感。灵感还存在于我们的梦中。而人类形态本身也将继续给予启发，如同它千百年来所做的一样（图 6.28 - 图 6.39）。

分形

过多设计师在设计中使用的微积分、几何学，和代数学中的线性并不能简单地应用到自然形态中。一个对于城市公共开放空间新几何形态的启发资源就是几何分形。分形在 1975 年被本华·曼德布罗特第一次提出来，它是一种自相似的形状；就是说，这种形状是有机发展的，看起来在无尽的对自身的复制中生长。每一个分支都反映了整体。生长秩序是内在固有的。这被叫作分形对称。分形对称的形状是曲线的，没有直线存在其中。它可以在树中，云中，雪花中，水晶中，河流排水系统中，珊瑚礁中，海岸线中，以及山脉中被发现。有些讽刺的是，新的分形秩序来自

图 6.26
二维码
QR Code

如果二维码设计应用到建筑立面或者铺装上，那么移动设备就可以读取它并且进入到一个有着场地相关信息的网页中。
If the QR ("quick response") code is designed in a façade or pavement, a mobile device can read it and access a web site with relevant information about the site, whether mundane or cultural. There is a current fad for imitating the QR code look in a paving pattern, which is visually appealing for combining randomness and geometric order in harmonious balance. (Credit: brdall)

图 6.27
推特墙
Twitter Wall

互动性将越来越多地增加公共空间的生气以及不同场地之间无形的连接。位于洛杉矶果树林这块触摸屏是加拿大旅游委员会在美国放置的若干触摸屏之一，这样路人可以点击查看照片也可以发表关于加拿大旅游地点的推特，并且和加拿大的旅游者进行互动。
Interactivity will increasingly animate public space and link distant places virtually. The touchscreen at The Grove in Los Angeles is one of several that the Canadian Tourism Commission has installed in the U. S. so that passers-by can touch on photos and twitter streams about Canadian destinations and interact with tourists in Canada. (Credit: Brand Canada Library, Canadian Tourism Commission)

图 6.28 - 图 6.29
自然的形式和色彩
Patterns and colors from nature

大自然在不同尺度上给我们的启发是无穷无尽的，就像鱼类，我们从雄性鱼的背鳍获得灵感，或者爬行动物类，我们从豹纹变色龙的头部获得启发。
The inspiration from Nature at all scales is endless, such as from fish, like the dorsal fin of a male dragonet, or reptiles, like the head of a panther chameleon. (Credits: © Hans Hillewaert / CC-BY-SA-3.0 and Tom Junek)

---

to generate concept design alternatives. Participatory design is an approach which is focused on process. It is not a design style. A strong advocate of the process is the Project for Public Spaces (www.pps.org)

An emerging trend is the more intense use of all media including social media to reach people all across the community and invite those with the most relevant and fresh ideas to participate. "Crowdsourcing" implies reaching beyond a narrow group of specific individuals to the wider community (crowd) through an open call. The goal is to create environments that are more responsive and appropriate to local cultural, emotional, spiritual and practical needs. People can stay involved throughout the implementation process through social networks and websites.

In the post-PC era foreseen by Steve Jobs, perhaps even owning hand-held devices will be superseded by the concept of free public access in public space, both cyberspace and physical space.

High-tech kiosks will someday act as digital information billboards where visitors can download an "application" specific to the venue and can post their digital mark or "tag" the kiosk. Like community bulletin boards, the snippets of digital information will become an ever-changing, evocative collage of information and encourage community participation. In particular, kiosks will provide on-site interactive displays to serve educational functions for a variety of age groups, and encourage visitors to add comments, submit inquiries or make recommendations for improved management. These information centers will be "refreshable", that is, updated for new content, and visitor-interactive, encouraging return visits. Refreshable content displays will loosely resemble the giant screens at sporting events, but the ambience of randomly generated art managed by visually-intelligent software will be worlds apart. At its best it will be bold, often celebratory, never clichéd, will reject commercialism, and will embrace, however cautiously, free expression.

Those who deplore this trend of technology as an intrusion that undermines the experience of Nature may be unduly pessimistic. Surely, Nature must maintain the upper hand, not be buried under technology. Just as surely, visitors will remain at liberty to experience immersion in Nature for as long as they want before electing to step out of it and connect with others. Indeed, experiencing a moment in Nature—whether savoring natural beauty or finding inspiration from Nature—and then sharing that moment immediately with a friend far away is a precious gift which earlier generations could not have imagined. To be sent an image or video clip of that moment is the next best thing to being there oneself, and surely also has power to make people feel better.

4. Design Hyperconnectivity

The very best examples of design will aspire to achieve something more than simply mimic the look of someone else's success. They will dare to design deliberately and deliberatively. There are too many good sources of design inspiration remaining underappreciated to force us to settle for imitation of another designer's work. Imitate nature, but not another designer. The inspiration is all around us, in natural forms—plants of course, but also insects, birds and butterflies, cellular structures, fractals, geology, even the constellations. Design inspiration can likewise be found in kindred art forms like jewelry, glassblowing, textiles, calligraphy, screensavers, graphic arts, artsy advertisements, dance and music, in industrial design and technology like circuit board design, and in our dreams. The human form itself will continue to inspire, as it has for millennia. (6.28-6.39)

Fractals

The linear preoccupation of calculus, geometry, and algebra and too many designers simply does not apply to natural forms. One source of inspiration for new geometries to organize urban public open space is fractal theory. First conceived by Benoit Mandelbrot in 1975, fractals are forms that are self-similar; that is, the form is organic, and seems to have grown by replicating itself endlessly. Each branch reflects the whole. The order is innate. It is called fractal symmetry. The forms are curves—nothing about them is linear. Fractal symmetry can be found in trees, clouds, snowflakes, crystals, riverine drainage systems, coral reefs, coastlines and mountain ranges. Ironically, the new order of fractals came out of the work of the chaos theorists. What they found, on the other side of chaos, was simplicity. There is something holistic, even Daoist, in this big idea, and it is worth celebrating in design.

The well-known design firm AECOM's Discovery Communications Garden in Silver Springs, Maryland, points the way. Its fractal-inspired intersecting arcs and circle segments allows a site narrative to unfold in layers, revealing content for this content-provider client both literally and metaphorically, as AECOM puts it. After all, the mission of the cable media company is to reveal the stories of Nature and culture.

Computers will aid in the layout and fabrication of complex fractal geometries, as they have with complex curving architectural forms conceived by architects like Frank Gehry.

The design of the outdoor realm will increasingly borrow from design indoors; art installations from inside museums will inspire outdoor counterparts—call them avatars. For example, French artist Serge Salat's installation "Beyond Infinity", which toured Chinese cities in 2011, blends mirrors, sculpture, light and music to alter the visitor's perception of space and evoke the cycle of day to night. The manipulation of outdoor sensory immersion will continue to be refined. Cross-sensory connectivity will intensify. Longing to be surrounded and enveloped, we are told, goes back to the womb, the primal immersion experience. Two art-forms are said to offer immersion—architecture and music. But this is being unfair to landscape design, especially in a disruptive world. Immersion in Nature and taking in the full multisensory experience have the power to restore emotional wellness and reduce stress. The more multisensory the experience, we are told, the more deeply we are touched and our spirits uplifted, and the more powerful our memory. Then what could be more natural and powerful than experiencing music in Nature? The convergence of sensory

于混沌理论家的工作成果。他们所发现的是，在混沌的另一面，是简易。在这个大概念中，有一些内容是整体化的，甚至是道学的，都值得在设计中被应用。

著名设计公司AECOM为探索频道（一个电视节目）所设计的马里兰州银泉市探索交流花园运用了这一概念。其中相互交叉的弧线和圆片段由分形启发而来，由此场地的故事被层层掀开，从实际和隐喻两方面为这个电视节目客户展现出AECOM所赋予场地的内容。毕竟，探索频道的任务就是揭示自然和文化的故事。

计算机将辅助复杂几何分形的布局和制作，就像他们辅助像弗兰克·盖里那样的建筑师所设计的复杂曲线建筑形状一样。

户外领域的设计将越来越多地借鉴室内设计，室内的艺术装置将启发室外的对景——就叫他们化身好了。比如说，法国艺术家谢尔杰·萨拉的作品"超越无限"，曾于2011年在中国城市巡展，它融合了镜面、雕塑、灯光和音乐来改变参观者对空间的认知，并且激发日与夜的循环。这种对户外感官沉浸体验的控制会继续被改善。多感官的交叉连通将会加强。我们渴望被包围被笼罩，我们被告知，回到子宫去吧，那是最原始的沉浸体验。据说两种艺术形式可以提供沉浸的感觉——建筑艺术和音乐。但

图6.30 - 图6.32
纺织品的形式
Patterns from textiles

来自于纺织品的启发同样无穷无尽，无论是传统的印尼蜡染印花还是西方现代形式都是由自然启发得出的，就像红蓝波浪或者蓝绿色玉簪的叶子。
The inspiration from textiles is endless, whether traditional Indonesian batik or Western modern textiles inspired from nature, such as red and blue waves or the leaves of a blue green hosta plant. (Credits: Erik DeGraaf, Alexandra Makarova, and lisann)

图6.33
琉璃艺术
Glassblower's Art

戴尔·齐鲁利位于拉斯韦加斯贝拉吉奥赌场内的艺术来源于植物世界，这可以给予公共空间中的设施以启发。
The art of Dale Chilluly in the Bellagio casino in Las Vegas, derived from the plant world, can inspire installations in public space. (Credit: John Saxenian)

图 6.34
珠宝设计的艺术
Jewelry Designer's Art

复杂的珠宝设计可以在相应的尺度上启发开放空间形式的设计。
The intricate world of jewelry can inspire forms in scale with open space design. (Credit: Mark Allyn)

图 6.35
万花筒
The Kaleidoscope

在万花筒里看到的永远变化的放射对称世界可以启发公共空间里大尺度的放射性设计。
The ever changing radially symmetrical world seen through a kaleidoscope can inspire large scale radial design in pubic space. (Credit: Anna Yakimova)

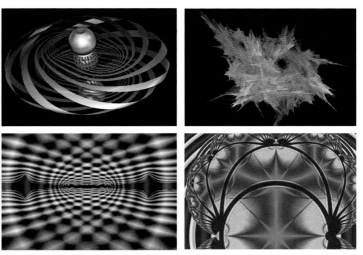

图 6.36 - 图 6.39
分形
Fractals

在很多的形式中,分形是另外一种获得灵感的丰富资源。现在分形可以通过电脑模型来设计,且通过电脑指引来建造。
In their many forms, fractals are another rich source of inspiration for forms that can now be designed by computer modeling and built by computer guided manufacturing. (Credits: Ruslan Zalivan, Jeremy Baumann, Ralph Langendam and Nevit Dilmen)

immersions—one visually spatial, the other acoustically spatial—works its own mysterious harmonics in what happens between them, within us. Acoustic sculptor and musician Ross Barrable brought his harp outdoors, heard the wind blowing through it, and reinvented the wind harp. This multisensory sculpture requires no human hand to turn even the softest breezes into deeply soothing harmonious music.

The interconnectedness of possibilities is truly infinite. The notion of connectedness can take an idea perfected in Chinese Taoist thought over the millennia—the correspondence of things and the unity of the world—and extend it to vast new realms. The current focus in China on the most basic of correspondences—the five elements, five virtues, five life phases, five colors, five fruits and the like—points the way to an updated and re-energized interest in making connections on the part of the most brilliant designers in our time.

5. Celebrating Heroes by Telling Their Stories

We are hard-wired for mythic stories. We have been this way since long before the legendary blind Homer enthralled his audiences with his recitation of the tales of Odysseus on the wine-dark sea, and Chinese opera still does when the Monkey King Sun Wukong (孙悟空) appears on the stage in a Beijing opera. It is not just that our children love to be read to, or that we watch sitcoms or go to movies or read novels, it is the way we think. Each sentence we speak or read tells a story, in linear fashion, has a beginning and an end. Storylines are what engage us, over and over. They are journeys from one place to another. Advertisers, so enamored of getting our attention, know this perfectly well: advertisements have storylines. Stories touch us somewhere deep in our psyches, and inspire us, move us, disgust us, galvanize us to action. Stories are the principal vehicle by which

是这对景观设计是不公平的,尤其在这个纷扰的世界上。沉醉于自然并且有完整的多感官体验可以重塑健康情感并且减小压力。我们被告知,多感官体验越多,我们被打动的越深,我们的精神越升华,我们的记忆越有力量。所以有什么能够比在大自然中欣赏音乐更加自然有力呢?感官沉浸的汇聚——一种是视觉空间的,另一种是听觉空间的——两者与我们一同产生了神秘的和谐之感。声学雕塑家和音乐人罗斯·巴若伯把他的竖琴带到了户外,他聆听风从竖琴间吹过的声音,然后重新做成了风竖琴。这件多感官雕塑并不需要通过手的演奏去把哪怕是最最轻柔的微风转变成深层舒缓的和谐音乐。

相互连通的可能性真是无穷尽的。千百年来中国道教中的一个想法很好地反映了对于连通的认识——事情间的相互对应和全世界的统一——并且将其延伸到了广大新的领域。目前在中国对于最基本的相互对应的关注——五个元素、五种美德、五个人生阶段、五种颜色、五种水果以及等等相似的——都体现出我们时

meaning is communicated. They are so fundamental that the question is why we have not layered public space with narrative.

So far, landscape architects who allude to a narrative in their urban public open space design are describing something that most visitors would overlook. In evaluating the use of "gentle narrative" in such design, one has to concur with Reuben Rainey:

> *One appreciates the subtlety, but one must ask, How apparent is this "narrative" to the general public?... [It] may be too subtle for many people's comprehension... [T]he narrative function...is very important, for it can tie the park to its particular region and help build a crucial sense of shared history...* [2]

In some flagship urban public open spaces, narrative will succeed in doing more. If interpreting the geological and ecological history of the site is a goal, it will take the form of subtle markers or "tagging" so people truly understand, rather than try to read the mind of the designer. Beyond that laudable goal, these spaces will be enriched by storylines that evoke a big cultural idea, an archetypal story. While there may be good reasons to celebrate international heroes, it ought to celebrate local heroes, local artists and writers, local talent of all kinds. But let us be clear. Narrative that succeeds in communicating with visitors will not be strident, like a rant, or verbose, like a lecture, or preachy, like a sermon. Outdoor space will not be "scripted", taken over or overrun by its message. Nor will it return to allegory or heroic statues so popular in earlier eras. Instead, the narrative will appropriately take the form of contemporary media, from events or performances to installations such as iconic imagery, murals by other means. It is enough to be suggestive, to engage the visitor, rather than overstate. Landscape architects will soon be getting more comfortable with this promising design opportunity than they have proven to be so far. (6.40-6.44)

Helping landscape architects to loosen up their thinking may have to come from the world of fantasy. At least one designer is tagging spaces and places globally by installing plaques that together weave a narrative of a parallel world of make-believe. Grandson of celebrated filmmakers and furniture designers Charles and Ray Eames, Eames Demetrios conceived the Kymaerica project, linking sites globally by installing plaques honoring events from an imaginary alternative world. According to Demetrios, "The Kymaerica project is like writing a book and putting every page in a different city, so where you are and finding your way there (or asking a friend to send you a picture of it if you are too far away)—all enrich the story". The sites for the plaques keep changing but may still include Beijing's 798 artist district.[3] If random spaces across the globe can be woven into a work of narrative fiction, perhaps commemorating a factual narrative linking places separated by geography is at least as plausible. One place can be linked to another for deep reasons. Indeed, if there can be twin towns and sister cities which have formed cultural ties across great distances, why not

代最杰出的设计师对于建立相互连通性的兴趣。这种兴趣是现代的，是重新给予了能量的。

5. 讲述英雄事迹来纪念英雄

我们着迷于神话故事，从史诗巨人盲人荷马在酒暗海中朗诵奥德赛的故事开始就被深深吸引，而在遥远的东方，美猴王孙悟空至今依然会出现在京剧的舞台上。这不仅仅是我们的孩子喜爱看的，也不仅仅是我们看的情景喜剧、电影或者读的小说，这是我们思考的方式。我们说的或者读的每一个句子都在以线性的方式讲述故事，有开始有结局。故事情节是一遍又一遍吸引我们的关键所在，他们就是从一个地方到另一个地方的旅行。专注于抓住人们注意力的广告策划们深谙此道：每一个广告都有它的故事情节。故事可以触碰我们内心深处，启发我们，感动我们，让我们感到厌恶，刺激我们去行动。故事是意味着连接两地的基础车辆。他们是最重要的根基，那么我们的问题就在于，为什么我们的公共空间设计还未有叙事性呢？

目前为止，游客们往往会忽视掉那些被设计师加在他们的城市公共开放空间中的叙事部分。当评价在这种设计中的"温和叙事"时，有一点是必须和鲁本·雷尼相一致：

> *"我们欣赏那些在细节中的叙事，但我们也必须要问，这样的叙事对大众是不是足够的醒目？……对于许多人的理解来说，它可能太过细节了……叙事的功能是非常重要的，因为它能……把公园和它所在的地域连在一起，并且帮助人们去感受那段分享的历史……"* [2]

在一些旗舰性质的城市公共开放空间，叙事性将在更多的方面有所建树。如果我们的目标是解释场地的地质和生态历史，那么就可以采用细节标注或者"标签"的形式来让人们真正地理解它，而不是尝试去读设计者的想法。除去这个令人称道的目标之外，故事情节还将使空间变得丰满起来，它们能够唤起一个大的文化想法，或者成为一类叙事的经典原型。我们有好的理由去纪念国际英雄，也有好的理由去纪念当地的英雄、艺术家、作家以及其他各路的人才。让我们来说清楚。与游客们达到成功沟通的叙事不会像咆哮一样刺耳，不会像讲座一样冗长，也不会像布道一样说教。户外空间不会像写好的剧本一样，被信息词句所接管或覆盖，也不会倒退回很早之前所流行的寓言或者英雄式雕像。反而，从活动或表演，到标志性的图像或者具有其他意义的大型壁画，叙事都将合理地应用当代的媒体形式。它将有足够的建设性，足够的引人，而不仅仅是夸大。景观设计师们将很快地比现在更适应这个有前景的设计机遇（图6.40 - 图6.44）。

也许只能通过幻想世界来帮助设计师放松他们的思想。至少有一位设计师通过安装标牌在全世界范围内陆续标注一些地点，这些标牌编织了一个令人信服的有关平行世界的故事。著名电影制作人和家具设计师查尔斯·埃姆斯和雷·埃姆斯的孙子埃姆斯·德米特里奥斯构想了凯迈瑞克项目，这个项目在世界范围内通过安装标牌来纪念幻想世界中的事件，从而把不同的场地连接起来。德米特里奥斯说，"凯迈瑞克项目就像是写一本书，然后把每一页都放在一个不同的城市。所以你在哪里以及如何寻找到达那里

图 6.40
传统墙壁艺术
Traditional Wall Art

在英格兰，布里斯托尔以街道艺术而闻名，空置建筑立面的窗户变为表达各种文化主题的全景画，化单调为生动。
In an area renowned for its street art in Bristol, England, covered windows on the public façade of a vacant building have been transformed into a panorama expressing a variety of cultural themes, and turned monotony into variety. (Credit: Patrick J Hanrahan)

图 6.42
LED 叙述墙
LED Narrative Wall

用多种变化和移动影像来叙述故事的景墙是未来城市开放空间的一个方向。
Walls with a multiplicity of changing and moving images that tell a story are the way of the future in urban open space. (Credit: James Thew)

图 6.41
新型墙壁艺术
Innovative Wall Art

面对一个小公园，并标志着巴黎圣丹尼斯街105号的是一幅由铝片构成的圣人文森特·保罗（1581-1660）的肖像。他是一位著名的人道主义者，帮助了许多穷人。肖像中逼真而又不生硬的姿势把传统表现与新材料融于一体。
Facing the small park marking the site at 105 rue du Faubourg Saint-Denis in Paris where Saint-Vincent de Paul (1581-1660), a famous humanitarian, helped the poor is his portrait composed of aluminum blades, an eloquent and not too assertive gesture fusing traditional representation and new materials. (Credit: Tangopaso)

图 6.43
脚下的魔术
Magic Underfoot

硬质铺装表面提供了大量的机会去叙述故事，无论是文字、地图还是图像。在这个画在布宜诺斯艾利斯人行道的立体绘画中，著名的艺术家朱利安·比弗展示了这种既丰富公共空间又包含轻松易懂文化含义的暂时性艺术。
Hardscape surfaces offer ample opportunity for narrative, whether text, map or graphic. In this trompe-l'oeil drawing on a sidewalk in Buenos Aires, noted artist Julian Beever showcases ephemeral art that enriches public space with easily understood cultural meaning. (Credit: Flickr user lrargerich's photo of a Julian Beever artwork)

sister urban public open spaces?

Imagine a Birthday Park using digital technology on a wall mural to celebrate, each day, people born on that day, using a collage of images of famous people, living and dead, and local people of all ages. Local residents can upload images to be included in the ever changing collage of images. Families can bring children born on that day to celebrate their day and make connections to others who share the same birthday.

Runnymede and its Global Connections

The future of narrative space that makes global connections will owe a substantial debt to this greenway corridor along the Thames southwest of London that has been open to the public since 1929. Runnymede witnessed the sealing of the Magna Carta in 1215, under a yew tree that is now 2500 years old. The Magna Carta is the earliest constitutional document to enunciate the principle of freedom under law. It has inspired people around the world. In addition to memorials placed by the British government, the American Bar Association donated a memorial in 1957, and will celebrate the 800th anniversary of the Magna Carta at Runnymede in 2015. In 1994 the Prime Minister of India planted an oak tree and placed this inscription there:

> *As a tribute to the historic Magna Carta, a source of inspiration throughout the world, and as an affirmation of the values of Freedom, Democracy and the Rule of Law which the People of India cherish and have enshrined in their Constitution.*

In 1965, the people of Britain gave the United States an acre (0.4 hectares) of ground at Runnymede in memory of President John F. Kennedy. Inscribed on a limestone slab in a beautiful font are Kennedy's words:

> *Let every Nation know, whether it wishes us well or ill, that we shall pay any price, bear any burden, meet any hardship, support any friend or oppose any foe, in order to assure the survival and success of liberty.* (6.45)

FDR and Vietnam Veterans Memorials, Washington DC

Perhaps the best example of outdoor space that uses traditional narrative means—sculpture and inscription—in a fresh way is the Franklin Delano Roosevelt Memorial on the Tidal Basin in Washington, DC, completed in 1997. It is the crowning achievement of master landscape architect Lawrence Halprin, indeed the first major memorial in Washington by a landscape architect. Its four outdoor rooms, like museum galleries open to the sky, each tell the story of one of Roosevelt's four terms as President. Pink granite walls and floor define the rooms; a crescendo of waterfalls provides acoustical immersion. The use of rock and water to dominate and define space channels the gardens of Suzhou as much as the allegorical gardens of the Italian Renaissance. Planting is minimal, mostly outside the four main spaces and along the corridors that separate them.

The narrative consists of a dance between sculpture and inscription, the work of sculptors Leonard Baskin, Neil Estern, Robert Graham, Thomas Hardy, and George Segal, and master stone-carver John Benson. Sculptures unflinchingly depict Roosevelt in his wheelchair, a lone citizen listening to the radio during a rousing "fireside chat", a breadline, and widowed Eleanor Roosevelt as UN delegate. The many inscriptions distill Roosevelt's most memorable speeches.

图 6.44
头顶的魔术
Magic Overhead

有天赋的艺术家从地面立体绘画中得到灵感创造出一个充满想象力的世界,丰富了公共空间中我们头顶的世界。
Talented artists can take as their inspiration the trompe l'oeil fresco to create a magical imaginary world to enrich overhead surfaces in public space. (Credit: Guido Bertolotti photo of Galliari brothers, *Apollo Guiding the Sun*, Villa Arconati, Milan Province, Italy, after 1750).

的路(或者因为你离得太远所以让你的朋友寄给你书的照片)——都使这个故事更加丰富。"³ 如果遍布全球的随机场所能被编连起来成为一部叙事小说作品,那么也许庆祝一个连接不同地理位置的故事也至少是言之有理的。一个地方与另一个地方的连接可以有很深刻的原因。事实上,我们已经有兄弟城或者姐妹城,它们虽然相距甚远但是用文化连接彼此。所以我们为什么不能有兄弟城市公共开放空间呢?

请想象一个有大型电子科技景墙来庆祝生日的主题公园。每一天,墙上将演示所有生于此日的人们的图像,无论是在世或者故去的名人,还是当地各个年龄层的人们。当地居民可以上传图片到这个永远变换的景墙上。家人可以带着孩子来到这里庆祝生日,并且和同一天出生的其他人们建立联系。

兰尼米德以及其全球性连接

叙事空间可以创造全球性的连接。它的未来将欠下这个沿伦敦泰晤士西南、1929 年对公众开放的绿道一笔重债。1215 年,兰尼米德一颗现今有 2500 年历史的紫杉见证了英国大宪章的签署。大宪章是最早的宪法文件,它确切阐述了在法律下自由的原则。它激励了全世界的人民。除去英国政府建立的纪念碑,美国律师协会也在 1957 年捐赠了一座纪念碑,并将在 2015 年庆祝兰尼米德大宪章签署 800 周年。1994 年印度总理种下一棵橡树,并题字于此:

(这棵树)象征了对兰尼米德大宪章的供奉,它是全世界精神的来源,(这个树)同样象征了对自由民主和法制价值观的肯定,这些是印度人民所珍惜的,并被载入我们的宪法。

1965 年,英国人民在兰尼米德赠予美国一块一英亩(0.4 公顷)的场地用来纪念约翰•肯尼迪总统。刻在一块石灰岩石板上的优美文字是肯尼迪的讲话:

要让每一个国家都知道,无论它们希望我们好还是不好,我们都将会付出任何代价,承受任何负担,面对任何困难,支持任何朋友或是反对任何敌人,来确保自由的存在和成功。(图 6.45)

图 6.45
我们都是兄弟
We are all Brothers

在900年前兰尼米德——一个沿着英国泰晤士河的绿道,英格兰国王在一颗紫杉树下签署了大宪章,给予人们在法律下的自由。如今那棵紫杉依然健在。几乎五十年以前,英国人民把这个神圣场地的一部分赠予了美国人民,用来纪念美国已逝总统肯尼迪。这里刻有肯尼迪最愤慨激昂的演说词。这说明任何地方的任何人都可以选择对其他人来表达这样的友好。
Nine hundred years ago at Runnymede, a greenway along the Thames River in England, the King of England signed the Magna Carta granting the people the principle of freedom under law, under a yew tree that still stands. Nearly fifty years ago the people of Britain gave a part of this sacred site to the people of the United States, to honor a fallen American President, John F. Kennedy, commemorated in his most rousing words. Any people anywhere can choose to make the same good-will gesture to another people. (Credit: wyrdlight)

图 6.46
文字的力量
The Power of Text

在各个领域用名言警句来纪念伟大领袖的方式同样可以增加公共空间的意义。在华盛顿的罗斯福纪念碑,著名景观大师劳伦斯·哈普林天才般地把富兰克林·罗斯福的名句"我憎恨战争"刻在下落的石块上,以告诫世人。
The spare use of memorable words to honor great leaders in all fields can add rich meaning to public space. In the FDR Memorial in Washington DC the great landscape architect Lawrence Halprin brilliantly fused a famous quotation from FDR—"I hate war"—with fallen stone blocks, each reinforcing the other. (Credit: Karen Nutini)

富兰克林·罗斯福纪念碑和越战老兵纪念碑,华盛顿

在华盛顿,恐怕用新手法表达传统叙事含义的最好户外空间案例就是富兰克林·罗斯福纪念碑了。它位于华盛顿潮汐湖畔,1997年完工,是景观大师劳伦斯·哈普林至高无上的成就,事实上也是华盛顿特区第一个由景观设计师完成的主纪念碑。四个户外房间就像博物馆的四个展室一样,分别讲述了罗斯福在总统任期内四个时期的故事。粉色大理石的墙壁和地面定义着每一个房间;水声渐强的瀑布给予人们听觉上的沉浸体验。用水和石来主导和界定空间的手法可追溯到苏州园林,就像是比喻园之于意大利文艺复兴。植物种植在四个主空间内是极少的,大部分都位于四个主空间的外围以及用于界定它们的廊道。

整体叙事由雕像和碑文之间的舞蹈组成。这些作品是由雕塑师里昂纳得·贝斯金,内尔·埃斯特恩,罗伯特·格兰汉姆,汤马斯·哈迪,乔治·西格尔和石刻大师约翰·本森完成的。雕像群毫不避讳地描绘了坐在轮椅上的罗斯福,一个正在收听激动人心的"炉边谈话"广播节目的孤独市民,一条等待救济的队伍和作为联合国代表的总统遗孀埃莉诺·罗斯福。许多碑文都提炼于罗斯福那些最难忘的演讲。在二十多个碑文中,最引人注目的是刻在一处破碎石堆上的"我憎恨战争"。

没有其他的总统纪念碑能够引用如此多的故事,能够如此缩小人民与领袖之间的距离。我们可以实实在在感受到对历史事件按时间顺序的编排,真真正正以一个新的身份漫步在花园中,去纪念这位在他的一生中无法漫步,只能困难地移动的伟人,正是他在他的有限人生中,使国家发生了巨大的改变。相反顺序的漫步同样有纪念意义。富兰克林·罗斯福纪念碑是深刻的、动感情的。它充满了移动性,充满了感人情怀。最主要的,就像哈普林

Among the twenty plus inscriptions, the most arresting is I HATE WAR inscribed on a broken pile of blocks.

No other presidential memorial invokes so much narrative or so reduces the separation of the people and the leader. There is indeed a feeling of choreography about the sequence, truly a stroll garden in a new guise, celebrating a man who could not stroll and was mobile only with difficulty for most of his life, and for the part of his life that really made a difference. The stroll can just as rewardingly be experienced backwards. The FDR Memorial is poignant and unsentimental, full of movement, and fully moving. It is above all, in Halprin's words, "experiential rather than purely visual—one which is evocative, involving and appropriate for all ages and all people".

One of the most moving uses of immutable text in any public space is the narrative of names chiseled into the polished black granite wall of Maya Lin's Vietnam Veterans Memorial. Honoring the veterans of that conflict, the memorial is inscribed simply with the names of every single serviceman and woman who died or remains unaccounted for, perhaps the first such instance to include so many names in one public monument, over 58,000 names, listed chronologically from 1959 to 1975. In its supremely eloquent low profile, its bell curve turned upside-down, the most visited site in our nation's capital is a high altar for emotional release and renewal. The wall's wedge shape in plan anchors the meaning of sacrifice to the place of memorialization, as the arms point to the Washington Monument and Lincoln Memorial. The form itself is laden with subliminal content. Some may see the V-plan as suggesting Victory, others might infer the wingspan of a B52 Bomber. But most often it is compared to an open wound, and so was not allowed to stand on its own without the rebuttal of figurative sculpture, Frederick Hart's Three Serviceman of war-weary GIs and Glenna Woodacre's Vietnam Women's Memorial, added for good measure, in the hope that heroism, patriotism and honor would thereby be unequivocally expressed. Figurative sculpture is old-style narrative. Most visitors are drawn to the more eloquent narrative and spiritual power of the names, like a scroll of text. It is tenth on the list of America's favorite architecture of the American Institute of Architects.

Born a decade after her mainland Chinese parents fled from China in 1949, architect and artist Maya Lin confides that she did not study Chinese culture until years after she designed the Memorial, but some observers feel that her design taps into something profoundly Asian. Her uncle Liang Sicheng（梁思成）founded the School of Architecture at Tsinghua University in Beijing in 1946, and her aunt Lin Huiyin（林徽因）was the first female architect in China. (6.46-6.49)

Proposed Memorials for Three Principles of the People and Three Men of the People

Somewhere in China and the United States two urban public open spaces could celebrate a truly great idea that happens to link the Peoples' Republic of China and the United States of America—the Three Principles of the People. Dr. Sun Zhongshan credited President Abraham Lincoln's immortal phrase *of the people, by the people and for the people* from the Gettysburg Address with inspiring his own Three Principles of the People articulated in 1921 and adopted by the Republic of China in 1924. President Lincoln in turn borrowed

所说的，"让人去体验而不是仅仅去看——唤起人们的记忆，让人们参与进来，并且对每个年龄层的人们都适用"。

在公共空间中对永久性文字最感人的一次使用，是林璎所设计的越南老兵纪念碑，那些刻在抛光黑色大理石上的名字。为了纪念那场战斗中的老兵，纪念碑上没有其他别的碑文，只是简单地刻上了每一个牺牲的或者失踪的男兵和女兵的名字，或许这是第一次在一个公共纪念碑上刻有如此多的名字。他们超过了58000个，从1959年至1975年按照时间顺序排列。如此有说服力的内容呈钟形曲线分布，显示着这首都访客量最多的景点是一个让人释放感情、重塑感情的圣坛。楔形墙壁的两边分别指向华盛顿纪念碑和林肯纪念堂，在这个纪念场地体现出牺牲的意义。这种形式本身充满了潜意识的内容。有些人把这个平面上的V字形看做是胜利的代表，有些人则会把它看做是B52型轰炸机的机翼。但是人们最多的还是把它比作成一个裸露的伤口，所以它并没有孤零零地独自展现，而是伴随着有反抗意义的人物雕像。弗雷德里克·哈特的"三个厌战的美国大兵"雕像和葛琳娜·伍德卡尔的"越南妇女纪念雕像"都毫不含糊表达了英雄主义、爱国主义和荣誉感。巨型雕像是一种老式的叙事手法。大部分的游客对那些名字更有兴趣，他们富有说服力，富有精神的力量，就像一卷课文。在美国建筑师协会的最受喜爱的美国建筑排名上，这个纪念碑排名第十。

建筑师和艺术家林璎出生在她的父母于1949年离开中国大陆的十年之后。她自己说她并没有学习中国文化，直到设计越战纪念碑的多年以后才开始。但是一些观察家感到她的设计探寻了某些很深层次的亚洲文化。她的姑父梁思成1946年在清华大学创办了建筑学院，她的姑姑林徽因则是中国第一位女性建筑师（图6.46 - 图6.49）。

图 6.47 - 图 6.48
越南老兵纪念碑
The Vietnam Veterans Memorial

牺牲士兵的名字按照时间先后顺序被强有力地刻在黑色大理石墙上，大理石面同样反射出被受触动的访问者的身影。纪念碑设计者林璎是中国第一位女建筑师林徽因和1946年在北京建立清华大学建筑学院的梁思成先生的侄女。
The names of the fallen soldiers listed chronologically are the powerful narrative engraved into the black granite wall which reflects moving images of visitors moved by the names of fellow soldiers and family members and fellow Americans. Its designer Maya Lin is the niece of Lin Huiyin, the first female architect in China, and Liang Sicheng, who founded the School of Architecture at Tsinghua University in Beijing in 1946. (Credits: Skyring at en.wikipedia and Thomas M. Paine)

图 6.49
朝鲜战争纪念碑，华盛顿
Korean War Veterans Memorial, Washington DC

从越南纪念碑有感而来，在这里有历史意义的图片用喷砂的形式喷到了抛光的黑色大理石墙上，体现出巡逻中的人们钢铁般的意志，走过附近的参观者，身影映在其中，与之共同优雅地叙述着故事。
Taking the idea of the Vietnam Veterans Memorial further, here historic photographic images are sandblasted onto the polished black granite wall, reflecting the stainless steel statues of squad men on patrol, and the faces of visitors, merging the viewer and the event in an elegant narrative.
(Credit: Thomas M. Paine)

图 6.50 - 图 6.51
伟大的公共空间超越了民族主义
Transcending Nationalism in Great Public Space

孙中山先生通过林肯总统的启发建立了三民主义。大约在这张纪念 1942 年这个伟大联系的邮票发行 80 年后，一个有价值的，超越民族主义和暴力的公共空间可以被用来纪念这两位伟人，并把他们不朽的名言肩并肩放在一起。
Dr. Sun Zhongshan credited Abraham Lincoln with inspiring his formulation of the Three Principles of the People (三民主义). Perhaps 80 years after a postage stamp honored this connection in 1942 a worthy public space transcending nationalism and violence could honor these two great men and their immortal words side by side. (Credits: public domain and Alexander Gardner (1821-1882)/unidentified photographer)

三民主义纪念碑提案

在中国和美国的某个地方，我们可以用这两处城市公共开放空间来庆祝一个把中美两国联系起来的真正伟大的理念——三民主义以及这个理念的三位缔造者。孙中山先生借鉴了亚伯拉罕·林肯总统在葛底斯堡演讲中不朽的名言"民有、民治、民享"来启发出他自己的三民主义。三民主义在 1921 年发表，1924 年被中华民国采纳。林肯总统同样地借用了直率的波士顿先验论牧师西奥多·帕克的言论，是他在 1850 年首先提出三民的言论。经历了 74 年后，西奥多·帕克·亚伯拉罕·林肯和孙中山，这三位均在 60 岁之前就故去的伟人被这世俗世界上最神圣和不朽的一个理念联系起来，这个理念就是：自治。他们之间的联系展示了一个伟大的理念如何传遍世界，挑战死亡。林肯给予帕克"民有、民治、民享的政府"这个想法一个新的诠释，受它的激发孙中山领导了早应完成的革命。我们称之为理念借用：在这个故事里，伟大的理念乘着伟大领导人的翅膀传遍各地，民主、勇气、友谊、和谐和统一相互交错在一起，延伸出一个充满无限希望的未来。庆祝中美两国这个特别的联系是"取之于人民，用之于人民"的体现，不管是对于美国人，还是中国人，还是每一个公园里的人们，它都将是同样受人喜爱的，甚至是感人的。两个公园应该被指定为兄弟公共开放空间（图 6.50 - 图 6.51）。

> 1850 年 5 月 29 日西奥多·帕克在波士顿举行的新英格兰反奴隶制会议上作了名为"美国人民的理念"的演讲。他提出"民主，就是政府是所有人民的政府，它来源于人民，它服务于人民，当然，永久的公正也应该是政府的原则……我把它称之为自由"。

6. 全球化和本土化

随着流动和迁徙的增加，每一个文化群体都会越来越多地了解其他的文化群体，同时也会越来越清楚地意识到自己文化的特殊之处。在鼓励批判性思维的地方，人们会在形成对好坏事物判断力的过程中得到自信，并且意识到没有任何一种文化可以垄断另外一种文化。即便是本地居民也会欢迎来自其他地方、其他文化的人们，或者是某一文化区域的成员分散到其他地区去，但他们保持着与家里的联系，这时纪念这一文化区域和当地历史的重要性就变得越来

his phrase specifically from the outspoken Boston Transcendentalist preacher Theodore Parker, who first uttered the phrase in 1850. Over a span of seventy-four years, Theodore Parker, Abraham Lincoln, and Sun Zhongshan, who all died before age sixty, are linked by one of the most sacred and immortal ideas in our secular world: self-government. Their connection shows how a big idea takes wing, crosses oceans and defies mortality. Lincoln gave a new birth of currency to Parker's notion about government of, by and for the people, and with it Dr. Sun inspired a long overdue revolution in the largest country on earth. Call it borrowed ideascape: in this story of the travels of a powerful idea on the wings of great leaders, democracy, courage, friendship, harmony, and community all intersect, extending into an infinitely hopeful future. It would be gratifying, even moving, to Americans and Chinese alike, each in a park of the people, by the people and for the people, to celebrate this extraordinary connection in China and the United States. The two parks should be designated sister public open spaces. (6.50-6.51)

> In "The American Idea", a speech delivered at the New England Anti—Slavery Convention in Boston, May 29, 1850, Theodore Parker uttered the words "A democracy,—that is a government of all the people, by all the people, for all the people; of course, a government of the principles of eternal justice...I will call it the idea of Freedom."

6. Globalization and Localization

As mobility and migration increase, each cultural group will both become more familiar with other groups and more conscious of what is special about their own cultural roots. In places that encourage critical thinking, people will gain confidence to form their own judgments about what is good and what is not, and to realize that no culture has a monopoly on either. The importance of celebrating place and local history will become ever more highly prized, even as local residents welcome people from other places and other cultural groups, and see members of their own cultural group disperse to other regions but stay in touch with home. Each country will have its own "diaspora" or migrant community, a vast group of followers in a vast cultural network, increasingly able to gain access to the internet and stay in touch with anyone anywhere anytime, in our increasingly technology-rich, and increasingly "flat" world.

The implication of this exciting new reality for design is twofold. First, standards will tend to improve everywhere, as information about best practices is slowly but surely disseminated and freely borrowed. Design in one place will continue to learn from design in other places. And yet, in the end, design is local, because construction is local, and use is local. Local and regional distinctiveness is to be savored, not just in places that had the good fortune to retain their local character and attract tourists to share in their glories, but in places which lost nearly everything, and seek to preserve what few relics remain as all the more precious and even sacred.

Second, design practiced by design firms networked across

cultures and time zones may have a competitive advantage in adding value to a project by tapping into the best of local and global, combining inspiration and best practices and local knowledge, maximizing resources of all kinds, even financial. The best answer to the shortcomings of hiring the celebrity designer or airlifting the overseas expert is to take the pulse of not only local style and local conditions, but also the longings of the people and feel, over an extended period, the unexpected resonances with place. The more that local people are personally consulted for their opinions, rather than relying on second-hand information and web sources, the better the results will be.

As China looks back over its traditional cultural landscapes with renewed pride, the proper way to honor that legacy in the proper design of urban public open space will continue to be refined and made more deeply reverential. This will mean avoidance of pure imitation or trivial transformation, and certainly rejection of cheap construction. The borrowing of Western forms, if it persists, will continue to evolve toward an Asian sensibility. Facilities will tread lightly on the land.

7. Rising Sea Levels and Green Infrastructure

Today's discourse on climate change will pale in comparison to what lies ahead. Today, in the West, we wring our hands on issues like invasive species: exotic plants imported into gardens adapt so well that they escape the garden and threaten to replace indigenous species. Other plants migrate on their own. Pests from afar which have no natural predator can threaten to wipe out native plant species. In overreacting to these signs of a disruptive world, we miss the larger point that the so-called natural boundaries of plant communities have never been truly stable, they have migrated south with glaciers in the past and they will migrate north with global warming in the future. Pests that destroy plants have come and gone. Species that have done well in the past at a particular latitude may not continue to do so as temperatures rise. We can regret all of that, but we do better to adjust expectations, and deal with reality. Sadly, some common tree species may become a rarity. Sadly, some precious species, rarer today than a century ago, may become rarer still, even die out altogether, unless enlightened botanists work to preserve samples and genetic material in controlled environments, perhaps at facilities within urban public open space. Biodiversity loss is of great concern to us all.

Indigenous species vulnerability is serious enough, but overshadowing all this is something even more serious for humans. Tomorrow's discourse will confront ever more frankly the challenges we face on every low-lying coastal waterfront in the world. The challenges are two-fold: first, sea levels are expected to rise 0.6 to 1.2 meters (2 to 4 feet) over the next eight decades. That is the baseline. On top of that, extreme weather and earthquake events are expected to produce greater flooding and tsunamis with greater frequency. The rest of the world is about to experience what Venice and New Orleans have been experiencing for hundreds of years, and what the Netherlands has been experiencing for thousands. From these two challenges,

越有价值。每一个国家都会有自己的"离散人群"或者移民团体，他们是在一个巨大文化网络中的一大群追随者，在这个科技越来越发达，越来越"平坦"的世界上，他们有越来越多的途径连接到互联网，在任何地方、任何时间与任何人保持联系。

这种令人兴奋的新现实对于设计的含义有两部分。第一，随着好的案例信息慢慢地但毫无疑问地被传播和被自由地借鉴，无论在任何地方，标准都将会被提高。一个地方的设计将持续地从其他地方的设计上借鉴学习。到了最后，设计就成了当地性的，因为施工是当地的，用途也是当地的。地方和区域的独特性将被欣赏，不仅仅是在那些有幸保留了当地古迹并吸引着游客来分享它们荣誉的地方，而且也在那些几乎失去了一切，努力寻求保护仅剩的少数遗址的地方，对于这些地方而言，所剩下的一切都是无比珍贵，甚至神圣的。

第二，那些跨越文化和时区的设计公司所作的设计案例会发掘最好资源，无论是当地的还是国际的，把设计灵感、最佳案例与当地知识相结合，最大化各种资源，甚至是经济资源。这些都使得设计公司在给设计主体增加价值上有竞争优势。对于什么是聘请知名设计师或空降海外专家的缺点的最佳答案是缺乏对当地的"号脉"，不仅仅是关于当地风格和当地条件的，更是关于当地人们的渴望，和在相当长的一段时间内与环境不经意间发生的共振。越多地征求当地居民他们意见，而不是依赖于第二手资料，设计的效果越好。

中国正带着恢复的自豪感回望它的传统文化景观，在恰当的城市公共开放空间设计中以恰当的方式来纪念文化遗产，这条路会越来越完善以及充满更深的敬意。这将意味着避免单纯的模仿和不重要的转型，以及对廉价建造的坚决制止。对西方形式的借鉴，如果它继续存在的话，将继续与东方的情感相融合。公共设施应尽量轻巧地建在土地上。

7. 海平面上升以及绿色基础设施

在我们面临的现实问题面前，中国对当今气候变化的论述也黯然失色。在当今的西方，我们因为物种入侵这样的问题感到苦恼：在特定花园引进的外来植物由于对环境适应的过于良好以至于从花园蔓延出去并带来取代乡土物种的威胁。其他植物靠自己的力量来迁徙。外来的有害昆虫因为没有任何自然天敌同样会带来摧毁当地植物物种的威胁。在对这个破坏性世界中这些现象的过度反应中，我们忽视了更大的重点，那就是植物种群所谓的自然边界从来没有真正的稳定过，在过去它们随着冰川运动向南移动，在未来它们将随着全球变暖向北移动。那些摧毁植物的有害昆虫来了又走。过去那些在特定纬度一直存活得很好的物种因为气温的上升并不一定在将来会一直保持良好。我们可以为所有的一切懊悔，但我们也要更好地调整期望，面对现实。悲哀的是，一些常见的树种可能会变得很稀有。同样悲哀的是，一些在现今已经比在过去稀有的珍贵物种，会变得少之又少，甚至彻底灭绝。除非有远见的植物学家在可控的环境中，也许是城市公共开放空间中的设施场所，去保存样本和基因材料。生物多样性的减少对我们所有人来说都是一个很严重的忧虑。

乡土物种的脆弱性已经足以严峻，但是令这些都相形见绌的是一些更加严峻的关系到人类的问题。今后的讨论将更加直接地面对沿海地区下沉给我们带来的挑战。我们面对的挑战分为两个部分。首先，在未来的 80 年中海平面将预计升高 0.6-1.2 米（2-4 英尺）。这是我们的基准线。在这之上，预计极端的天气情况和地震将造成更严重、更频繁的洪水和海啸灾害。世界上的其他地区将体验到上百年来威尼斯和新奥尔良所一直体验的，经历到上千年来荷兰所一直经历的。这两种挑战将使我们的铁路、建筑、道路、桥梁、水道管线、电路管线、排水系统，以及公共开放空间和其他绿色基础设施变得更加脆弱，如果没有大量的补救和重建措施，它们终将会达到崩溃的临界点，或早或晚，那将是全球性的崩溃。

我们要如何应对这些新的认识还并不清晰，但是如果社区将要是"有弹性"[4]的，那么工程解决办法就是必须在一切可能的地方用绿色基础设施代替硬质结构。将地面层建筑设计为"自由空域"从而来应对高水位浪潮的再设计将成为沿海建筑的规范。目前，这样的做法只被某些沿海地区关注，比如三角洲城市，大量堤坝、泵站、海堤、海闸和其他设施会有选择性的使用。根据有远见的设计师和科学家的研究，这种选择是有优先权的，即优先于软质、多孔设施、滞洪区、雨水花园、水渠、人工湿地、滞洪广场、屏障土丘、小岛和珊瑚礁。这其中的一些设施比起常规的海平面上升，更适合于应对严重的洪涝和飓风风浪。滞洪广场或"氢气公园"通常是干燥的绿道地区，它们用于调节洪水。这些富有创造力的设计可以把由气候变化而导致的悲观预测转变为一些有希望的东西。这类设计的唯一限制就是支撑设计连贯完成的经济来源。

8. 设计寿命

长寿在中国文化中深受尊敬，传统设计主题中也充满了各式各样的对它的象征。许多含有长寿象征的园子都存在了好几代人的时间，因为它们好的设计寓意被人尊敬。当今的设计师会享受到同样长时间的声望么？设计师们必须迎接的巨大挑战之一就是，他们的设计是否会随着时间"衰老"？就像人一样，有些人会很优雅地衰老，有些人则不是。有些人不用很大的花销，有些人则要求做高昂费用的手术来保持年轻的样貌。开放空间的设计也是如此。

其中一个案例就是作者观察了 40 年的波士顿历史城区中心。在过去的 40 年中，市政厅广场的命运展示了一个大胆的、没有和当地气候相协调的设计是如何不适应所在的区域。设计时没有人研究过环境因素对设计的影响，尤其是在冬天会造成恶劣大风，以及在大范围倾斜的砖铺地上会形成冰层的潜在危险。相比于获奖的锡耶纳扇形广场，市政厅广场并没有经受住时间的考验。花岗岩的台阶已经不能再被使用，地砖也已经松动。曾经前卫的几何线条现在也成了碍眼之物。而没有人会这么认为如今的扇形广场。市政厅广场毫无必要地替代了斯柯雷广场，后者的尺度与罗马备受敬意的纳沃纳广场相似。

随着我们进入了一个资源紧缺的时代，我们所建的东西必须要可以持续更长的时间，而不是在短短几年后就需要重建，这就要求我们更仔细地研究它的环境影响。当然，我们也会期待建造质量的提高，这同样会帮助延长设计的寿命。但是，即使是最高质量的建造也无法完全抵御由于温度升高和降低对材料所造成的膨胀收缩的损害，甚至在更冷的气候中，比如结冰，对材料连接处的破坏。

the increasing vulnerability of subways, buildings, roadways, bridges, water supply lines, electric conduits, storm water drainage systems, in addition to public open space and other green infrastructure, will reach the breaking point unless massive remedial measures and reconstruction proceed. Sooner or later, they will have to, globally.

How we transition to this new realization is not at all clear, but engineering solutions will have to replace hard structure with green infrastructure wherever feasible, if communities are to be "resilient".[4] Redesigning the ground level of buildings as a "freeboard" zone to be safe from high-water flows will become the coastal norm. The first livable floor will be elevated to what is now thought of as the second floor level. So far, the dialogue seems to focus on selective coastal conditions like delta cities and selective application of massive levees and dikes, pumping stations, sea walls, sea gates, and other structures, but there is a preference, among visionary designers and scientists, for softer, porous structures, detention basins, raingardens, canals, artificial wetlands, wet plazas, barrier dunes, islets, and reefs. Some of these elements seem more appropriate to address severe flooding and storm surges than a general rise in sea level. Wet plazas or "hydric parks" are usually dry greenway areas that are designed to accommodate flooding. The opportunity for imaginative design to transform this gloomy forecast of the consequences of climate change into something hopeful is limited only by the financial resources available to fund implementation.

8. Design Longevity

Longevity is revered in Chinese culture, symbolized in numerous ways in traditional design motifs. Many gardens where these symbols can be found have lasted for generations, because their design genius has been revered. Will today's designers enjoy the same longevity of their reputation? The great challenge which designers must take into account along with everything else is, will their designs age well? Some people age well, others do not. Some age well without great expense, others require expensive surgery to stay young looking. The same is true of open space design.

One example that the author has been observing for four decades is in the heart of the historic downtown of Boston. The fate of City Hall Plaza over the past four decades shows how bold design unattuned to local climate can betray a community. No one studied the environmental impacts of the design, specifically its potential to generate harsh gusty winds each winter, and dangerous sheets of ice on vast sloping brick surfaces. Once considered an award-winning homage to the Piazza del Campo in Siena, City Hall Plaza has not aged well. Granite steps are out of step, bricks are loosened. Its once bold geometry is now considered an eyesore. No one says that of the Campo itself. City Hall Plaza needlessly replaced Scollay Square, in scale close to Rome's much-admired Piazza Navona.

As we move into an era of resource scarcity, what we build will need to last longer and therefore be studied more carefully for its environmental impacts, rather than need to be rebuilt in a matter of a few years. Of course, we can anticipate that construction quality will improve, which will help designs to last longer. However, even the best quality construction cannot overcome the damaging effects of expansion and contraction of materials as temperatures rise and fall, or worse in colder climates, as waters freeze, expand in joints, and crack them. A wonderfully precise geometric composition that photographs so well for design publications can deteriorate into an embarrassment in just a few years. That reality suggests that large plazas with bold geometry are doomed to look shabby. Likewise, the use of bold color in materials like fiberglass or other plastic is likely to fade dramatically in several years. That kind of design is not very forward-thinking or resource-efficient. The true measure of success ought to be how well the project looks in year five. If local officials and the designers whom they hire take this issue seriously, they will come up with forms that age well because they are more pliant, and look well even after undergoing the stresses of expansion and contraction and soil movement. And they will use materials and colors that do not disappoint us by fading from constant exposure to direct sun. Longevity has always mattered, now more than ever.

■ **Conclusion**

This book on urban public open space has examined what has been done, what is being done, and what should be done to honor the people's needs for urban public open space excellence. As said in the Introduction, this book has played an advocacy role. It has taken a bold stand on what is at stake, and what can make a difference in peoples' lives beyond anything that has been achieved before. It has served as a guidebook, providing a framework for the all-important visionary part at the outset of the process, on the road to implementation, before an urban master plan has been created, let alone a design contract signed. The details that should guide the analysis of water supply, natural resources, demographics, transportation, market feasibility, finance mechanisms, available resources, applicable laws and codes, product research, facility design, cost estimates, construction drawings and park administration are left to manuals, handbooks, web sites and databases. But without an overarching framework to put those details in perspective, and without the involvement of a creative multi-disciplinary design team led by landscape architects, it is possible to "miss the forest through the trees."

It is hoped that this book will make a difference in how cities choose to approach the challenge of providing urban open space to honor the people's most basic and most aspirational needs. There can be no more glorious work. There can be no more urgent need than to do so wisely, and to avoid needless waste. Building on a tradition extending back to the oldest cities from Athens to Hangzhou and blossoming in public spaces of all types over the past 150 years, all across the globe, tomorrow's leaders have it within their power to provide these fundamental benefits in ways that will inspire us and move us as never before. Where cities have heart, the people have hope.

在设计杂志中被拍摄的一个完美精密的几何构图会在短短几年内变得面目全非。这个事实表明那些使用大胆几何造型的大型广场们注定了会变得残破。同样的，那些具有鲜亮颜色的材料，比如玻璃纤维或者其他塑料材料也会在几年内严重的褪色。这类设计并没有很好的前瞻性或是节省资源。判断一个设计是否成功的真正方法是它在五年后是否看起来依旧很好。如果当地官员和设计者很认真的对待这个问题，他们会在设计中使用经得起时间考验的形式，因为它们更柔韧，并且在经历了膨胀收缩和土壤移动后依然看起来很好。他们同样会用那些不会在强烈阳光照射下很快褪色的材料和颜色。设计寿命一直是重要的，现在比过去更甚。

■ **总结**

这本书仔细检验了人们对优秀城市公共开放空间的需求中我们已经完成的、正在进行的以及应该被做的事情。正如序言里所说，这本书扮演的是一个提倡者的角色。什么是濒临危险的，什么是可以给人们生活带来前所未有的影响的，本书都大胆提出了它的看法。它作为一本指导书，在设计过程开始，完善过程中，总平面图完成前，以及签订合同这些所有重要的方面都给出了框架。关于设计的细节，比如如何分析给水排水、自然资源、人口统计、公共交通、市场可行性、经济机制、可利用资源、相关法律法规、产品研究、设施设计、经济预算、施工图、公园管理等等，就交给专有手册、指南、网站和数据库了。但是如果没有一个把这些细节统筹在一起的大框架，没有一个景观设计师领导的富有创造力的多学科团队的投入，那么我们很可能会"看到树却无法领会它身后的森林"。

提供可以满足人们最基本、最渴望的需求的城市开放空间一直是我们的挑战。希望这本书可以在城市应对这个挑战上有所帮助。我们可以做到没有更铺张的作品，没有不理智的赶工，没有不需要的浪费。未来的领导者将在传统、古老如雅典和杭州般的城市里开启新的建设，他们将用他们自己的力量，以激励我们和触动我们的方式来满足这些基本的需求，而这些建设将在150多年来遍布全球的各种公共空间中绽放新的设计，在有心的城市里，人们才会有希望。

注释
1. 梅丽莎·厄坎，"继续从拉斯维加斯中学习：30年后罗伯特·凡杜里和丹尼斯·斯科特·布朗担纲罪恶之称建筑界"，2014年2月11日
2. 鲁宾·瑞内，'肢体性'与'叙事性'：哈格里夫斯的城市公园．摘自"过程：建筑128：哈格里夫斯：景观作品"（1996），36
3. http://www.discoverkymaerica.com/
4. 唐纳德·沃特森，米歇尔·亚当斯．《洪水设计：根据洪水和气候变化而具有弹性的建筑，景观和城市设计》（霍伯肯，新泽西：约翰，威利公司，2011）

NOTES
1. Melissa Urcan. "Still Learning From Las Vegas: Robert Venturi and Denise Scott Brown take on Sin City architecture three decades later", www.american suburbx.com/2011/04/interview-venturi-scott-brown-still.html, 11 February 2014
2. Reuben Rainey, "'Physicality' and 'Narrative': the Urban Parks of Hargreaves Associates", in *Process: Architecture 128: Hargreaves: Landscape Works* (1996), 36
3. http://www.discoverkymaerica.com/
4. Donald Watson FAIA and Michele Adams PE, *Design for Flooding: Architecture, Landscape and Urban Design for Resilience to Flooding and Climate Change* (Hoboken, NJ: John Wiley & Sons Inc., 2011), for a survey of issues and tentative solutions which focus on flooding rather than sea level rise.

附录 A | Appendix A

城市公共开放空间设计准则
Urban Public Open Space Supplementary Guidelines

本篇所列准则并不强制要求遵守，但强烈建议在大部分的场合使用。这些准则旨在避免常犯的错误，防止疏漏。除非设计师有充足理由证明本准则不适用于具体项目基地和使用者，否则不该拒绝遵守。

基地分析

向地区与国家机构收集以下人文因素资料：
- 土地利用模式
- 历史及考古
- 使用者需求
- 交通网路及可达性
- 安全
- 法规
- 水电

向地区与国家机构收集以下环境因素资料：
- 气候—日照与风向分布
- 地理—土壤、水文与地形
- 植栽与野生生物
- 风景

功能设置（参见城市公共开放空间活动表，表4.1）

活动

提供功能性舞台，并且能在没有表演活动时供闲坐使用。
在活动期间提供暂时性场地。

饮食

在餐饮区提供饮水器（视具体条件而定）、洗手间、电话以增加餐饮设施。
分散设置足够的垃圾桶。

园内资讯及指示牌

在出口处以及主要交通节点附近设置园区地图。
考虑设置一个简单、清楚的区域地图。

通道、交通以及联系

与公有道路的竖向高差应小于1米。
至少有两侧应为市政道路。
在远离人行入口处提供维修车辆专用的出入口。

These guidelines supplement those presented in the main text. They are strongly recommended for most situations. They are intended to avoid making common mistakes or omitting something easily overlooked. The designer must have a very good reason to justify why the current situation (this site, and its users) would not benefit from these guidelines, before rejecting them.

Site Analysis

Collect all available local and national agency information on the following human and cultural factors:
- Land-use Patterns
- Historic and Archeological
- User needs
- Circulation - Transportation network and Accessibility
- Security
- Codes/Zoning
- Utilities

Collect all available local and national agency information on the following environmental factors:
- Climate - Sun Patterns and Wind Patterns
- Geology - Soil, Hydrology and Topography
- Vegetation and Wildlife
- Views

Programming (see also the Urban Public Open Space Activity Table, Table 4.1)

Events

Provide a functional stage and allow it to be used for sitting between performances or events.
Provide a place for temporary concessions on event days.

Food

Provide drinking fountains (if available in China), restrooms, and telephones to augment the facilities for eating.
Distribute enough trash receptacles.

Information and Signs

Place a map for the Public Open Space near the entrance and at key nodes.
Consider providing a simple, clear map of the neighborhood.

Access, Circulation and Linkages

Grade change from public right of way should be less than 1 meter.

Minimum of two sides should front on public roads.

Provide separate entry for maintenance vehicles away from the main pedestrian park entry.

Prevent interior vehicular circulation by using gates or bollards.

Provide lots of seating along the main circulation pedestrian corridor.

Provide direct access to the play area, restroom and sports fields.

In softscape parks, align meandering pathways to pass by or through a variety of natural settings.

Provide ramps instead of stairs where possible.

Provide adequate parking to minimize parking problems on residential and arterial streets, but avoid providing excess parking.

Provide adequate access for fire, emergency and maintenance vehicles in parks and open space.

Avoid use of color or differentiated paving patterns to direct foot traffic; it does not work.

Barriers

Avoid use of peripheral walls unless exterior conditions are noisy because of high volume truck traffic, rail, or industrial use.

Avoid smooth materials for walls (to discourage graffiti).

Provide 1 meter high fence around tot lots.

Provide high-quality low-visibility fencing around active recreation facilities.

Spatial Quality

Provide an open dominant space with good visibility, preferably large contiguous turf areas if justified by the level of usage.

Provide at least one flat area.

Provide subspaces that are each distinct and distinguishable.

Provide subspaces which are not so tight that a second person must invade the personal space of the first person (in China this may be less important).

Seating

Provide seating at rate of 1 meter per 10 square meters (1 linear foot per 30 square feet) of plaza space.

Avoid too many benches to intimidate lonely single users; provide secondary seating in form of planters, low retaining walls, steps, but primary seating capacity should exceed secondary capacity.

Set benches back from circulation paths so that pedestrians do not disturb bench sitters.

Provide some benches that allow a sitter to watch the people moving through the park or along the adjacent sidewalks.

Provide perpendicular pairings or concave groupings of benches to encourage social interaction; facing benches discourage social interaction so avoid spacing facing benches closer than 6 meters.

Locate benches against shrubs, wall or tree to provide a back for sense of security.

Provide benches next to various facilities – tennis courts, tot lot, recreation building – to enable conversation among bench users. Examples: at the park entry, at regular intervals along the main circulation path, along the park perimeter away from the street, alone and grouped to support conversation and gathering, for viewing activities or pleasant views, and for direct supervision of

使用门栏以及桩柱防止园内车辆过往。

主要人行廊道应配备大量座椅。

提供通往儿童游乐区、洗手间以及体育场地的直达路径。

在软景公园里，设计蜿蜒小径时倚就自然景观，让道路经过或穿过自然景观。

条件许可的情况下，尽量使用斜坡取代阶梯。

提供足量的停车位以解决周遭住宅区巷道及主干道的停车问题，但是不宜设置过量车位。

在公园以及开放空间里提供足够的通道供消防车、救护车以及维修车辆使用。

避免使用不同的颜色或是铺面样式引导步行方向，这办法没有效果。

区隔

对城市中心以外的公园，避免使用外围围墙，除非园外的情况为嘈杂、大量卡车经过、近铁路或是工业用途。

避免使用平滑材料为墙面（减少涂鸦）。

在儿童游乐场外设置 1 米高的围篱。

在动态活动场地周围设置高品质、低能见度的围篱。

空间品质

在主要开放空间提供好视野，如果有相应使用要求的话最好是设置大量连续的草皮。

提供至少一个平坦区域。

提供各具特色独立不同的子空间。

子空间范围必须够大让第二位使用者不会干扰第一位使用者的个人空间。（在中国，这也许不那么重要）

座椅

每 10 平方米的广场设置 1 米长的长凳。（即每 30 平方英尺的广场需设置 1 英尺长的长凳）（纽约）

避免过多的长凳惊吓到孤单的使用者；提供如植栽槽、低矮挡土墙、阶梯等次级座椅。但是主要座椅的量应该要大于次级座椅。

长凳设置于交通步道后面，以避免座椅使用者被行人打扰。

一些长凳的位置应当便于人们坐在长凳上可以看在公园内或在附近人行道行走的人。

将座椅排列成"L"形或是使用凹弧形长凳。这可以鼓励交流。面对面配置的长凳会妨碍交流，所以避免距离小于 6 米的面对面长凳。

使座椅背对灌木丛、墙面或是树丛，给人以依靠和安全感。

在各式设施旁提供长凳——网球场、儿童游乐区、休闲娱乐馆——以使长凳使用者得以交谈。比如在园区入口、主交通步道各区段上、公园四周远离街道处，单独以及集中放置长凳以促进对话交谈与聚会，并观赏活动和美景，更能直接看护小孩。

座椅应该被放置在可以获得最大夏日浓荫及冬日太阳之处。

特殊用途区域

提供园内遮蔽处（雨天使用）。

提供独立的桌椅让人们可以在自然景观中进食、阅读或学习。

提供野餐桌。

在儿童游乐区内的秋千以及滑梯下使用特殊合成软垫。

提供一些玩水的条件，有自然河床者为最佳。

考虑设置一个专门为宠物狗使用的区域。

儿童游乐区至少需要离街道或是停车场 15 米（50 英尺）远，少于 7.5

米（25英尺）远的游乐场地周围需要有1米（英尺）高的钢管围篱。儿童游乐场地处主要交通路线的附近，同时靠近野餐区群与开放草坪区。

运动场地应该坐落于园区边缘以利用增加可见度来加强安全性。其至少应距离街道4.5至6米（15至20英尺）并使用低矮堆坡或是低矮景观缓冲以加强效果。

为体育场地安排适当方位，比如，体育场地长轴以南北向为宜。

所有的城市公园、市区公园、大型公园、社区公园都应该设置洗手间。

大型公园和绿道的休息区应该坐落于林间小径上的适当位置。休息区应该包含脚踏车架、饮水器（视具体条件而定）、荫蔽处以及野餐设施。

在所有长凳与桌子附近摆放跟街道家具样式符合的垃圾桶。同时儿童游乐区、运动设施与所有使用量高的区域也都需要提供垃圾桶。

在儿童游乐区、野餐区、洗手间、运动设施附近设置饮水器（视具体条件而定）。

硬景边缘与街道家具间应净空0.6米（2英尺）。

提供容易行走的铺面供行动不便者、穿着高跟鞋者以及学龄前儿童使用。使用光滑度低的材料，连续铺设至桌椅处。

社交

通过风格鲜明的子空间设计来设计开放空间。使人们约在那里见面的时候能轻易描述会面地点。

提供容许多样化摆设的座椅，有助于人们社交活动，且不受陌生人打扰。

为不同的使用群体提供可能的社交活动区域。

空间设计应避免不同使用群体间可能的冲突。

舒适及印象

利用植栽和细节创造一个色彩、质感、形状、气味丰富多样化的美感环境。

将儿童游乐区设置在其他活动节点附近。

在园内提供多种路径选择，同时提供多处出入口。

允许人们跟园内计划展示的公共艺术互动。

为了视觉上与嗅觉上的吸引力，放置喷泉或是其他水景。

设计社区以及区域公园时需考虑到适当的夜间使用。为了安全以及休闲活动需提供夜间照明。防止光透出场外造成光污染。

可持续性

使用原生耐旱植物，减少浇灌。

考虑降低铺面面积。

考虑使用透水、高反射铺装，减少强光。

选择使用对环境危害最小的施工方法。

树木种植位置应考虑其最终的树体大小，这样不用时常修剪。

使用节能照明设施，并且不要溢光至园外。灯柱不能相邻太近。

优先使用低能量材料及再生材料。

有些装饰性水景并未做到循环用水，应避免使用。因为一旦水景关闭，就会又脏又丑。

野生生物以及自然区域

考虑野生生态保育及复育。

考虑自然区域如湿地与林地的保育及复育。

children.

Position benches to maximize shade in the summer and sun in the winter.

Special Use Areas

Provide park shelter (for rainy weather).

Provide somewhat isolated tables and benches or chairs, where a person can eat, read, or study in a natural setting.

Provide picnic tables.

Provide special synthetic cushioned paving under swings and slides in children's play areas.

Provide some provision for water play, with a natural stream-bed the most desirable.

Consider the creation of a dogs-only section in the park.

Place play areas a minimum of 15 meters (50 feet) from the street or parking lot. Play areas closer than 7.5 meters (25 feet) should be surrounded by a 1 meter (3 feet) high tubular steel fence.

Locate play areas near the main circulation route and near group picnic areas and open lawn areas.

Sports courts should be located along the edges of the park to maximize visibility for security. Provide minimum separation from the street of 4.5 to 6 meters (15 to 20 feet), enhanced by a low berm or low landscape buffer.

Provide for the optimum orientation of sports fields. For example, sports courts should be oriented with the long axis north-south.

Provide restroom facilities in all Civic Plazas, Downtown Parks, Large Parks and Neighborhood Parks.

Provide rest areas in Large Parks and Greenways along trails where appropriate. Rest areas should include bike racks, drinking fountains (if available in China), shade and picnic facilities.

Provide trash receptacles, matching site furniture near all benches and tables, and at play areas and sports facilities and all high-use areas.

Place drinking fountain (if available in China) near children's play area, group picnic areas, restrooms and sports facilities.

Provide a 0.6 meter (2 feet) clearance between hardscape edges and site furnishings.

Provide paving that is easy for handicapped users, users wearing high heels and preschool children, that is low glare, and is continuous to seating or tables.

Sociability

Design the urban open space with subareas that are distinct, enabling people to describe easily a planned meeting area.

Provide seating that allows a variety of arrangements, both to support socializing and to permit use without intrusion by strangers.

Provide a potential social gathering area for each user group.

Design the space to avoid possible conflict between different user groups.

Comfort and Image

Create a rich and varied aesthetic environment, with a range of colors, textures, shapes, and smells, by planting and detailing.

Locate children's play areas near other activity areas.

Provide a choice of routes within the park, with multiple entrance/exit points.

Allow people to interact with planned public art.

Include a water fountain or other water feature in the design, for visual and aural attraction.

Design community and regional parks for night use, as appropriate. Provide lighting at night for safety, and anticipated recreational uses, with no spillage off-site.

Sustainability

Consider native drought-tolerant plants to minimize irrigation.

Consider reduced paving areas.

Consider permeable, highly reflective (high albedo) paving as well as glare reduction.

Select a construction method that minimizes environmental damage.

Plant trees in locations that will allow them to grow to their full size without drastic pruning.

Provide low energy use lighting, with no spillage off-site. Light pole fixtures should not be spaced too close together.

Prefer low embedded energy and recycled materials.

Minimize use of ornamental water fountains which do not use recirculated water, and which look ugly and dirty when they are shut off.

Wildlife and Natural Areas

Consider wildlife habitat preservation or restoration.

Consider preservation or restoration of natural areas such as wetlands and woodland.

Grading

Do not grade turfed slopes steeper than 5:1, because they cannot be easily mowed.

Crown playing fields such as baseball, softball and soccer, at 1.5% minimum, preferably 2%.

Provide cross slope on walkways not to exceed 2% .

Grade hard court surfaces at a slope of 1%.

Grade the park site to provide topographical relief. Consider using berms in some parts of the park site.

Balance cut and fill in grading the park.

Drainage

Provide a play area catch basin within each play area and slope the play area subgrade at 1% minimum toward the catch basin.

For swales in planted or turf area, ensure a minimum flow line slope of 2%.

Do not drain planted areas or turf areas across a paved area or walkway.

Planting

Provide a 3.7 meter (12-foot) minimum clearance between the tree trunk and the edge of hardscape.

Provide a 6 meter (20-foot) minimum clearance between canopy trees or between trees and other vertical site improvements in turf areas unless project manager approves a differing width. Locate individual plants far enough apart to avoid thinning or pruning.

Consider widest possible variety of indigenous species, not just the short list of commonly used ones.

Avoid species requiring irrigation or high maintenance.

竖向

不要让草坡坡度大于 5：1，这样不方便割草。

有路冠的游乐场地如棒球、垒球及足球场最低要有 1.5% 的坡度，2% 为佳。

步道应有不超过 2% 的横坡。

硬铺面场地表面应有 1% 坡度。

通过竖向设计使地形起伏，场地部分地方应堆坡处理。

保持公园土方平衡。

排水

在每个游乐区设置集水区，而且将游乐区路基往集水区方向至少倾斜 1%。

有植物覆盖或草坪覆盖的低洼地应确保有最低 2% 坡度的出水线。

不要将水从有植被或草坪覆盖的区域排向有铺装的区域以及人行道。

植栽

树干以及硬景边缘间应有 3.7 米（12 英尺）净空。

树冠应间隔至少 6 米（20 英尺）。树木以及其他在草坪区的直立构造物也应相隔至少 6 米（20 英尺），除非项目经理批准不同的宽度。各植栽间应有足够的空间以避免枝叶过于繁茂需要修剪。

考虑使用越多种类的原生植物越好，而不是只使用常用的那寥寥数种。

避免使用需要高度维护以及灌溉的植物。

提供多样化的日光与树荫。

使用深根树种，使用抗病虫害树种。

规避常见施工错误

1．地基太浅、不够夯实、排水粗略。

2．地基内忽略土建纺织物。

3．表层土没有经过毒素、酸度、孔隙度和肥沃度测试。

4．公用工程之间没有协作，导致铺管混乱以及地面管盖准线不对。

5．硬景竖向设计时平地和坡地设计不合逻辑，看起来像横裂纹带，而不是平缓的。

6．硬景在延伸段节点、材料替换过渡段节点施工混乱粗糙。

7．硬景排水点入口排布错乱，比如不是在低点。

8．石头铺装饰面过于平滑（湿的抛光面行走不安全）或过于粗糙（卵石和压碎石不利于女子穿高跟鞋行走）。

9．人行交通过于频繁地区的硬质空间用木散步道或木平台。（木材不能持久，会不规则的腐坏，影响游人）

10．道路准线不稳定不平整。

11．道路铺装在连接处不平整，导致人们摔跤跌倒。

12．阶梯梯步或踏面尺寸不统一。

13．石墙只在混凝土墙和水泥墙外贴一层薄片，容易剥落，或看起来廉价（特别是有锯痕的地方），而不是全部贴石。

14．挡土墙底部没有预留输水孔，大雨后没法疏导排水。

15．木柱、长凳以及其他场地露天木件不够持久，比如一两年之后就腐坏（通常应该持续 5 到 10 年）。

16．钢件由于在焊接点没有做充足的防锈处理，不到一年就生锈了。

17．屋顶、长椅、矮墙凳等所用的玻璃材料没有打扫，积累尘土和潮气。

18．未验证的不实用的材料或细节暴露在日照下 1-2 年引起退色。

19．雕塑等公共艺术处于低修护状态或不持久。

20．新替换和修补的材料与原材料不统一。

这些不足对使用者的安全和健康造成威胁，在西方，由于疏忽造成个人伤害会引起官司。有些也是不可持续的设计。

维护

需要有足够维护植栽的工作人员以及器材。如不可得，改使用低维护植栽。

放置足够的垃圾桶并设定适当的垃圾收集安排以避免垃圾过满。

草坪的灌溉安排须合理，确保草地在午餐时间是干燥的，种植灌木以及草花的可坐式植栽槽也一样。

Sources

1. Jan Gehl, *Cities for People* (Washington DC: Island Press, 2010)
2. Leonard J. Hopper, ed., *Landscape Architecture Graphic Standards* (Hoboken: John Wiley, 2007)
3. Kevin Lynch, *Site Planning* (Cambridge, MA: MIT Press, 1971)
4. Clare Cooper Marcus and Carolyn Francis ed., *People Places, Design Guidelines for Urban Open Space* (Wiley, New York, 1998)
5. Website of Project for Public Places, New York, USA (www.pps.org)
6. *Park Design Guidelines,* City of Sacramento, California, Department of Parks and Recreation, Landscape Architecture Section, revised 6/26/2001
7. San Francisco [CA] Parks Maintenance Standards: http://sf-recpark.org/ftp/uploadedfiles/wcm_recpark/Mowing_Schedule/SFParkMSManual.pdf
8. AGER Group

Provide sun and shade variety.

Use tree species with deep root systems and resistance to diseases and pests.

Common Construction Errors to Avoid

1. Subbase that is too shallow, poorly compacted, poorly drained
2. Omission of geotextile in subbase
3. Topsoil that has not been tested for toxins, acidity, porosity, and fertility
4. Lack of coordination among utilities to avoid disorderly arrangement and alignment of conduit access covers on ground
5. Hardscape grading in illogical planes and slopes, perhaps looking like a broken-back, rather than smooth
6. Hardscape with haphazardly constructed expansion joints, material change joints
7. Hardscape with misplaced drain inlets, for example not at low points
8. Stone paving surface treatment that is either too smooth (polished surfaces are unsafe to walk on when wet) or too rough (cobblestone or crushed stone pebbles are brutal on women's high-heel shoes)
9. Boardwalk or wood deck instead of paving in high-foot-traffic hardscape areas (wood is not durable; it rots irregularly causing discomfort to visitors)
10. Path alignment that is wobbly not smooth
11. Path paving with uneven settlement at joints that causes people to trip and fall
12. Stairway risers or treads that are not of uniform dimension
13. Stone walls that are a thin veneer over concrete brick or block masonry that falls off too easily or just looks cheap (especially where there are exposed saw cuts), rather than solid stone all the way through)
14. Lack of weep holes at bottom of retaining walls to let groundwater escape after heavy rains
15. Wood posts, benches or other built-in wood furnishings that are not durable, such that they rot after a year or two of exposure (they should last at least 5 to 10 years)
16. Steel that rusts within a year due to improper or inadequate rust protection coating especially at joints due to welding
17. Glass material for roofs, benches or sitting walls left uncleaned and collecting dirt and moisture
18. Unproven impractical materials or details that fade from exposure to strong sunlight or fail after 1 or 2 years
19. Public art such as sculpture that is not durable or is kept in poor repair
20. Replacement or repair using different materials than original ones. Many of these deficiencies pose a safety or health risk to users and in the West personal injury caused by such negligence could lead to legal action. Some are unsustainable practices.

Maintenance

Staff and equipment for maintaining plantings should be adequate for the proposed planting. If there is uncertainty about the availability of maintenance, planting should be low-maintenance.

Follow a litter collection schedule to prevent the overflowing of litter receptacles. Determine the best schedule for watering lawns, as well as shrubs and flowers in planters that double as seats, but allow these areas to be dry during lunchtime.

附录 B | Appendix B

可持续景观的评价体系
Sustainable Sites Initiative™ (SITES™) Rating System for Sustainable Landscapes

Prerequisite 1.1 Limit development of soils designated as farmland
Prerequisite 1.2 Protect floodplain functions
Prerequisite 1.3 Preserve wetlands
Prerequisite 1.4 Preserve threatened or endangered species and their habitats
Credit 1.5 Select brownfields or greyfields for redevelopment
Channel development to urban areas to reduce pressure on undeveloped land, reduce resource consumption and restore ecosystem services to damaged sites.
Credit 1.6 Select sites within existing communities
Encourage site development within existing communities to reduce pollution and development impacts, support local economy and improve human health.
Credit 1.7 Select sites that encourage non-motorized transportation and use of public transit
Encourage site development that is accessible by pedestrians and bicyclists and near public transit to reduce pollution and improve human health.

Prerequisite 2.1 Conduct a pre-design site assessment and explore opportunities for site sustainability
Prerequisite 2.2 Use an integrated site development process
Credit 2.3 Engage users and other stakeholders in site design
Engage with site users and other stakeholders in meaningful participation during the site design process to identify needs and to supplement professional expertise with local knowledge.

Prerequisite 3.1 Reduce potable water use for landscape irrigation by 50 percent from established baseline
Credit 3.2 Reduce potable water use for landscape irrigation by 75 percent or more from established baseline.
Limit or eliminate the use of potable water, natural surface water (such as lakes, rivers, and streams), and groundwater withdrawals for landscape irrigation. Encourage alternative irrigation methods and water conservation strategies.
Credit 3.3 Protect and restore riparian, wetland, and shoreline buffers
Preserve and enhance riparian, wetland, and shoreline buffers to improve flood control and water-quality, stabilize soils, control erosion, and provide wildlife corridors and habitat.
Credit 3.4 Rehabilitate lost streams, wetlands, and shorelines
Rehabilitate ecosystem functions and values of any streams, wetlands, or shorelines that have been artificially modified, using stable geomorphological and vegetative methods.
Credit 3.5 Manage stormwater on site
Replicate the hydrologic condition (infiltration, runoff, and evapotranspiration) of the site based on historic, natural, and undeveloped ecosystems in the region.
Credit 3.6 Protect and enhance on-site water resources and receiving water quality

必要项 1.1 限制开发被指定为农田的土地
必要项 1.2 保护泄洪平原功能
必要项 1.3 保留湿地
必要项 1.4 保护面临生存威胁物种和濒危物种以及其栖息地
加分项 1.5 选择棕色土地或者灰色土地再开发
合理引导通向城市区域的发展应来减少对未开发土地的压力，减少资源消耗，修复受损区域的生态系统。
加分项 1.6 在现有的社区内选择场地
鼓励现有社区内的场地开发，以减少污染和开发的影响，支持地方经济以及改善人类健康。
加分项 1.7 鼓励使用非机动车交通或者公共交通的场地
鼓励以人行、自行车和附近公共交通方式来进入场地的开发，以减少污染和改善人类健康。

必要项 2.1 评估预先设计的场地，并且考察场地可持续性的机会
必要项 2.2 使用完整的场地开发程序
加分项 2.3 让使用者和其他利益相关者参与到设计过程中
在场地设计过程中使场地使用者和相关利益者有意义地参与进来，以便更好地了解需求和提供与当地情况相关的专业技术。

必要项 3.1 在设定基准时减少 50% 用于绿化灌溉的饮用水使用
加分项 3.2 在设定基准时减少 75% 用于绿化灌溉的饮用水使用
限制或者减少饮用水，天然地表水（比如湖、河和溪）和抽取地下水在绿化灌溉上的使用。鼓励替代的灌溉方式和节水措施。
加分项 3.3 保护和恢复河岸、湿地和海岸线的缓冲区
维护和加强河岸、湿地和海岸线的缓冲区以增强防洪能力和提高水质，达到固土、控制水土流失和保护野生动物栖息地的目的。
加分项 3.4 恢复失去的溪流、湿地和海岸线
使用稳定地形和植被的方法，恢复被人为改动过溪流、湿地和海岸线的生态系统功能和价值。
加分项 3.5 管理场地中的雨水
基于该地区历史的、自然的和未开发的生态系统情况，模拟场地的水文条件（渗透、地表径流和蒸发量）
加分项 3.6 保护和提高现场的水资源和承受水体的质量
阻止或减少与承受水体相关的常见雨水污染物和具体水域污染物的产生、流动和传播。承受水体包括地表水、地下水和合流式下水道或雨水系统。
加分项 3.7 根据降雨/雨水的特点进行设计，融于景观设施中
以一种美观的方式在视觉上和实际上将降雨特点融入场地中。
加分项 3.8 维护水景的同时节约水资源和其他资源

设计和维护水景时，尽量不用或者最小化饮用水或者其他自然地表水和地下水来作为水源。

必要项 4.1 控制和管理在场地中已知的入侵植物
必要项 4.2 使用合适的非入侵植物
必要项 4.3 建立一个土壤维护计划
加分项 4.4 设计和施工中减少对土体的破坏
限制对健康土壤的破坏以保护土层，维持土壤结构以及储存于土壤中现有的水分、有机物和营养物质。
加分项 4.5 保存所有被指定为特殊物种的植物
识别和保存所有被本地、州或者联邦政府指定为特殊物种的植物。
加分项 4.6 保存或复原场地中适当的植物生物量
维持或建立区域性适当的植物生物量以支持由现场植被提供的生态系统服务效益。
加分项 4.7 使用本土植物
种植合适的原产于场地生态区域的植物。
加分项 4.8 保存原产于该生态区域的植物群落
保存原产于场地生态区域的植物群落，这样有助于区域植物群的多样性，并且为本地的野生动物提供了栖息地。
加分项 4.9 恢复原产于该生态区域的植物群落
适当恢复原产于场地生态区域的植物群落，这样有助于区域植物群的多样性，并且为本地的野生动物提供了栖息地。
加分项 4.10 使用植被以尽量减少建筑采暖要求
将植被种植于建筑物周围的关键位置以减少能量消耗和与室内取暖控制相关联的成本。
加分项 4.11 使用植被以尽量减少建筑制冷要求
将植被种植于建筑物周围的关键位置以减少能量消耗和与室内制冷控制相关联的成本。
加分项 4.12 减少城市热岛效应
使用植被和反光材料以减少热岛效应和最小化对小气候、人类和野生动物栖息地的影响。
加分项 4.13 减小灾难性野火发生的风险
设计、建造和维护场地，管理燃料以减小在现场以及临近景区发生灾难性野火的风险。

必要项 5.1 不使用濒危树种木材
加分项 5.2 维持现有结构、硬质景观和景观设施
维持现有结构、硬质景观和景观设施（例如，挡土墙和长凳以其现有的形式增加到现有建筑物存量的生命周期中，节约资源，减少浪费。）
加分项 5.3 对拆建和拆除进行设计
通过设计促进再利用，避免将有用的材料送往垃圾填埋场。
加分项 5.4 再利用回收的材料和植物
再利用回收的材料和合适的植物以节约资源，以及避免将有用的材料送往垃圾填埋场。
加分项 5.5 使用可再生材料
使用可再生材料以减少原材料的使用，以及避免将有用的材料送往垃圾填埋场。
加分项 5.6 使用经过认证的木材
购买认证过的木材以鼓励规范性森林管理，这是对环境和社会负责。
加分项 5.7 使用当地材料
减少运输的能源使用；加强对当地提取、制造和生长的材料、植物

Prevent or minimize generation, mobilization, and transport of common stormwater pollutants and watershed-specific pollutants of concern to receiving waters, including surface water and groundwater, and combined sewers or stormwater systems.
Credit 3.7 Design rainwater/stormwater features to provide a landscape amenity
Integrate visually and physically accessible rainwater/stormwater features into the site in an aesthetically pleasing way.
Credit 3.8 Maintain water features to conserve water and other resources
Design and maintain water features created in the landscape with minimal or no make-up water from potable sources or other natural surface or subsurface water resources.

Prerequisite 4.1 Control and manage known invasive plants found on site
Prerequisite 4.2 Use appropriate, noninvasive plants
Prerequisite 4.3 Create a soil management plan
Credit 4.4 Minimize soil disturbance in design and construction
Limit disturbance of healthy soil to protect soil horizons and maintain soil structure, existing hydrology, organic matter and nutrients stored in the soil.
Credit 4.5 Preserve all vegetation designated as special status
Identify and preserve all vegetation designated as special status by local, state, or federal entities.
Credit 4.6 Preserve or restore appropriate plant biomass on site
Maintain or establish regionally appropriate vegetative biomass to support the ecosystem service benefits provided by vegetation on-site.
Credit 4.7 Use native plants
Plant appropriate vegetation that is native to the ecoregion of the site.
Credit 4.8 Preserve plant communities native to the ecoregion
Preserve plant communities native to the ecoregion of the site to contribute to regional diversity of flora and provide habitat for native wildlife.
Credit 4.9 Restore plant communities native to the ecoregion
Restore appropriate plants and plant communities native to the ecoregion of the site to contribute to regional diversity of flora and provide habitat for native wildlife.
Credit 4.10 Use vegetation to minimize building heating requirements
Place vegetation in strategic locations around buildings to reduce energy consumption and costs associated with indoor climate control for heating.
Credit 4.11 Use vegetation to minimize building cooling requirements
Place vegetation and/or vegetated structures in strategic locations around buildings to reduce energy consumption and costs associated with indoor climate control.
Credit 4.12 Reduce urban heat island effects
Use vegetation and reflective materials to reduce heat islands and minimize effects on microclimate and on human and wildlife habitat.
Credit 4.13 Reduce the risk of catastrophic wildfire
Design, build, and maintain sites to manage fuels to reduce the risk of catastrophic wildfire both on-site and in adjacent landscapes.

Prerequisite 5.1 Eliminate the use of wood from threatened tree species
Credit 5.2 Maintain on-site structures, hardscape, and landscape amenities
Maintain existing structures, hardscape, and landscape amenities (e.g., retaining walls and benches in their existing form to extend the life cycle of existing building stock, conserve resources, and reduce waste.
Credit 5.3 Design for deconstruction and disassembly
Design to facilitate reuse and avoid sending useful materials to the landfill.
Credit 5.4 Reuse salvaged materials and plants
Reuse salvaged materials and appropriate plants to conserve resources and avoid sending useful materials to the landfill.

Credit 5.5 Use recycled content materials
Use materials with recycled content to reduce the use of virgin materials and avoid sending useful materials to the landfill.
Credit 5.6 Use certified wood
Purchase certified lumber to encourage exemplary forest management that is both environmentally and socially responsible.
Credit 5.7 Use regional materials
Reduce energy use for transportation; increase demand for materials, plants, and soils that are extracted, manufactured, or grown within the region to support the use of local resources; and promote a regional identity.
Credit 5.8 Use adhesives, sealants, paints, and coatings with reduced VOC emissions
Select paints, sealants, adhesives, coatings, and other products used in site development that contain reduced amounts of volatile organic compounds (VOCs) to reduce harmful health effects associated with air pollution.
Credit 5.9 Support sustainable practices in plant production
Purchase plants from providers that reduce resource consumption and waste.
Credit 5.10 Support sustainable practices in materials manufacturing
Support sustainable practices in materials manufacturing by purchasing materials from manufacturers whose practices increase energy efficiency, reduce resource consumption and waste, and minimize negative effects on human health and the environment.

Credit 6.1 Promote equitable site development
During construction of the site, ensure that the project provides economic or social benefits to the local community.
Credit 6.2 Promote equitable use of the site
During site use, ensure that the project provides economic or social benefits to the local community.
Credit 6.3 Promote sustainability awareness and education
Interpret on-site features and processes to promote understanding of sustainability in ways that positively influence user behavior on-site and beyond.
Credit 6.4 Protect and maintain unique cultural and historical places
Protect and maintain cultural and historical locations, attributes and artifacts to enhance a site's sense of place and meaning.
Credit 6.5 Provide for optimum site accessibility, safety, and wayfinding
Promote site use by increasing user's ability to understand and safely access outdoor spaces.
Credit 6.6 Provide opportunities for outdoor physical activity
Provide on-site opportunities that encourage outdoor physical activity to improve human health.
Credit 6.7 Provide views of vegetation and quiet outdoor spaces for mental restoration
Provide visual and physical connections to the outdoors to optimize the mental health benefits of site users.
Credit 6.8 Provide outdoor spaces for social interaction
Provide outdoor gathering spaces of various sizes and orientations to accommodate groups, for the purpose of building community and improving social ties.
Credit 6.9 Reduce light pollution
Reduce light pollution by minimizing light trespass on-site for the purpose of reducing sky-glow, increasing nighttime visibility and minimizing negative effects on nocturnal environments and human health and functioning.

Prerequisite 7.1 Control and retain construction pollutants
Prerequisite 7.2 Restore soils disturbed during construction

和土壤的需求，以支持当地资源的利用；促进区域性的一致。
加分项5.8 使用减少挥发性有机化合物排放的黏合剂、密封剂、油漆和涂料
在选择油漆、密封剂、黏合剂和涂料以及其他在场地开发所需的产品时，为了减少对健康有害的空气污染，选择含有较少挥发性有机化合物（VOC等）的产品。
加分项5.9 在植物采买中支持可持续的做法
从减少资源消耗和浪费的供方购买植物。
加分项5.10 在材料制造上支持可持续的做法
在增加能源效率、减少资源消耗和浪费以及减少对人类健康和环境负面影响的材料供应商中购买材料，在材料制造上支持可持续的做法。

加分项6.1 促进公平的场地开展
在场地的建设过程中，确保该项目为当地社区提供经济或社会效益。
加分项6.2 促进公平的场地使用
在场地的使用中，确保该项目为当地社区提供经济或社会效益。
加分项6.3 促进可持续发展的意识和教育
解读现场的特点和过程来促进对可持续发展的理解，这种方式能够积极地影响使用者在场地以及今后的行为。
加分项6.4 保护和维持有独特文化和历史的地点
保护和维持具有文化和历史的场地、标志和雕像，以提高场地的地方认同感和意义。
加分项6.5 提供最佳的场地可达性、安全性和便利性。
通过提高使用者对到达户外空间的理解能力和安全性来促进场地的使用。
加分项6.6 提供户外体育活动的机会
提供现场的活动机会，鼓励户外体育活动，以提高人们的健康水平。
加分项6.7 提供植物景色和安静的户外空间让人们得到精神上的恢复
在视觉上和身体上提供与户外的联系来增加对场地使用者精神健康的益处。
加分项6.8 提供社交联络的户外空间
提供不同大小和针对不同人群的户外集会空间，来达到建设社区和改善社会关系的目的。
加分项6.9 减少光污染
通过减少场地的光入侵来减少光污染，这样做的目的是减低天色，增加夜间能见度和减少对夜间环境和对人类健康的负面影响。

必要项7.1 控制和保留建筑污染物
必要项7.2 复原施工期间破坏的土壤
加分项7.3 复原早期开发中被破坏的土壤
复原早期开发中被破坏的表层土和下层土区域的土壤功能，重建场地的自然能力来支持健康植物、生物群落、蓄水和渗透的能力。
加分项7.4 转移废弃物中的建筑和拆建材料
从垃圾填埋场和焚烧炉燃料中转移由场地开发而产生的施工和拆建材料。如果可能的话在现场回收和／或再利用这些材料，或者把它们运回到生产制造过程中，或者运到其他施工场地，或建筑材料再利用市场，以支持零浪费场地和减少材料的向下循环。
加分项7.5 再利用或循环使用施工期间产生的植被、岩石和土壤
转移施工期间产生的废弃植被、土壤和矿物／岩石废料以实现零废物场地。

加分项 7.6 减少施工过程中温室气体的产生和向地方排放空气污染物
使用施工设备来减少地方空气污染物的排放和温室气体的产生。

必要项 8.1 制定场地可持续维护的计划
必要项 8.2 提供可回收物品的存储和收集
加分项 8.3 回收场地运营和维护期间产生的有机物
设计回收植被修剪碎料和食品垃圾（如果可行的话），来产生混合肥和植物覆盖层来支持营养物循环，改善土壤健康和减少运往垃圾场的材料和运费。
加分项 8.4 减少所有景观和外部运营的户外能耗
选择节能的户外装置和设备来减少场地使用和运营时的能源消耗和成本。
加分项 8.5 使用可再生能源满足景观用电需求
使用来自于可再生能源的电力来降低场地运营中温室气体排放和减少空气污染、栖息地破坏和以化石燃料为基础的能源污染。
加分项 8.6 减少暴露于环境烟草烟雾（ETS）
减少场地使用者暴露于环境烟草烟雾（比如二手烟），从而改善人类健康。
加分项 8.7 减少温室气体的产生以及暴露于景观维护所产生的空气污染物
降低、避免或者消除此类景观维护设备的使用，这些设备使场地和附近建筑物使用者暴露于空气污染物中，并且产生温室气体。
加分项 8.8 减少废气排放和促进节油车辆的使用
促进减少废气排放或/和具有高燃料效率车辆的使用，以及减少车辆使用带来的污染和对土地开发的影响。

加分项 9.1 监视可持续设计在实际中的表现
监视和记录可持续设计实践的实际表现，以评估其随后的性能，并提升人们对长期可持续性作用的认识。
加分项 9.2 在场地设计中进行创新
鼓励和奖励在上述评价系统中有卓越表现的可持续创新实践，以及/或并未在此评价体系中具体说明的创新表现。

来源：http://www.sustainablesites.org/pilot/

Credit 7.3 Restore soils disturbed by previous development
Restore soil function in areas of previously disturbed topsoils and subsoils to rebuild the site's ability to support healthy plants, biological communities, water storage and infiltration.
Credit 7.4 Divert construction and demolition materials from disposal
Divert construction and demolition (C&D) materials generated by site development from disposal in landfills and combusting in incinerators. Recycle and/or reuse C&D materials on-site, when possible, or redirect these materials back to the manufacturing process, other construction sites, or building materials reuse markets to support a net zero-waste site and minimize down-cycling of materials.
Credit 7.5 Reuse or recycle vegetation, rocks, and soil generated during construction
Divert from disposal vegetation, soils, and mineral/rock waste generated during construction to achieve a net zero-waste site.
Credit 7.6 Minimize generation of greenhouse gas emissions and exposure to localized air pollutants during construction
Use construction equipment that reduces emissions of localized air pollutants and greenhouse gas emissions.

Prerequisite 8.1 Plan for sustainable site maintenance
Prerequisite 8.2 Provide for storage and collection of recyclables
Credit 8.3 Recycle organic matter generated during site operations and maintenance
Design for recycling of vegetation trimmings and, where applicable, food waste to generate compost and mulch to support nutrient cycling, improve soil health, and reduce transportation costs and materials going to landfills.
Credit 8.4 Reduce outdoor energy consumption for all landscape and exterior operations
Select energy efficient outdoor fixtures and equipment to reduce energy consumption and costs associated with site use and operations.
Credit 8.5 Use renewable sources for landscape electricity needs
Use electricity from renewable sources to reduce the greenhouse gas emissions associated with site operations and minimize air pollution, habitat destruction, and pollution from fossil fuel-based energy production.
Credit 8.6 Minimize exposure to Environmental Tobacco Smoke (ETS)
Minimize exposure of site users to Environmental Tobacco Smoke (i.e. secondhand smoke) to improve human health.
Credit 8.7 Minimize generation of greenhouse gases and exposure to localized air pollutants during landscape maintenance activities
Reduce, avoid, or eliminate the use of landscape maintenance equipment that exposes site and adjacent building users to localized air pollutants and generates greenhouse gas emissions.
Credit 8.8 Reduce emissions and promote the use of fuel efficient vehicles
Promote the use of vehicles that have reduced emissions and/or high fuel-efficiency to reduce pollution and land development impacts from automobile use.

Credit 9.1 Monitor performance of sustainable design practices
Monitor and document sustainable design practices to evaluate their performance over time and improve the body of knowledge on long-term sustainability.
Credit 9.2 Innovation in site design
To encourage and reward innovative sustainable practices for exceptional performance above requirements and/or innovative performance in sustainable sites categories not specifically addressed by the Sustainable Sites Initiative Guidelines and Performance Benchmarks.

Source: http://www.sustainablesites.org/pilot/

图书在版编目（CIP）数据

有心的城市 /（美）潘德明（Thomas M.Paine）著. —北京：中国建筑工业出版社，2014.8
ISBN 978-7-112-17151-4

Ⅰ.①有… Ⅱ.①潘… Ⅲ.①城市规划-建筑设计-汉、英 Ⅳ.①TU984

中国版本图书馆CIP数据核字（2014）第186768号

本书通过对城市和个人关系的细致观察，针对城市开放空间在东西方城市发展演变过程中所扮演的角色进行了系统化的剖析，将开放空间在城市中的地位上升到"心"的高度，生动地再现了开放空间在城市中扮演的重要角色。全书共六章，讲述了公园的起源，城市开放空间的重要性、设计规划原则与方法以及未来公共空间的发展等内容。本书对于社会公众、专业规划设计人员以及城市发展的决策者了解和改进城市开放空间品质将具有重要的启迪作用。

责任编辑：杜　洁　兰丽婷　田启铭
责任校对：陈晶晶　刘梦然

有心的城市

（美）潘德明（Thomas M.Paine）　著

*

中国建筑工业出版社出版、发行（北京西郊百万庄）
各地新华书店、建筑书店经销
北京嘉泰利德公司制版
北京盛通印刷股份有限公司印刷

*

开本：787×1092毫米　1/16　印张：15　字数：650千字
2015年1月第一版　2015年1月第一次印刷
定价：99.00元
ISBN 978-7-112-17151-4
（25927）

版权所有　翻印必究
如有印装质量问题，可寄本社退换
（邮政编码　100037）

Advance Praise

Cities with Heart is written by a Harvard educated landscape architect who has lived and worked in China. He cares deeply about the quality of life in fast-growing cities. *Cities with Heart* will reflect upon various cases of city parks and makes the persuasive case that excellent urban open space is important to the current quality of life and future sustainability. This book is the place to start to understand and get familiar with emerging best practices globally.

—**Wang Shi, Founder and Chairman, Vanke**

After reading this book, never have I so wanted to grab my walking shoes and get an around-the-world travel ticket. Its pages are filled with the pictures and stories of virtually all the world's brilliantly designed and wonderfully maintained urban public spaces and landscapes. To read *Cities with Heart* is to have a revelation; not only do all the world's problems seem to melt away, but Tom Paine then moves on to describe and prescribe how more of these great parks can be built today, even in the wildly burgeoning metropolises of Asia and Africa today. A marvelous combination of "what is" and "what if."

—**Peter Harnik, Director, Center for City Park Excellence, The Trust for Public Land Washington, D.C., USA. Author of Urban Green: Innovative Parks for Resurgent Cities**

Parks and green space matter deeply for the lives of people in cities—indeed for the life of the city itself—as Tom Paine explains with authority and conviction. He's given us three essential books in one: a global overview of urban parks, a convincing case for investment in open space excellence, and a much needed roadmap to creating cities with heart. Without question, the road we must follow.

—**Helaine Kaplan Prentice, ASLA, Center for Community Innovation, University of California, Berkeley. Author of Suzhou: Shaping an Ancient City for the New China**

With brilliance and wide-ranging vision, Tom Paine presents a captivating portrait of the ways open space and aesthetically-designed landscapes nourish our souls in a variety of places from dense urban centers to local neighborhoods. Anyone with an eye to their environment, but certainly urban planners, architects, and developers, should read this enlightening book to see the direction we must take–one that, as Paine beautifully illustrates, can make our cities truly livable.

—**Arthur B. Weissman, Ph.D. Author of In the Light of Humane Nature: Human Values, Nature, the Green Economy, and Environmental Salvation, and President and CEO of Green Seal, Inc.**

Cities with Heart is an interesting book on inner city parks. Mr. Paine's profusely illustrated book has a breadth and scope which communicates to readers the importance of the urban environment and outdoor space—its history, planning principles, design principles, design guidelines—and offers a vision for the future. There is much to excite the curiosity of anyone entering the field and for everyone else interested in how urban spaces can improve urban life.

—**Carol R. Johnson, Founder of CRJA Landscape Architects. Author of A Life in the Landscape**

Cities with Heart is highly informative and richly attractive. Paine has been instrumental in community and landscape planning and design in China projects since 1976. Anyone active in the fields of urban open space planning and design or interested in these fields will be well rewarded by this well illustrated volume's comprehensive and illuminating coverage. Paine shows how lifeless plazas and other urban spaces can be beneficially revitalized, and helps us understand, quite clearly, why open space, attractive and happily functional, matters for our urban futures.

—**Roy B. Mann, ASLA, NAEP, landscape architect and planner, founder and principal of The Rivers Studio, Austin Texas. Author of Rivers in the City and manager of urban open space and riverside projects in the U.S., Canada, Mexico, Argentina, and Jordan**

For those who care about cities and livability, Thomas M. Paine's book *Cities with Heart* provides easy-to-use planning and design principles, with 332 illustrations and photos, to help us achieve better design for 5 types of urban open space—civic plaza, downtown park, large park, greenway corridor, and neighborhood park. In addition, the historical survey of city open space in Asia and the subject of future vision he brings in are wonderful subjects for tutorial discussion.

—**Rachel Lee, Lecturer, Department of Landscape Architecture at Tunghai University, Taichung, Taiwan.**